D0221308

Statistical
Computing
with R

Chapman & Hall/CRC
Computer Science and Data Analysis Series

The interface between the computer and statistical sciences is increasing, as each discipline seeks to harness the power and resources of the other. This series aims to foster the integration between the computer sciences and statistical, numerical, and probabilistic methods by publishing a broad range of reference works, textbooks, and handbooks.

SERIES EDITORS
David Madigan, Rutgers University
Fionn Murtagh, Royal Holloway, University of London
Padhraic Smyth, University of California, Irvine

Proposals for the series should be sent directly to one of the series editors above, or submitted to:

Chapman & Hall/CRC
23-25 Blades Court
London SW15 2NU
UK

Published Titles

Bayesian Artificial Intelligence
Kevin B. Korb and Ann E. Nicholson

Pattern Recognition Algorithms for Data Mining
Sankar K. Pal and Pabitra Mitra

Exploratory Data Analysis with MATLAB®
Wendy L. Martinez and Angel R. Martinez

Clustering for Data Mining: A Data Recovery Approach
Boris Mirkin

Correspondence Analysis and Data Coding with Java and R
Fionn Murtagh

R Graphics
Paul Murrell

Design and Modeling for Computer Experiments
Kai-Tai Fang, Runze Li, and Agus Sudjianto

Semisupervised Learning for Computational Linguistics
Steven Abney

Statistical Computing with R
Maria L. Rizzo

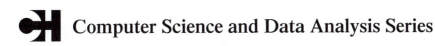

Computer Science and Data Analysis Series

Statistical Computing with R

Maria L. Rizzo

Bowling Green State University
Bowling Green, Ohio, U.S.A.

Chapman & Hall/CRC
Taylor & Francis Group

Boca Raton   London   New York

Chapman & Hall/CRC is an imprint of the
Taylor & Francis Group, an **informa** business

Chapman & Hall/CRC
Taylor & Francis Group
6000 Broken Sound Parkway NW, Suite 300
Boca Raton, FL 33487-2742

Library of Congress Cataloging-in-Publication Data

Rizzo, Maria L.
    Statistical computing with R / Maria L. Rizzo.
        p. cm. -- (Chapman & Hall/CRC computer science and data analysis series)
    Includes bibliographical references and index.
    ISBN-13: 978-1-58488-545-0 (alk. paper)
    ISBN-10: 1-58488-545-9 (alk. paper)
    1. Mathematical statistics--Data processing. 2. Statistics--Data processing. 3. R
(Computer program language) I. Title. II. Series.

QA276.45.R3R59 2007
519.50285'5133--dc22   2007034218

Visit the Taylor & Francis Web site at
http://www.taylorandfrancis.com

and the CRC Press Web site at
http://www.crcpress.com

Contents

List of Tables

List of Figures

Preface

This book is an introduction to statistical computing and computational statistics. Computational statistics is a rapidly expanding area in statistical research and applications. It includes computationally intensive methods in statistics, such as Monte Carlo methods, bootstrap, MCMC, density estimation, nonparametric regression, classification and clustering, and visualization of multivariate data. Gentle [113] and Wegman [295] describe *computational statistics* as computationally intensive methods in statistics. *Statistical computing*, at least traditionally, focused on numerical algorithms for statistics (see e.g. Thisted [269]). Generally a book has only one of these terms in the title; for example, Givens and Hoeting's *"Computational Statistics"* [121] includes classical statistical computing topics in optimization, numerical integration, density estimation and smoothing, as well as the Monte Carlo and MCMC methods of computational statistics. We chose the title *"Statistical Computing with R"* for this book, which is both computational statistics and statistical computing, and perhaps emphasizes Monte Carlo and resampling methods more than the title would suggest.

R is a statistical computing environment based on the S language. The software is free under the terms of the Free Software Foundation's GNU General Public License. It is available for a wide variety of platforms including among others Linux, Windows, and MacOS. See http://www.r-project.org/ for a description. All examples in the text are implemented in R.

This book is designed for graduate students or advanced undergraduates with preparation in calculus, linear algebra, probability and mathematical statistics. The text will be suitable for an introductory course in computational statistics, and may also be used for independent study. In addition, because of the computational nature of the material, this book serves as an excellent tutorial on the R language, providing examples that illustrate programming concepts in the context of practical computational problems. The text does not assume previous expertise in any particular programming language.

The presentation will focus on implementation rather than theory, but the connection to the mathematical ideas and theoretical foundations will be made clear. The first chapter provides an overview of computational statistics and a brief introduction to the R statistical computing environment. The second chapter is a summary and review of some basic concepts in probability and classical statistical inference. Each of the remaining chapters covers a topic in computational statistics.

The selection of topics includes the traditional core material of computational statistics: simulating random variables from probability distributions, Monte Carlo integration and variance reduction methods, Monte Carlo and MCMC methods, bootstrap and jackknife, density estimation, and visualization of multivariate data. Although R includes random generators for the commonly used probability distributions, there is instructive value in studying the algorithms for generating them. Research problems often involve distributions that are non-standard, generalized, or not implemented. Methods for generating mixtures and multivariate data are also covered. The text concludes with a chapter on numerical methods in R.

A large number of examples and exercises are included. All examples are fully implemented in the R statistical computing environment, and the R code for examples in the book can be downloaded from the author's web site at `personal.bgsu.edu/~mrizzo`. In an effort to keep the material self-contained, most examples and exercises use datasets available in the R distribution (base plus recommended packages), or simulated data. Some functions and datasets in contributed packages available on CRAN are used, which can be installed by functions provided in R.

Books in print have a long lifetime, while software is constantly evolving. By the time this book is in a reader's hands, one or more newer versions of R will have been released. Every effort has been made to check the code samples under the current version of R; comments, suggestions, and corrections are always welcome.

Acknowledgements

This book was inspired at least in part by the excellent statistical computing package R, and the author would like to acknowledge the team of developers for continuing to support and improve this software.

I would like to thank several reviewers who made invaluable suggestions and comments, especially Jim Albert, Hua Fang, Herb McGrath, Xiaoping Shen, and Gábor Székely. I would also like to acknowledge the contribution of my students who used a preliminary draft of the text at Ohio University and provided much helpful feedback, with special thanks to Roxana Hritcu, Nihar Shah, and Jinfei Zhang. Editor Bob Stern, Project Editor Marsha Hecht, and Project Coordinator Amber Donley of Taylor & Francis / CRC Press have been very helpful throughout the entire project. Finally, I would like to thank my family for their constant support and encouragement.

Maria L. Rizzo
Department of Mathematics and Statistics
Bowling Green State University

Chapter 1

Introduction

1.1 Computational Statistics and Statistical Computing

Computational statistics and statistical computing are two areas within statistics that may be broadly described as computational, graphical, and numerical approaches to solving statistical problems. *Statistical computing* traditionally has more emphasis on numerical methods and algorithms, such as optimization and random number generation, while *computational statistics* may encompass such topics as exploratory data analysis, Monte Carlo methods, and data partitioning, etc. However, most researchers who apply computationally intensive methods in statistics use both computational statistics and statistical computing methods; there is much overlap and the terms are used differently in different contexts and disciplines. Gentle [113] and Givens and Hoeting [121] use "computational statistics" to encompass all the relevant topics that should be covered in a modern introductory text, so that "statistical computing" is somewhat absorbed under this more broad definition of computational statistics. On the other hand, journals and professional organizations seem to use both terms to cover similar areas. Some examples are the International Association for Statistical Computing (IASC), part of the International Statistical Insititute, and the Statistical Computing section of the American Statistical Association.

This book encompasses parts of both of these subjects, because a first course in computational methods for statistics necessarily includes both. Some examples of topics covered are described below.

Monte Carlo methods refer to a diverse collection of methods in statistical inference and numerical analysis where simulation is used. Many statistical problems can be approached through some form of Monte Carlo integration. In parametric bootstrap, samples are generated from a given probability distribution to compute probabilities, gain information about sampling distributions of statistics such as bias and standard error, assess the performance of procedures in statistical inference, and to compare the performance of competing methods for the same problem. Resampling methods such as the ordinary bootstrap and jackknife are nonparametric methods that can be applied when the distribution of the random variable or a method to simulate it directly is unavailable. The need for Monte Carlo analysis also arises because in many problems, an asymptotic approximation is unsatisfactory or intractable. The convergence to the limit distribution may be too slow, or we require results for finite samples; or the asymptotic distribution has unknown parameters. Monte Carlo methods are covered in Chapters 5, 6, 7, 8, and 9. The first tool needed in a simulation is a method for generating psuedo random samples; these methods are covered in Chapter 3.

Markov Chain Monte Carlo (MCMC) methods are based on an algorithm to sample from a specified target probability distribution that is the stationary distribution of a Markov chain. These methods are widely applied for problems arising in Bayesian analysis, and in such diverse fields as computational physics and computational finance. Markov Chain Monte Carlo methods are covered in Chapter 9.

Several special topics also deserve an introduction in a survey of computationally intensive methods. Density estimation (Chapter 10) provides a nonparametric estimate of a density, which has many applications in addition to estimation ranging from exploratory data analysis to cluster analysis. Computational methods are essential for the visualization of multivariate data and reduction of dimensionality. The increasing interest in massive and streaming data sets, and high dimensional data arising in applications of biology and engineering, for example, demand improved and new computational approaches for multivariate analysis and visualization. Chapter 4 is an introduction to methods for visualization of multivariate data. A review of selected topics in numerical methods for optimization and numerical integration is presented in Chapter 11.

Many references can be recommended for further reading on these topics. Gentle [113] and the volume edited by Gentle, et al. [114] have thorough coverage of topics in computational statistics. Givens and Hoeting [121] is a recent graduate text on computational statistics and statistical computing. Martinez and Martinez [192] is an accessible introduction to computational statistics, with numerous examples in Matlab. Texts on statistical computing include the classics by Kennedy and Gentle [161] and Thisted [269], and a more recent survey of methods in statistical computing is covered in Kundu and Basu [165]. For statistical applications of numerical analysis see Lange [168] or Monahan [202]. Books that primarily cover Monte Carlo methods or resampling methods include Davison and Hinkley [63], Efron and Tibshirani [84], Hjorth [143], Liu [179], and Robert and Casella [228]. On density estimation see Scott [244] and Silverman [252].

1.2 The R Environment

The R environment is a suite of software and programming language based on S, for data analysis and visualization. "What is R" is one of the frequently asked questions included in the online documentation for R. Here is an excerpt from the R FAQ [217]:

> R is a system for statistical computation and graphics. It consists of a language plus a run-time environment with graphics, a debugger, access to certain system functions, and the ability to run programs stored in script files.

The home page of the R project is `http://www.r-project.org/`, and the current R distribution and documentation are available on the Comprehensive R Archive Network (CRAN). The CRAN master site is at TU Wien, Austria, `http://cran.R-project.org/`. The R distribution includes the base and recommended packages with documentation. A help system and several reference manuals are installed with the program.

R is based on the S langauge. Some details about differences between R and S are given in the R FAQ [147]. Venables and Ripley [278] is a good resource for applied statistics with S, Splus, and R. Other references on the S language include [24, 41, 42, 277].

An excellent starting point is the manual *Introduction to R* [279]. Some introductory books using R include Dalgaard [62] and Verzani [280]. On programming methods see Chambers [41], and Venables and Ripley [277, 278]. Other texts that feature Splus, S, and/or R may also be helpful (see e.g. Crawley [57] or Everitt and Hothorn [88]). Albert [5] is an introductory text on Bayesian computation. On statistical models see Faraway [90, 91], Fox, [97], Harrell [131], and Pinhiero and Bates [211]. Many more references can be found through links on the R project home page.

Programming is discussed as needed in the chapters that follow. In this text, new functions or programming methods are explained in remarks called "R notes" as they arise. Readers are always encouraged to consult the R help system and manuals [147, 279, 217]. For platform specific details about installation and interacting with the graphical user interface the best resource is the R manual [218] and current information at `www.r-project.org`.

In the remainder of this chapter, we cover some basic information aimed to help a new user get started with R. Topics include basic syntax, using the online help, datasets, files, scripts, and packages. There is a brief overview of basic graphics functions. Also see Appendix B on working with data frames.

1.3   Getting Started with R

R has a command line interface that can be used interactively or in batch mode. Commands can be typed at the prompt in the R Console window, or submitted by the `source` command (see Section 1.8). For example, we can evaluate the standard normal density $\phi(x) = \frac{1}{\sqrt{2\pi}} e^{-x^2/2}$ at $x = 2$ by typing the formula or (more conveniently) the `dnorm` function:

```
> 1/sqrt(2*pi) * exp(-2)
[1] 0.05399097
> dnorm(2)
[1] 0.05399097
```

In the example above, the command prompt is `>`. The `[1]` indicates that the result displayed is the first element of a vector.

A command can be continued on the next line. The prompt symbol changes whenever the command on the previous line is not complete. In the example below, the plot command is continued on the second line, as indicated by the prompt symbol changing to `+`.

```
> plot(cars, xlab="Speed", ylab="Distance to Stop",
+ main="Stopping Distance for Cars in 1920")
```

Whenever a statement or expression is not complete at the end of a line, the parser automatically continues it on the next line. No special symbol is needed to end a line. (A semicolon can be used to separate statements on a single line, although this tends to make code harder to read.) A group of statements can be gathered into a single (compound) expression by enclosing them in curly braces `{ }`.

To cancel a command, a partial command, or a running script use Ctrl-C, or in the Windows version of the R GUI, press the escape key (Esc). To exit the R system type the command `q()` or close the R GUI.

The usual assignment operator is `<-`. For example, `x <- sqrt(2 * pi)` assigns the value of $\sqrt{2\pi}$ to the symbol `x`.

Commands entered at the command prompt in the R console are automatically echoed to the console, but assignment operations are silent. Some objects have print methods so that the output displayed is not necessarily the entire object, but a summarized report. Compare the effect of these commands. The first command displays a sequence (0.0 0.5 1.0 1.5 2.0 2.5 3.0), but does not store it. The second command stores the sequence in x, but does not display it.

```
seq(0, 3, 0.5)
x <- seq(0, 3, 0.5)
```

TABLE 1.1: R Syntax and Commonly Used Operators

Description	R symbol	Example
Comment	#	#this is a comment
Assignment	<-	x <- log2(2)
Concatenation operator	c	c(3,2,2)
Elementwise multiplication	*	a * b
Exponentiation	^	2^1.5
x mod y	x %% y	25 %% 3
Integer division	%/%	25 %/% 3
Sequence from a to b by h	seq	seq(a,b,h)
Sequence operator	:	0:20

Syntax

Below are some help topics on R operators and syntax. The ? invokes the help system for the indicated keyword.

```
?Syntax
?Arithmetic
?Logic
?Comparison   #relational operators
?Extract      #operators on vectors and arrays
?Control      #control flow
```

Symbols or labels for functions and variables are case-sensitive and can include letters, digits, and periods. Symbols cannot contain the underscore character and cannot start with a digit. Many symbols are already defined by the R base or recommended packages. To check if a symbol is already defined, type the symbol at the prompt. The symbols q, t, I, T, and F, for example, are used by R. Note that whenever a package is loaded, other symbols may now be defined by the package.

```
> T
[1] TRUE
> t
function (x) UseMethod("t") <environment: namespace:base>
> g
Error: Object "g" not found
```

Here we see that both T and t are already defined, but g is not yet defined by R or by the user. Nothing prevents a user from assigning a new value to predefined symbols such as t or T, but it is a bad programming practice in general and can lead to unexpected results and programming errors.

Most new R users have some experience with other programming environments and languages such as C, MATLAB, or SAS. Some operations and

features are common to all these languages. A brief list summarizing R syntax
for some of these common elements is shown in Table 1.1. For more details see
the help topic `Syntax`. Some of the functions common to most development
environments are listed in Table 1.2.

Most arithmetic operations are vectorized. For example, `x^2` will square
each of the elements of the vector `x`, or each entry of the matrix `x` if `x` is a
matrix. Similarly, `x*y` will multiply each of the elements of the vector `x` times
the corresponding element of `y` (generating a warning if the vectors are not
the same length). Operators for matrices are described in Table 1.3.

TABLE 1.2:   Commonly Used Functions

Description	R symbol
Square root	sqrt
$\lfloor x \rfloor$, $\lceil x \rceil$	floor, ceiling
Natural logarithm	log
Exponential function e^x	exp
Factorial	factorial
Random Uniform numbers	runif
Random Normal numbers	rnorm
Normal distribution	pnorm, dnorm, qnorm
Rank, sort	rank, sort
Variance, covariance	var, cov
Std. dev., correlation	sd, cor
Frequency tables	table
Missing values	NA, is.na

TABLE 1.3:   R Syntax and Functions for Vectors and Matrices

Description	R symbol	Example
Zero vector	numeric(n)	x <- numeric(n)
	integer(n)	x <- integer(n)
	rep(0,n)	x <- rep(0,n)
Zero matrix	matrix(0,n,m)	x <- matrix(0,n,m)
i^{th} element of vector a	a[i]	a[i] <- 0
j^{th} column of a matrix A	A[,j]	sum(A[,j])
ij^{th} entry of matrix A	A[i,j]	x <- A[i,j]
Matrix multiplication	%*%	a %*% b
Elementwise multiplication	*	a * b
Matrix transpose	t	t(A)
Matrix inverse	solve	solve(A)

1.4 Using the R Online Help System

For documentation on a topic, type `?topic` or `help(topic)` where "topic" is the name of the topic for which you need help. For example, `?seq` will bring up documentation for the sequence function. In some cases, it may be necessary to surround the topic with quotation marks.

```
> ?%%
Error: syntax error, unexpected SPECIAL in " ?%%"
```

The second version (below) produces the help topic.

```
> ?"%%"
```

On most systems Html help is also available by the command `help.start()`; in Windows also try the Help menu, Html help. This command displays Help in a web browser, with hyperlinks. The Html help system has a search engine.

Another way to search for help on a topic is `help.search()`. This and the search engine in Html help may help locate several relevant topics. For example, if we are searching for a method to compute a permutation,

```
help.search("permutation")
```

produces two results: `order` and `sample`. We can then consult the help topics for `order` and `sample`. The help topic for `sample` shows that x is sampled without replacement (a permutation of the elements of vector x) by:

```
sample(x)          #permutation of all elements of x
sample(x, size=k)  #permutation of k elements of x
```

(If the goal was to *count* permutations, and evaluate $\frac{n!}{(n-k)!}$, we want `?Special`, a list of special functions including `factorial` and `gamma`.)

Many help files end with executable examples. The examples can be copied and pasted at the command line. To run all the examples associated with topic, use `example(topic)`. See e.g. the interesting set of examples for `density`. To run all the examples for `density`, type `example(density)`. To see one example, open the help page, copy the lines and paste them at the command prompt.

```
help(density)
# copy and paste the lines below from the help page
```

```
# The Old Faithful geyser data
d <- density(faithful$eruptions, bw = "sj")
d
plot(d)
```

A list of available data sets in the base and loaded packages is displayed by
`data()`, and documentation on a loaded data set is displayed by the associated
help topic For example, *help(faithful)* displays the Old Faithful geyser
data help topic. If a package is installed but not yet loaded, specify the name
of the package. For example, *help("geyser", package = MASS)* displays
help for the dataset `geyser` without loading the package `MASS` [278].

R note 1.1 *Data sets in the base package can be accessed without explicitly*
*loading them via **data**. Data sets in other packages can be loaded by the **data***
function. For example,
```
data("geyser", package = "MASS")
```
loads geyser data from the MASS package.

1.5   Functions

The syntax for a function definition is

```
function( arglist ) expr
return(value)
```

Many examples of functions are documented in the chapter "Writing your
own functions" of the manual [279].

Here is a simple example of a user-defined R function that "rolls" n fair
dice and returns the sum.

```
sumdice <- function(n) {
    k <- sample(1:6, size=n, replace=TRUE)
    return(sum(k))
}
```

The function definition can be entered by several methods.

1. Typing the lines at the prompt, if the definition is short.

2. Copy from an editor and paste at the command prompt.

3. Save the function in a script file and source the file.

Note that the R GUI provides an editor and toolbar for submitting code.
Once the user-defined function is entered in the workspace, it can be used like
other R functions.

```
#to print the result at the console
> sumdice(2)
[1] 9

#to store the result rather than print it
a <- sumdice(100)

#we expect the mean for 100 dice to be close to 3.5
> a / 100
[1] 3.59
```

The value returned by an R function is the argument of the `return` statement or the value of the last evaluated expression. The `sumdice` function could be written as

```
sumdice <- function(n)
    sum(sample(1:6, size=n, replace=TRUE))
```

Functions can have default argument values. For example, `sumdice` can be generalized to roll s-sided dice, but keep the default as 6-sided. The usage is shown below.

```
sumdice <- function(n, sides = 6) {
    if (sides < 1) return (0)
    k <- sample(1:sides, size=n, replace=TRUE)
    return(sum(k))
}

> sumdice(5)        #default 6 sides
[1] 12
> sumdice(n=5, sides=4)   #4 sides
[1] 14
```

1.6   Arrays, Data Frames, and Lists

Arrays, data frames, and lists are some of the objects used to store data in R. A matrix is a two dimensional array. A data frame is not a matrix, although it can be represented in a rectangular layout like a matrix. Unlike a matrix, the columns of a data frame may be different types of variables. Arrays contain a single type.

Data Frames

A data frame is a list of variables, each of the same length but not necessarily of the same type. In this section we will discuss how to extract values of variables from a data frame.

Example 1.1 (Iris data)

The Fisher `iris` data set gives four measurements on observations from three species of iris. The first few cases in the `iris` data are shown below.

	Sepal.Length	Sepal.Width	Petal.Length	Petal.Width	Species
1	5.1	3.5	1.4	0.2	setosa
2	4.9	3.0	1.4	0.2	setosa
3	4.7	3.2	1.3	0.2	setosa
4	4.6	3.1	1.5	0.2	setosa

The `iris` data is an example of a data frame object. It has 150 cases in rows and 5 variables in columns. After loading the data, variables can be referenced by $name (the column name), by subscripts like a matrix, or by position using the [[ ]] operator. The list of variable names is returned by names. Some examples with output are shown below.

```
> names(iris)
[1] "Sepal.Length" "Sepal.Width"  "Petal.Length" "Petal.Width"
[5] "Species"
> table(iris$Species)

    setosa versicolor  virginica
        50         50         50
> w <- iris[[2]]    #Sepal.Width
> mean(w)
[1] 3.057333
```

Alternately, the data frame can be attached and variables referenced directly by name. If a data frame is attached, it is a good practice to `detach` it when it is no longer needed, to avoid clashes with names of other variables.

```
> attach(iris)
> summary(Petal.Length[51:100]) #versicolor petal length
   Min. 1st Qu.  Median    Mean 3rd Qu.    Max.
   3.00    4.00    4.35    4.26    4.60    5.10
```

If we only need the `iris` data temporarily, we can use `with`. The syntax is in this example would be

```
with(iris, summary(Petal.Length[51:100]))
```

Suppose we wish to compute the means of all variables, by species. The first four columns of the data frame can be extracted with `iris[,1:4]`. Here the missing row index indicates that all rows should be included. The by function easily computes the means by species.

```
> by(iris[,1:4], Species, mean)
Species: setosa
Sepal.Length  Sepal.Width Petal.Length  Petal.Width
      5.006        3.428        1.462        0.246
------------------------------------------------------
Species: versicolor
Sepal.Length  Sepal.Width Petal.Length  Petal.Width
      5.936        2.770        4.260        1.326
------------------------------------------------------
Species: virginica
Sepal.Length  Sepal.Width Petal.Length  Petal.Width
      6.588        2.974        5.552        2.026

> detach(iris)
```

◇

R note 1.2 *Although `iris$Sepal.Width`, `iris[[2]]`, and `iris[ ,2]` all produce the same result, the $ and [[ ]] operators can only select one element, while the [ ] operator can select several. See the help topic **Extract**.*

Arrays and Matrices

An array is a multiply subscripted collection of a single type of data. An array has a dimension attribute, which is a vector containing the dimensions of the array.

Example 1.2 (Arrays)

Different arrays are shown. The sequence of numbers from 1 to 24 is first a vector without a dimension attribute, then a one dimensional array, then used to fill a 4 by 6 matrix, and finally a 3 by 4 by 2 array.

```
x <- 1:24                    # vector
dim(x) <- length(x)          # 1 dimensional array
matrix(1:24, nrow=4, ncol=6) # 4 by 6 matrix
x <- array(1:24, c(3, 4, 2)) # 3 by 4 by 2 array
```

The $3 \times 4 \times 2$ array defined by the last statement is displayed below.

```
, , 1
      [,1] [,2] [,3] [,4]
[1,]    1    4    7   10
[2,]    2    5    8   11
[3,]    3    6    9   12


, , 2
      [,1] [,2] [,3] [,4]
[1,]   13   16   19   22
[2,]   14   17   20   23
[3,]   15   18   21   24
```

The array x is displayed showing x[, , 1] (the first 3×4 elements) followed by x[, , 2] (the second 3×4 elements).   ◇

A matrix is a doubly subscripted array of a single type of data. If A is a matrix, then A[i, j] is the ij-th element of A, A[, j] is the j-th column of A, and A[i ,] is the i-th row of A. A range of rows or columns can be extracted using the : sequence operator. For example, A[2:3, 1:4] extracts the 2×4 matrix containing rows 2 and 3 and columns 1 through 4 of A.

Example 1.3 (Matrices)

The statements

```
A <- matrix(0, nrow=2, ncol=2)
A <- matrix(c(0, 0, 0, 0), nrow=2, ncol=2)
A <- matrix(0, 2, 2)
```

all assign to A the 2×2 zero matrix. Matrices are filled in column major order by default; that is, the row index changes faster than the column index. Thus,

```
A <- matrix(1:8, nrow=2, ncol=4)
```

stores in A the matrix

$$\begin{bmatrix} 1 & 3 & 5 & 7 \\ 2 & 4 & 6 & 8 \end{bmatrix}.$$

If necessary, use the option byrow=TRUE in matrix to change the default.   ◇

Example 1.4 (Iris data: Example 1.1, cont.)

We can convert the first four columns of the iris data to a matrix using as.matrix.

```
> x <- as.matrix(iris[,1:4]) #all rows of columns 1 to 4

> mean(x[,2])              #mean of sepal width, all species
[1] 3.057333
> mean(x[51:100,3])    #mean of petal length, versicolor
[1] 4.26
```

It is possible to convert the matrix to a three dimensional array, but arrays (and matrices) are stored in "column major order" by default. For arrays, "column major" means that the indices to the left are changing faster than indices to the right. In this case it is easy to convert the matrix to a $50 \times 3 \times 4$ array, with the species as the second dimension. This works because in the data matrix, by column major order, the iris species changes faster than the variable name (column).

```
> y <- array(x, dim=c(50, 3, 4))
> mean(y[,,2])  #mean of sepal width, all species
[1] 3.057333
> mean(y[,2,3]) #mean of petal length, versicolor
[1] 4.26
```

It is somewhat more difficult to produce a $50 \times 4 \times 3$ array of iris data, with species as the third dimension. Here is one approach. First the matrix is sliced into three blocks of 50 observations each, corresponding to the three species. Then the three blocks are concatenated into a vector length 600, so that species is changing the most slowly, and observation (row) is changing fastest. This vector then fills a $50 \times 4 \times 3$ array.

```
> y <- array(c(x[1:50,], x[51:100,], x[101:150,]),
+ dim=c(50, 4, 3))

> mean(y[,2,])  #mean of sepal width, all species
[1] 3.057333

> mean(y[,3,2]) #mean of petal length, versicolor
[1] 4.26
```

This array is provided in R as the data set `iris3`.   ◇

Lists

A list is an ordered collection of objects. The members of a list (the components) can be different types. Lists are more general than data frames; in fact, a data frame is a list with class "data.frame". A list can be created by the `list()` function.

Lists are frequently used to return several results of a function in a single object. Several classical hypothesis tests that return class `htest` are a good

example. See e.g. the help topic for t.test or chisq.test. Refer to the
"Value" section of the documentation. The value returned is a list containing
the test statistic, p-value, etc. The components of a list can be referenced by
name using $ or by position using [[ ]].

Example 1.5 (Named list)

The Wilcoxon rank sum test is implemented in the function wilcox.test.
Here the test is applied to two normal samples with different means.

```
w <- wilcox.test(rnorm(10), rnorm(10, 2))
> w      #print the summary

        Wilcoxon rank sum test

data:  rnorm(10) and rnorm(10, 2)
W = 2, p-value = 4.33e-05
alternative hypothesis:
true location shift is not equal to 0

> w$statistic        #stored in object w
W 2
> w$p.value
[1] 4.330035e-05
```

Try unlist(w) and unclass(w) to see more details.   ◇

Some examples of functions in this book that return a named list can be
found in Examples 7.14 on page 205, 10.12 on page 305, and 11.17 on page 349.

Example 1.6 (A list of names)

Below we create a list to assign row and column names in a matrix. The
first component for row names will be NULL in this case because we do not
want to assign row names.

```
a <- matrix(runif(8), 4, 2)    #a 4x2 matrix
dimnames(a) <- list(NULL, c("x", "y"))
```

Here is the 4×2 matrix with column names (type a to display it).

```
            x          y
[1,] 0.88009604 0.6583918
[2,] 0.32964955 0.1385332
[3,] 0.61625490 0.1378254
[4,] 0.08102034 0.1746324
```

```
# if we want row names
> dimnames(a) <- list(letters[1:4], c("x", "y"))
> a
          x         y
a 0.88009604 0.6583918
b 0.32964955 0.1385332
c 0.61625490 0.1378254
d 0.08102034 0.1746324

# another way to assign row names
> row.names(a) <- list("NE", "NW", "SW", "SE")
> a
           x         y
NE 0.88009604 0.6583918
NW 0.32964955 0.1385332
SW 0.61625490 0.1378254
SE 0.08102034 0.1746324
```

◇

1.7   Workspace and Files

The workspace in R contains data and other objects. User defined objects created in a session will persist until R is closed. If the workspace is saved before quitting R, the objects created during the session will be saved. It is not necessary to save the workspace for the examples and code here.

The ls command will display the names of objects in the current workspace. One or more objects can be removed from the workspace by the rm or remove command. For more information consult the R documentation.

Note that saving objects in the workspace can lead to unexpected results and serious hidden programming errors. For example, in the following, suppose that the programmer intended to randomly generate the value of b, but accidentally omitted the code.

```
y <- runif(100, 0, b)
```
Now, if an object named b happens to be found in the workspace, and the value of b produces a valid expression in runif, no error will be reported. An error will occur, but the programmer will not realize that it has occurred.

It is recommended that the user occasionally check what is stored in the workspace, and remove unneeded objects. The entire list of objects returned by ls() can be removed (without warning!) by rm(list = ls()).

In general, it is probably a bad practice to save functions in the workspace, because the user may forget that certain objects exist and these objects are either not documented at all or only through comments. It is a better idea to save functions in scripts and data in files. Collections of functions and data sets can also be organized and documented in packages. (See Sections 1.8 and 1.9 below.)

The Working Directory

Many scripts and data sets are provided, and many will be created by users. It is convenient to create a folder or directory with a short path name to store these files. In the examples, we assume that the files are located in /Rfiles, which will be created by the user. Any other name or path can be used.

Although it is not necessary to specify the working directory, sometimes it may be convenient to do so. A user can get or set the current working directory by the commands `getwd` and `setwd`. To set the working directory to "/Rfiles", for example, the command is `setwd("/Rfiles")`. Windows users can make this change the default by editing the Properties (Start in) in the Windows shortcut to R-GUI. More information about startup options for R can be found in the help topic `Startup`.

Reading Data from External Files

Often data to be analyzed is stored in external files. Typically, data is stored in plain text files, delimited by white space such as tabs or spaces, or by special characters such as commas.

Univariate data from an external file can be read into a vector by the `scan` command. If the file contains a data frame or a matrix, or is csv (comma separated values) format, use the `read.table` function. The `read.table` function has many options to support different file formats. Here are a few simple examples that refer to data files in Hand, et al. [126]. The data files currently are available at `http://www.stat.ncsu.edu/sas/sicl/data/` or at `http://www.stat.ucla.edu/data/`. To download, do not save the web page. Instead copy the data into a local text editor and save as plain text. Windows users note the unix style forward slashes in the path name below. See the *R for Windows FAQ* [225].

```
forearm <- scan("/Rfiles/forearm.dat") #a vector
x <- read.table("/Rfiles/irises.dat") #a data frame

> dim(x)
[1] 50 12

#get the fourth variable in the data frame
x <- read.table("/Rfiles/irises.dat")[[4]] #a vector
```

```
#read and coerce to matrix
x <- as.matrix(read.table("/Rfiles/irises.dat"))
```

The version of the iris data in [126] is given in a 50 by 12 array, with the variables in columns 1:4, 5:8, and 9:12 corresponding to the four measurements on each of the three species. Note that many of the data files from [126] are divided in groups by horizontal white space only (see e.g. the Tibetan skulls data), so they may require reformatting before reading into a data frame.

The help topic for `read.table` also contains documentation for `read.csv` and `read.delim`, for reading comma-separated-values (.csv) files and text files with other delimiters. Also see Appendix B.3.4 for an example with .csv format.

R note 1.3 *By default,* `read.table` *will convert character variables to factors. To prevent conversion of character data to factors, set* `as.is = TRUE` *(also see the* `colClasses` *argument of* `read.table`*).*

One of the recommended R packages included with the distribution is the `foreign` package, which provides several utility functions for reading files in Minitab, S, SAS, SPSS, Stata, and other formats. For details type `help(package = foreign)`.

1.8 Using Scripts

R scripts are plain text files containing R code. Once code is saved in a script, all of it can be submitted via the source command, or part of it can be executed by copy and paste (to the console).

To save R commands in a file, prepare the file with a plain text editor and save with extension .R. The Windows R GUI provides an integrated text editor. The File menu contains commands "New Script", "Open Script", "Source R code", etc. If a script editor is open, more commands for submitting the code are provided under the Edit menu and on the toolbar.

There are many other GUI's available for preparing and submitting scripts in R. Currently a list of several appears at the URL `www.sciviews.org/_rgui`. The `RWinEdt` package [177] is particularly nice for Windows users who like WinEdt.

The `source` command loads and executes the commands in the script. It is not necessary to close the file, and in fact, it may be convenient to keep it open for editing. Save changes before source-ing the file. For example, if "/Rfiles/example.R" is a file containing R code, the command

```
source("/Rfiles/example.R")
```

will enter all lines of the file at the command prompt and execute the code. Windows users should use the unix style forward slashes above or double backslashes like the command below.

```
source("\\Rfiles\\example.R")
```

Recent commands can be recalled using the up-arrow key. To edit your source file and run it again (after saving), simply use the up-arrow to recall your source command and press Enter.

Note that by default, evaluations of expressions are not printed at the console when a script is running. Use the print command within a script to display the value of an expression.

Thus, in interactive mode, an expression and its value are both printed

```
> sqrt(pi)
[1] 1.772454
```

but from a script it is necessary to use print(sqrt(pi)).

Alternately, set options in the source statement to control how much is printed. By setting echo=TRUE the statements and evaluation of expressions are echoed to the console. To see evaluation of expressions but not statements, leave echo=FALSE and set print.eval=TRUE. The examples are below.

```
source("/Rfiles/example.R", echo=TRUE)
source("/Rfiles/example.R", print.eval=TRUE)
```

1.9   Using Packages

The R installation consists of the base and several recommended packages. Type library() to see a list of installed packages. A package must be installed and loaded to be available. Base packages are automatically loaded. Other packages can be installed and loaded as needed.

Several of the recommended packages are used in this text. Some contributed packages are also used. The R system provides an interface to install contributed packages from CRAN as needed (see install.packages; in the Windows GUI see the Packages menu). A frequent error is the 'Object not found' error, which can occur when a symbol is used from a package that is not available. If this error occurs, check spelling, then check that the package containing the object is loaded.

To load an installed package use the library or require command. For example, to load the recommended package boot, type library(boot) at the command prompt. If the package is loaded, the help system for the package is also loaded. The package can also be loaded via the Packages

TABLE 1.4: Some Basic Graphics Functions in R (`graphics`) and Other Packages

Method	in (graphics)	in (package)
Scatter plot	plot	
Add regression line to plot	abline	
Add reference line to plot	abline	
Reference curve	curve	
Histogram	hist	truehist (MASS)
Bar plot	barplot	
Plot empirical CDF	plot.ecdf	
QQ Plot	qqplot	qqmath (lattice)
Normal QQ plot	qqnorm	
QQ normal ref. line	qqline	
Box plot	boxplot	
Stem plot	stem	

menu in the GUI. Typing the command `help(package=boot)` will bring up a window showing the contents of the package, whether or not the package is loaded. Once the package is loaded, typing `?boot` will bring up the help topic for the `boot` function in the `boot` package (if not loaded, use `help(boot, package=boot)`).

A complete list of all available packages is provided on the CRAN web site. A list of available packages is also included in the R FAQ [147]. Type `installed.packages()` to see a list of all the installed packages.

1.10 Graphics

The R `graphics` package contains most of the commonly used graphics functions. In this section, for reference, some of the graphics functions and options or parameters are listed. Examples of graphics and the R code used to produce them appear throughout the text. See Murrell [204] for many more examples. Maindonald and Braun [184]), and Venables and Ripley [278] also have many examples of graphics in R.

Table 1.4 lists some basic 2D graphics functions in R (`graphics`) and other packages. Several examples using the graphics functions in Table 1.4 are given throughout the text. See Table 4.1 and the examples of Chapter 4 for more 2D graphics functions and some 3D visualization methods. Also see the gallery of graphics at `http://addictedtor.free.fr/graphiques/`.

Colors, plotting symbols, and line types

In most plotting functions, colors, symbols, and line types can be specified using col, pch, and lty. The size of a symbol is specified by cex. Available plotting characters are shown in the manual [279, Ch. 12], which includes this example for displaying plotting characters in a legend.

```
plot.new()      #if a plot is not open
legend(locator(1), as.character(0:25), pch=0:25)
#then click to locate the legend
```

The example above can be used to display line types, by substituting lty for pch. The following produces a display of colors.

```
legend(locator(1), as.character(0:8), lwd=20, col=0:8)
```

Other colors and color palettes are available. For example,

```
plot.new()
palette(rainbow(15))
legend(locator(1), as.character(1:15), lwd=15, col=1:15)
```

puts a 15 color rainbow palette into effect and displays the colors. Use colors() to see the vector of named colors.

The figures in this text have been drawn in black and white. Where color palettes would normally be used, we have substituted a grayscale palette. In these cases, on screen it is better to substitute one of the pre-defined color palettes or a custom palette. To define a color palette, refer to ?palette, and to use a defined color palette, see the topic ?rainbow (the topics rainbow, heat.colors, topo.colors, and terrain.colors are documented on the same page.)

A table of plotting characters is produced by show.pch() (Hmisc). A utility to display available colors in R is show.colors() in the DAAG package [184]. Also see show.col() in the Hmisc package [132].

Setting the graphical parameter par(ask = TRUE) has the effect that the graphics device will wait for user input before displaying the next plot; e.g. the message "Waiting to confirm page change ... " appears, and in the GUI the user should click on the graphics window to display the next screen. To turn off this behavior, type par(ask = FALSE).

Chapter 2

Probability and Statistics Review

In this chapter we briefly review without proofs some definitions and concepts in probability and statistics. Many introductory and more advanced texts can be recommended for review and reference. On introductory probability see e.g. Bean [23], Ghahramani [118], or Ross [232]. Mathematical statistics and probability books at an advanced undergraduate or first year graduate level include e.g. DeGroot and Schervish [64], Freund (Miller and Miller) [201], Hogg, McKean and Craig [146] or Larsen and Marx [170]. Casella and Berger [39] or Bain and Englehart [16] are somewhat more advanced. Durrett [77] is a graduate probability text. Lehmann [172] and Lehmann and Casella [173] are graduate texts in statistical inference.

2.1 Random Variables and Probability

Distribution and Density Functions

The cumulative distribution function (cdf) of a random variable X is F_X defined by

$$F_X(x) = P(X \leq x), \qquad x \in \mathbb{R}.$$

In this book $P(\cdot)$ denotes the probability of its argument. We will omit the subscript X and write $F(x)$ if it is clear in context. The cdf has the following properties:

1. F_X is non-decreasing.

2. F_X is right-continuous; that is,

$$\lim_{\epsilon \to 0^+} F_X(x + \epsilon) = F_X(x), \quad \text{for all } x \in \mathbb{R}.$$

3. $\lim_{x \to -\infty} F_X(x) = 0$ and $\lim_{x \to \infty} F_X(x) = 1.$

A random variable X is continuous if F_X is a continuous function. A random variable X is discrete if F_X is a step function.

Discrete distributions can be specified by the probability mass function (pmf) $p_X(x) = P(X = x)$. The discontinuities in the cdf are at the points where the pmf is positive, and $p(x) = F(x) - F(x^-)$.

If X is discrete, the cdf of X is

$$F_X(x) = P(X \le x) = \sum_{\{k \le x : p_X(k) > 0\}} p_X(k).$$

Continuous distributions do not have positive probability mass at any single point. For continuous random variables X the probability density function (pdf) or density of X is $f_X(x) = F'_X(x)$, provided that F_X is differentiable, and by the fundamental theorem of calculus

$$F_X(x) = P(X \le x) = \int_{-\infty}^{x} f_X(t)dt.$$

The joint density of continuous random variables X and Y is $f_{X,Y}(x,y)$ and the cdf of (X,Y) is

$$F_{X,Y}(x,y) = P(X \le x; Y \le y) = \int_{-\infty}^{y} \int_{-\infty}^{x} f_{X,Y}(s,t)dsdt.$$

The marginal probability densities of X and Y are given by

$$f_X(x) = \int_{-\infty}^{\infty} f_{X,Y}(x,y)dy; \qquad f_Y(y) = \int_{-\infty}^{\infty} f_{X,Y}(x,y)dx.$$

The corresponding formulas for discrete random variables are similar, with sums replacing the integrals. In the remainder of this chapter, for simplicity $f_X(x)$ denotes either the pdf (if X is continuous) or the pmf (if X is discrete) of X.

The set of points $\{x : f_X(x) > 0\}$ is the *support set* of the random variable X. Similarly, the bivariate distribution of (X,Y) is supported on the set $\{(x,y) : f_{X,Y}(x,y) > 0\}$.

Expectation, Variance, and Moments

The mean of a random variable X is the *expected value* or mathematical expectation of the variable, denoted $E[X]$. If X is continuous with density f, then the expected value of X is

$$E[X] = \int_{-\infty}^{\infty} x f(x)dx.$$

If X is discrete with pmf $f(x)$, then

$$E[X] = \sum_{\{x : f_X(x) > 0\}} x f(x).$$

(The integrals and sums above are not necessarily finite. We implicitly assume that $E|X| < \infty$ whenever $E[X]$ appears in formulas below.) The expected value of a function $g(X)$ of a continuous random variable X with pdf f is defined by

$$E[g(X)] = \int_{-\infty}^{\infty} g(x)f(x)dx.$$

Let $\mu_X = E[X]$. Then μ_X is also called the first moment of X. The rth *moment* of X is $E[X^r]$. Hence if X is continuous,

$$E[X^r] = \int_{-\infty}^{\infty} x^r f_X(x)dx.$$

The *variance* of X is the second central moment,

$$Var(X) = E[(X - E[X])^2].$$

The identity $E[(X - E[X])^2] = E[X^2] - (E[X])^2$ provides an equivalent formula for variance,

$$Var(X) = E[X^2] - (E[X])^2 = E[X^2] - \mu_X^2.$$

The variance of X is also denoted by σ_X^2. The square root of the variance is the *standard deviation*. The reciprocal of the variance is the *precision*.

The expected value of the product of continuous random variables X and Y with joint pdf $f_{X,Y}$ is

$$E[XY] = \int_{-\infty}^{\infty} \int_{-\infty}^{\infty} xy f_{X,Y}(x,y)dxdy.$$

The *covariance* of X and Y is defined by

$$\begin{aligned} Cov(X,Y) &= E[(X - \mu_X)(Y - \mu_Y)] \\ &= E[XY] - E[X]E[Y] = E[XY] - \mu_X\mu_Y. \end{aligned}$$

The covariance of X and Y is also denoted by σ_{XY}. Note that $Cov(X, X) = Var(X)$. The product-moment *correlation* is

$$\rho(X,Y) = \frac{Cov(X,Y)}{\sqrt{Var(X)Var(Y)}} = \frac{\sigma_{XY}}{\sigma_X \sigma_Y}.$$

Correlation can also be written as

$$\rho(X,Y) = E\left[\left(\frac{X - \mu_X}{\sigma_X}\right)\left(\frac{Y - \mu_Y}{\sigma_Y}\right)\right].$$

Two variables X and Y are *uncorrelated* if $\rho(X,Y) = 0$.

Conditional Probability and Independence

In classical probability, the conditional probability of an event A given that event B has occurred is

$$P(A|B) = \frac{P(AB)}{P(B)},$$

where $AB = A \cap B$ is the intersection of events A and B. Events A and B are independent if $P(AB) = P(A)P(B)$; otherwise they are dependent. The joint probability that both A and B occur can be written

$$P(AB) = P(A|B)P(B) = P(B|A)P(A).$$

If random variables X and Y have joint density $f_{X,Y}(x,y)$, then the conditional density of X given $Y = y$ is

$$f_{X|Y=y}(x) = \frac{f_{X,Y}(x,y)}{f_Y(y)}.$$

Similarly the conditional density of Y given $X = x$ is

$$f_{Y|X=x}(y) = \frac{f_{X,Y}(x,y)}{f_X(x)}.$$

Thus, the joint density of (X, Y) can be written

$$f_{X,Y}(x,y) = f_{X|Y=y}(x)f_Y(y) = f_{Y|X=x}(y)f_X(x).$$

Independence

The random variables X and Y are *independent* if and only if

$$f_{X,Y}(x,y) = f_X(x)f_Y(y)$$

for all x and y; or equivalently, if and only if $F_{X,Y}(x,y) = F_X(x)F_Y(y)$, for all x and y.

The random variables $X_1, \ldots, X_d$ are *independent* if and only if the joint pdf f of $X_1, \ldots, X_d$ is equal to the product of the marginal density functions. That is, $X_1, \ldots, X_d$ are independent if and only if

$$f(x_1, \ldots, x_d) = \prod_{j=1}^{d} f_j(x_j)$$

for all $x = (x_1, \ldots, x_d)^T$ in $\mathbb{R}^d$, where $f_j(x_j)$ is the marginal density (or marginal pmf) of X_j.

The variables $\{X_1, \ldots, X_n\}$ are a *random sample* from a distribution F_X if $X_1, \ldots, X_n$ are independently and identically distributed with distribution F_X. In this case the joint density of $\{X_1, \ldots, X_n\}$ is

$$f(x_1, \ldots, x_n) = \prod_{i=1}^{n} f_X(x_i).$$

If X and Y are independent, then $Cov(X, Y) = 0$ and $\rho(X, Y) = 0$. However, the converse is not true; uncorrelated variables are not necessarily independent. The converse is true in an important special case: if X and Y are normally distributed then $Cov(X, Y) = 0$ implies independence.

Properties of Expected Value and Variance

Suppose that X and Y are random variables, and a and b are constants. Then the following properties hold (provided the moments exist).

1. $E[aX + b] = aE[X] + b$.
2. $E[X + Y] = E[X] + E[Y]$.
3. If X and Y are independent, $E[XY] = E[X]E[Y]$.
4. $Var(b) = 0$.
5. $Var[aX + b] = a^2 Var(X)$.
6. $Var(X + Y) = Var(X) + Var(Y) + 2Cov(X, Y)$.
7. If X and Y are independent, $Var(X + Y) = Var(X) + Var(Y)$.

If $\{X_1, \ldots, X_n\}$ are independent and identically distributed (iid) we have

$$E[X_1 + \cdots + X_n] = n\mu_X, \qquad Var(X_1 + \cdots + X_n) = n\sigma_X^2,$$

so the sample mean $\overline{X} = \frac{1}{n}\sum_{i=1}^{n} X_i$ has expected value μ_X and variance σ_X^2/n. (Apply properties 2, 7, and 5 above.)

The conditional expected value of X given $Y = y$ is

$$E[X|Y = y] = \int_{-\infty}^{\infty} x f_{X|Y=y}(x)dx,$$

if $F_{X|Y=y}(x)$ is continuous.

Two important results are the conditional expectation rule and the conditional variance formula:

$$E[X] = E[E[X|Y]] \qquad (2.1)$$
$$Var(X) = E[Var(X|Y)] + Var(E[X|Y]). \qquad (2.2)$$

See e.g. Ross [233, Ch. 3] for a proof of (2.1, 2.2) and many applications.

2.2 Some Discrete Distributions

Some important discrete distributions are the "counting distributions." The counting distributions are used to model the frequency of events and waiting time for events in discrete time, for example. Three important counting distributions are the binomial (and Bernoulli), negative binomial (and geometric), and Poisson.

Several discrete distributions including the binomial, geometric, and negative binomial distributions can be formulated in terms of the outcomes of Bernoulli trials. A Bernoulli experiment has exactly two possible outcomes, "success" or "failure." A Bernoulli random variable X has the probability mass function

$$P(X = 1) = p, \qquad P(X = 0) = 1 - p,$$

where p is the probability of success. It is easy to check that $E[X] = p$ and $Var(X) = p(1 - p)$. A sequence of Bernoulli trials is a sequence of outcomes $X_1, X_2, \ldots$ of iid Bernoulli experiments.

Binomial and Multinomial Distribution

Suppose that X records the number of successes in n iid Bernoulli trials with success probability p. Then X has the Binomial(n, p) distribution [abbreviated $X \sim \text{Bin}(n, p)$] with

$$P(X = x) = \binom{n}{x} p^x (1 - p)^{n-x} = \frac{n!}{x!(n-x)!} p^x (1 - p)^{n-x}, \quad x = 0, 1, \ldots, n.$$

The mean and variance formulas are easily derived by observing that that the binomial variable is an iid sum of n Bernoulli(p) variables. Therefore

$$E[X] = np, \qquad Var(X) = np(1 - p).$$

A binomial distribution is a special case of a multinomial distribution. Suppose that there are $k+1$ mutually exclusive and exhaustive events $A_1, \ldots, A_{k+1}$ that can occur on any trial of an experiment, and each event occurs with probability $P(A_j) = p_j$, $j = 1, \ldots, k + 1$. Let X_j record the number of times that event A_j occurs in n independent and identical trials of the experiment. Then $X = (X_1, \ldots, X_k)$ has the multinomial distribution with joint pdf

$$f(x_1, \ldots, x_k) = \frac{n!}{x_1! x_2! \ldots x_{k+1}!} p_1^{x_1} p_2^{x_2} \ldots p_{k+1}^{x_{k+1}}, \quad 0 \le x_j \le n, \qquad (2.3)$$

where $x_{k+1} = n - \sum_{j=1}^{k} x_j$.

Geometric Distribution

Consider a sequence of Bernoulli trials, with success probability p. Let the random variable X record the number of failures until the first success is observed. Then

$$P(X = x) = p(1 - p)^x, \qquad x = 0, 1, 2, \ldots. \qquad (2.4)$$

A random variable X with pmf (2.4) has the Geometric(p) distribution [abbreviated $X \sim \text{Geom}(p)$]. If $X \sim \text{Geom}(p)$, then the cdf of X is

$$F_X(x) = P(X \leq x) = 1 - (1-p)^{\lfloor x \rfloor + 1}, \qquad x \geq 0,$$

and otherwise $F_X(x) = 0$. The mean and variance of X are given by

$$E[X] = \frac{1-p}{p}; \qquad Var[X] = \frac{1-p}{p^2}.$$

Alternative formulation of Geometric distribution

The geometric distribution is sometimes formulated by letting Y be defined as the number of trials until the first success. Then $Y = X + 1$, where X is the random variable defined above with pmf (2.4). Under this model, we have $P(Y = y) = p(1-p)^{y-1}$, $y = 1, 2, \ldots$, and

$$E[Y] = E[X+1] = \frac{1-p}{p} + 1 = \frac{1}{p};$$

$$Var[Y] = Var[X+1] = Var[X] = \frac{1-p}{p^2}.$$

However, as a counting distribution, or frequency model, the first formulation (2.4) given above is usually applied, because frequency models typically must include the possibility of a zero count.

Negative Binomial Distribution

The negative binomial frequency model applies in the same setting as a geometric model, except that the variable of interest is the number of failures until the rth success. Suppose that exactly X failures occur before the rth success. If $X = x$, then the rth success occurs on the $(x+r)^{th}$ trial. In the first $x + r - 1$ trials, there are $r - 1$ successes and x failures. This can happen $\binom{x+r-1}{r-1} = \binom{x+r-1}{x}$ ways, and each way has probability $p^r q^x$. The probability mass function of the random variable X is given by

$$P(X = x) = \binom{x+r-1}{r-1} p^r q^x, \qquad x = 0, 1, 2, \ldots. \tag{2.5}$$

The negative binomial distribution is defined for $r > 0$ and $0 < p < 1$ as follows. The random variable X has a negative binomial distribution with parameters (r, p) if

$$P(X = x) = \frac{\Gamma(x+r)}{\Gamma(r)\Gamma(x+1)} p^r q^x, \qquad x = 0, 1, 2, \ldots, \tag{2.6}$$

where $\Gamma(\cdot)$ is the complete gamma function defined in (2.8). Note that (2.5) and (2.6) are equivalent when r is a positive integer. If X has pmf (2.6) we will write $X \sim \text{NegBin}(r, p)$. The special case $\text{NegBin}(r = 1, p)$ is the $\text{Geom}(p)$ distribution.

Suppose that $X \sim \text{NegBin}(r, p)$, where r is a positive integer. Then X is the iid sum of r $\text{Geom}(p)$ variables. Therefore, the mean and variance of X given by

$$E[X] = r\frac{1-p}{p}, \qquad Var[X] = r\frac{1-p}{p^2},$$

are simply r times the mean and variance of the $\text{Geom}(p)$ variable in (2.4). These formulas are also valid for all $r > 0$.

Note that like the geometric random variable, there is an alternative formulation of the negative binomial model that counts the number of trials until the rth success.

Poisson Distribution

A random variable X has a Poisson distribution with parameter $\lambda > 0$ if the pmf of X is

$$p(x) = \frac{e^{-\lambda}\lambda^x}{x!}, \quad x = 0, 1, 2, \ldots.$$

If $X \sim \text{Poisson}(\lambda)$ then

$$E[X] = \lambda; \qquad Var(X) = \lambda.$$

A useful recursive formula for the pmf is $p(x + 1) = p(x)\frac{\lambda}{x+1}$, $x = 0, 1, 2, \ldots$. The Poisson distribution has many important properties and applications (see e.g. [124, 158, 233]).

Examples

Example 2.1 (Geometric cdf)

The cdf of the geometric distribution with success probability p can be derived as follows. If $q = 1 - p$, then at the points $x = 0, 1, 2, \ldots$ the cdf of X is given by

$$P(X \leq x) = \sum_{k=0}^{x} pq^k = p(1 + q + q^2 + \cdots + q^x) = \frac{p(1 - q^{x+1})}{1 - q} = 1 - q^{x+1}.$$

Alternately, $P(X \leq x) = 1 - P(X \geq x + 1) = 1 - P(\text{first x+1 trials are failures}) = 1 - q^{x+1}$.   ◇

Example 2.2 (Mean of the the Poisson distribution)

If $X \sim \text{Poisson}(\lambda)$, then

$$E[X] = \sum_{x=0}^{\infty} x \frac{e^{-\lambda}\lambda^x}{x!} = \lambda \sum_{x=1}^{\infty} \frac{e^{-\lambda}\lambda^{x-1}}{(x-1)!} = \lambda \sum_{x=0}^{\infty} \frac{e^{-\lambda}\lambda^x}{x!} = \lambda.$$

The last equality follows because the summand is the Poisson pmf and the total probability must sum to 1.   ◇

2.3 Some Continuous Distributions

Normal Distribution

The normal distribution with mean μ and variance σ^2 [abbreviated $N(\mu, \sigma^2)$] is the continuous distribution with pdf

$$f(x) = \frac{1}{\sqrt{2\pi}\sigma} \exp\left\{ -\frac{1}{2}\left(\frac{x-\mu}{\sigma}\right)^2 \right\}, \qquad -\infty < x < \infty.$$

The standard normal distribution $N(0,1)$ has zero mean and unit variance, and the standard normal cdf is

$$\Phi(z) = \int_{-\infty}^{z} \frac{1}{\sqrt{2\pi}} e^{-t^2/2}\, dt, \qquad -\infty < z < \infty.$$

The normal distribution has several important properties. We summarize some of these properties, without proof. For more properties and characterizations see [156, Ch. 13], [210], or [270].

A linear transformation of a normal variable is also normally distributed. If $X \sim N(\mu, \sigma)$ then the distribution of $Y = aX + b$ is $N(a\mu + b, a^2\sigma^2)$. It follows that if $X \sim N(\mu, \sigma)$, then

$$Z = \frac{X - \mu}{\sigma} \sim N(0, 1).$$

Linear combinations of normal variables are normal; if $X_1, \ldots, X_k$ are independent, $X_i \sim N(\mu_i, \sigma_i^2)$, and $a_1, \ldots, a_k$ are constants, then

$$Y = a_1 X_1 + \cdots + a_k X_k$$

is normally distributed with mean $\mu = \sum_{i=1}^{k} a_i \mu_i$ and variance $\sigma^2 = \sum_{i=1}^{k} a_i^2 \sigma_i^2$.

Therefore, if $X_1, \ldots, X_n$ is a random sample $(X_1, \ldots, X_n$ are iid) from a $N(\mu, \sigma^2)$ distribution, the sum $Y = X_1 + \cdots + X_n$ is normally distributed with $E[Y] = n\mu$ and $Var(Y) = n\sigma^2$. It follows that the sample mean $\overline{X} = Y/n$ has the $N(\mu, \sigma^2/n)$ distribution if the sampled distribution is normal. (In case the sampled distribution is not normal, but the sample size is large, the Central Limit Theorem implies that the distribution of Y is approximately normal. See Section 2.5)

Gamma and Exponential Distributions

A random variable X has a gamma distribution with shape parameter $r > 0$ and rate parameter $\lambda > 0$ if the pdf of X is

$$f(x) = \frac{\lambda^r}{\Gamma(r)} x^{r-1} e^{-\lambda x}, \qquad x \geq 0, \tag{2.7}$$

where $\Gamma(r)$ is the complete gamma function, defined by

$$\Gamma(r) = \int_0^\infty t^{r-1} e^{-t} dt, \qquad r \neq 0, -1, -2, \ldots. \tag{2.8}$$

Recall that $\Gamma(n) = (n-1)!$ for positive integers n.

The notation $X \sim \text{Gamma}(r, \lambda)$ indicates that X has the density (2.7), with shape r and rate λ. If $X \sim \text{Gamma}(r, \lambda)$ then

$$E[X] = \frac{r}{\lambda}; \qquad Var(X) = \frac{r}{\lambda^2}.$$

Gamma distributions can also be parameterized by the scale parameter $\theta = 1/\lambda$ instead of the rate parameter λ. In terms of (r, θ) the mean is $r\theta$ and the variance is $r\theta^2$. An important special case of the gamma distribution is $r = 1$, which is the exponential distribution with rate parameter λ. The Exponential(λ) pdf is

$$f(x) = \lambda e^{-\lambda x}, \qquad x \geq 0.$$

If X is exponentially distributed with rate λ [abbreviated $X \sim \text{Exp}(\lambda)$], then

$$E[X] = \frac{1}{\lambda}; \qquad Var(X) = \frac{1}{\lambda^2}.$$

It can be shown that the sum of iid exponentials has a gamma distribution. If $X_1, \ldots, X_r$ are iid with the Exp(λ) distribution, then $Y = X_1 + \cdots + X_r$ has the Gamma(r, λ) distribution.

Chisquare and t

The Chisquare distribution with ν degrees of freedom is denoted by $\chi^2(\nu)$. The pdf of a $\chi^2(\nu)$ random variable X is

$$f(x) = \frac{1}{\Gamma(\nu/2) 2^{\nu/2}} x^{(\nu/2)-1} e^{-x/2}, \qquad x \in \mathbb{R}, \ \nu = 1, 2, \ldots,.$$

Note that $\chi^2(\nu)$ is a special case of the gamma distribution, with shape parameter $\nu/2$ and rate parameter $1/2$. The square of a standard normal variable has the $\chi^2(1)$ distribution. If $Z_1, \ldots, Z_\nu$ are iid standard normal then $Z_1^2 + \cdots + Z_\nu^2 \sim \chi^2(\nu)$. If $X \sim \chi^2(\nu_1)$ and $Y \sim \chi^2(\nu_2)$ are independent, then $X + Y \sim \chi^2(\nu_1 + \nu_2)$. If $X \sim \chi^2(\nu)$, then

$$E[X] = \nu, \qquad Var(X) = 2\nu.$$

The Student's t distribution [256] is defined as follows. Let $Z \sim N(0,1)$ and $V \sim \chi^2(\nu)$. If Z and V are independent, then the distribution of

$$T = \frac{Z}{\sqrt{V/\nu}}$$

has the Student's t distribution with ν degrees of freedom, denoted $t(\nu)$. The density of a $t(\nu)$ random variable X is given by

$$f(x) = \frac{\Gamma(\frac{\nu+1}{2})}{\Gamma(\frac{\nu}{2})} \frac{1}{\sqrt{\nu\pi}} \frac{1}{\left(1 + \frac{x^2}{\nu}\right)^{(\nu+1)/2}}, \qquad x \in \mathbb{R}, \, \nu = 1, 2, \ldots$$

The mean and variance of $X \sim t(\nu)$ are given by

$$E[X] = 0, \quad \nu > 1; \qquad Var(X) = \frac{\nu}{\nu - 2}, \quad \nu > 2.$$

In the special case $\nu = 1$ the $t(1)$ distribution is the standard Cauchy distribution. For small ν the t distribution has "heavy tails" compared to the normal distribution. For large ν, the $t(\nu)$ distribution is approximately normal, and $t(\nu)$ converges in distribution to standard normal as $\nu \to \infty$.

Beta and Uniform Distributions

A random variable X with density function

$$f(x) = \frac{\Gamma(\alpha + \beta)}{\Gamma(\alpha)\Gamma(\beta)} x^{\alpha-1}(1-x)^{\beta-1}, \qquad 0 \leq x \leq 1, \, \alpha > 0, \, \beta > 0. \qquad (2.9)$$

has the Beta(α, β) distribution. The constant in the beta density is the reciprocal of the beta function, defined by

$$B(\alpha, \beta) = \int_0^1 t^{\alpha-1}(1-t)^{\beta-1} dt = \frac{\Gamma(\alpha)\Gamma(\beta)}{\Gamma(\alpha + \beta)}.$$

The continuous uniform distribution on (0,1) or Uniform(0,1) is the special case Beta(1,1).

The parameters α and β are shape parameters. When $\alpha = \beta$ the distribution is symmetric about $1/2$. When $\alpha \neq \beta$ the distribution is skewed, with the direction and amount of skewness depending on the shape parameters. The mean and variance are

$$E[X] = \frac{\alpha}{\alpha + \beta}; \qquad Var(X) = \frac{\alpha\beta}{(\alpha + \beta)^2(\alpha + \beta + 1)}.$$

If $X \sim \text{Uniform}(0, 1) = \text{Beta}(1, 1)$, then $E[X] = \frac{1}{2}$ and $Var(X) = \frac{1}{12}$.

In Bayesian analysis, a beta distribution is often chosen to model the distribution of a probability parameter, such as the probability of success in Bernoulli trials or a binomial experiment.

Lognormal Distribution

A random variable X has the Lognormal(μ, σ^2) distribution [abbreviated $X \sim \text{LogN}(\mu, \sigma^2)$] if $X = e^Y$, where $Y \sim N(\mu, \sigma^2)$. That is, $\log X \sim N(\mu, \sigma^2)$. The lognormal density function is

$$f_X(x) = \frac{1}{x\sqrt{2\pi}\sigma} e^{-(\log x - \mu)^2/(2\sigma^2)}, \qquad x > 0.$$

The cdf can be evaluated by the normal cdf of $\log X \sim N(\mu, \sigma^2)$, so the cdf of $X \sim \text{LogN}(\mu, \sigma^2)$ is given by

$$F_X(x) = \Phi\left(\frac{\log x - \mu}{\sigma}\right), \qquad x > 0.$$

The moments are

$$E[X^r] = E[e^{rY}] = \exp\left\{r\mu + \frac{1}{2}r^2\sigma^2\right\}, \qquad r > 0. \tag{2.10}$$

The mean and variance are

$$E[X] = e^{\mu + \sigma^2/2}, \qquad Var(X) = e^{2\mu + \sigma^2}(e^{\sigma^2} - 1).$$

Examples

Example 2.3 (Two-parameter exponential cdf)

The two-parameter exponential density is

$$f(x) = \lambda e^{-\lambda(x - \eta)}, \qquad x \geq \eta, \tag{2.11}$$

where λ and η are positive constants. Denote the distribution with density function (2.11) by $\text{Exp}(\lambda, \eta)$. When $\eta = 0$ the density (2.11) is exponential with rate λ.

The cdf of the two-parameter exponential distribution is given by

$$F(x) = \int_\eta^x \lambda e^{-\lambda(t - \eta)}\, dt = \int_0^{x-\eta} \lambda e^{-\lambda u}\, du = 1 - e^{-\lambda(x - \eta)}, \qquad x \geq \eta.$$

In the special case $\eta = 0$ we have the cdf of the $\text{Exp}(\lambda)$ distribution,

$$F(x) = 1 - e^{-\lambda x}, \qquad x \geq 0.$$

◇

Example 2.4 (Memoryless property of the exponential distribution)

The exponential distribution with rate parameter λ has the memoryless property. That is, if $X \sim \text{Exp}(\lambda)$, then

$$P(X > s + t | X > s) = P(X > t), \qquad \text{for all } s, t \geq 0.$$

The cdf of X is $F(x) = 1 - \exp(-\lambda x)$, $x \geq 0$ (see Example 2.3). Therefore, for all $s, t \geq 0$ we have

$$
\begin{aligned}
P(X > s + t | X > s) &= \frac{P(X > s + t)}{P(X > s)} = \frac{1 - F(s + t)}{1 - F(s)} \\
&= \frac{e^{-\lambda(s+t)}}{e^{-\lambda s}} = e^{-\lambda t} = 1 - F(t) \\
&= P(X > t).
\end{aligned}
$$

The first equality is simply the definition of conditional probability, $P(A|B) = P(AB)/P(B)$.

◇

2.4 Multivariate Normal Distribution

The bivariate normal distribution

Two continuous random variables X and Y have a bivariate normal distribution if the joint density of (X, Y) is the bivariate normal density function, which is given by

$$
\begin{aligned}
f(x, y) = &\frac{1}{2\pi\sigma_1\sigma_2\sqrt{1 - \rho^2}} \exp\Bigg\{ -\frac{1}{2(1 - \rho^2)} \Bigg[\left(\frac{x - \mu_1}{\sigma_1}\right)^2 \\
&- 2\rho\left(\frac{x - \mu_1}{\sigma_1}\right)\left(\frac{y - \mu_2}{\sigma_2}\right) + \left(\frac{y - \mu_2}{\sigma_2}\right)^2 \Bigg] \Bigg\},
\end{aligned} \qquad (2.12)
$$

$(x, y) \in \mathbb{R}^2$. The parameters are $\mu_1 = E[X]$, $\mu_2 = E[Y]$, $\sigma_1^2 = Var(X)$, $\sigma_2^2 = Var(Y)$, and $\rho = Cor(X, Y)$. The notation $(X, Y) \sim \text{BVN}(\mu_1, \mu_2, \sigma_1^2, \sigma_2^2, \rho)$ indicates that (X, Y) have the joint pdf (2.12). Some properties of the bivariate normal distribution (2.12) are:

1. The marginal distributions of X and Y are normal; that is $X \sim N(\mu_1, \sigma_1^2)$ and $Y \sim N(\mu_2, \sigma_2^2)$.

2. The conditional distribution of Y given $X = x$ is normal with mean $\mu_2 + \rho\sigma_2/\sigma_1(x - \mu_1)$ and variance $\sigma_2^2(1 - \rho^2)$.

3. The conditional distribution of X given $Y = y$ is normal with mean $\mu_1 + \rho\sigma_1/\sigma_2(y - \mu_2)$ and variance $\sigma_1^2(1 - \rho^2)$.

4. X and Y are independent if and only if $\rho = 0$.

Suppose $(X_1, X_2) \sim \text{BVN}(\mu_1, \mu_2, \sigma_1^2, \sigma_2^2, \rho)$. Let $\mu = (\mu_1, \mu_2)^T$ and

$$\Sigma = \begin{bmatrix} \sigma_{11} & \sigma_{12} \\ \sigma_{21} & \sigma_{22} \end{bmatrix},$$

where $\sigma_{ij} = Cov(X_i, X_j)$. Then the bivariate normal pdf (2.12) of (X_1, X_2) can be written in matrix notation as

$$f(x_1, x_2) = \frac{1}{(2\pi)|\Sigma|^{1/2}} \exp\left\{ -\frac{1}{2}(x - \mu)^T \Sigma^{-1}(x - \mu) \right\},$$

where $x = (x_1, x_2)^T \in \mathbb{R}^2$.

The multivariate normal distribution

The joint distribution of continuous random variables $X_1, \ldots, X_d$ is multivariate normal or d-variate normal, denoted $N_d(\mu, \Sigma)$, if the joint pdf is given by

$$f(x_1, \ldots, x_d) = \frac{1}{(2\pi)^{d/2}|\Sigma|^{1/2}} \exp\left\{ -\frac{1}{2}(x - \mu)^T \Sigma^{-1}(x - \mu) \right\}, \qquad (2.13)$$

where Σ is the $d \times d$ nonsingular covariance matrix of $(X_1, \ldots, X_d)^T$, $\mu = (\mu_1, \ldots, \mu_d)^T$ is the mean vector, and $x = (x_1, \ldots, x_d)^T \in \mathbb{R}^d$.

The one-dimensional marginal distributions of a multivariate normal variable are normal with mean μ_i and variance σ_i^2, $i = 1, \ldots, d$. Here σ_i^2 is the ith entry on the diagonal of Σ. In fact, all of the marginal distributions of a multivariate normal vector are multivariate normal (see e.g. Tong [273, Sec. 3.3]).

The normal random variables $X_1, \ldots, X_d$ are independent if and only if the covariance matrix Σ is diagonal.

Linear transformations of multivariate normal random vectors are multivariate normal. That is, if C is an $m \times d$ matrix and $b = (b_1, \ldots, b_m)^T \in \mathbb{R}^m$, then $Y = CX + b$ has the m-dimensional multivariate normal distribution with mean vector $C\mu + b$ and covariance matrix $C\Sigma C^T$.

Applications and properties of the multivariate normal distribution are covered by Anderson [8] and Mardia et al. [188]. Refer to Tong [273] for properties and characterizations of the bivariate normal and multivariate normal distribution.

2.5 Limit Theorems

Laws of Large Numbers

The *Weak Law of Large Numbers* (WLLN) or (LLN) states that the sample mean converges *in probability* to the population mean. Suppose that $X_1, X_2 \ldots$ are independent and identically distributed (iid), $E|X_1| < \infty$ and $\mu = E[X_1]$. For each n let $\overline{X}_n = \frac{1}{n} \sum_{i=1}^{n} X_i$. Then $\overline{X}_n \to \mu$ in probability as $n \to \infty$. That is, for every $\epsilon > 0$,

$$\lim_{n \to 0} P(|\overline{X}_n - \mu| < \epsilon) = 1.$$

For a proof, see e.g. Durrett [77].

The *Strong Law of Large Numbers* (SLLN) states that the sample mean converges *almost surely* to the population mean μ. Suppose that $X_1, X_2, \ldots$ are pairwise independent and identically distributed, $E|X_1| < \infty$ and $\mu = E[X_1]$. For each n let $\overline{X}_n = \frac{1}{n} \sum_{i=1}^{n} X_i$. Then $\overline{X}_n \to \mu$ almost surely as $n \to \infty$. That is, for every $\epsilon > 0$,

$$P(\lim_{n \to 0} |\overline{X}_n - \mu| < \epsilon) = 1.$$

For Etemadi's proof see Durrett [77].

Central Limit Theorem

The first version of the Central Limit Theorem was proved by de Moivre in the early 18th century for random samples of Bernoulli variables. The general proof was given independently by Lindeberg and Lévy in the early 1920's.

THEOREM 2.1 (The Central Limit Theorem) *If $X_1, \ldots, X_n$ is a random sample from a distribution with mean μ and finite variance $\sigma^2 > 0$, then the limiting distribution of*

$$Z_n = \frac{\overline{X} - \mu}{\sigma/\sqrt{n}}$$

is the standard normal distribution.

See Durrett [77] for the proofs.

2.6 Statistics

Unless otherwise stated, $X_1, \ldots, X_n$ is a random sample from a distribution with cdf $F_X(x) = P(X \leq x)$, pdf or pmf $f_X(x)$, mean $E[X] = \mu_X$ and variance σ_X^2. The subscript X on F, f, μ, and σ is omitted when it is clear in context. Lowercase letters $x_1, \ldots, x_n$ denote an observed random sample.

A statistic is a function $T_n = T(X_1, \ldots, X_n)$ of a sample. Some examples of statistics are the sample mean, sample variance, etc. The sample mean is $\overline{X} = \frac{1}{n} \sum_{i=1}^n X_i$, and sample variance is

$$S^2 = \frac{1}{n-1} \sum_{i=1}^n (X_i - \overline{X})^2 = \frac{\sum_{i=1}^n X_i^2 - n\overline{X}^2}{n-1}.$$

The sample standard deviation is $S = \sqrt{S^2}$.

The empirical distribution function

An estimate of $F(x) = P(X \leq x)$ is the proportion of sample points that fall in the interval $(-\infty, x]$. This estimate is called the empirical cumulative distribution function (ecdf) or empirical distribution function (edf). The ecdf of an observed sample $x_1, \ldots, x_n$ is defined by

$$F_n(x) = \begin{cases} 0, & x < x_{(1)}, \\ \frac{i}{n}, & x_{(i)} \leq x < x_{(i+1)}, \ i = 1, \ldots, n-1, \\ 1, & x_{(n)} \leq x, \end{cases}$$

where $x_{(1)} \leq x_{(2)} \leq \cdots \leq x_{(n)}$ is the ordered sample.

A *quantile* of a distribution is found by inverting the cdf. The cdf may not be strictly increasing, however, so the definition is as follows. The q quantile of a random variable X with cdf $F(x)$ is

$$X_q = \inf_x \{x : F(x) \geq q\}, \qquad 0 < q < 1.$$

Quantiles can be estimated by the inverse ecdf of a random sample or other function of the order statistics. Methods for computing sample quantiles differ among statistical packages R, SAS, Minitab, SPSS, etc. (see Hyndman and Fan [148] and the `quantile` help topic in R).

R note 2.1 *The default method of estimation used in the R* `quantile` *function assigns cumulative probability* $(k-1)/(n-1)$ *to the* k^{th} *order statistic. Thus, the empirical cumulative probabilities are defined*

$$0, \frac{1}{n-1}, \frac{2}{n-1}, \ldots, \frac{n-2}{n-1}, 1.$$

Note that this set of probabilities differs from the usual assignment $\{k/n\}_{k=1}^n$ *of the ecdf.*

Bias and Mean Squared Error

A statistic $\hat{\theta}_n$ is an *unbiased* estimator of a parameter θ if $E[\hat{\theta}_n] = \theta$. An estimator $\hat{\theta}_n$ is *asymptotically unbiased* for θ if

$$\lim_{n \to \infty} E[\hat{\theta}_n] = \theta.$$

The *bias* of an estimator $\hat{\theta}$ for a parameter θ is defined $bias(\hat{\theta}) = E[\hat{\theta}] - \theta$.

Clearly $\overline{X}$ is an unbiased estimator of the mean $\mu = E[X]$. It can be shown that $E[S^2] = \sigma^2 = Var(X)$, so the sample variance S^2 is an unbiased estimator of σ^2. The maximum likelihood estimator of variance is

$$\hat{\sigma}^2 = \frac{1}{n} \sum_{i=1}^{n} (X_i - \overline{X})^2,$$

which is a biased estimator of σ^2. However, the bias $-\sigma^2/n$ tends to zero as $n \to \infty$, so $\hat{\sigma}^2$ is asymptotically unbiased for σ^2.

The mean squared error (MSE) of an estimator $\hat{\theta}$ for parameter θ is

$$MSE(\hat{\theta}) = E[(\hat{\theta} - \theta)^2].$$

Notice that for an unbiased estimator the MSE is the equal to the variance of the estimator. If $\hat{\theta}$ is biased for θ, however, the MSE is larger than the variance. In fact, the MSE can be split into two parts,

$$\begin{aligned} MSE(\hat{\theta}) = E[\hat{\theta}^2 - 2\theta\hat{\theta} + \theta^2] &= E[\hat{\theta}^2] - 2\theta E[\hat{\theta}] + \theta^2 \\ &= E[\hat{\theta}^2] - (E[\hat{\theta}])^2 + (E[\hat{\theta}])^2 - 2\theta E[\hat{\theta}] + \theta^2 \\ &= Var(\hat{\theta}) + (E[\hat{\theta}] - \theta)^2, \end{aligned}$$

so the MSE is the sum of variance and squared bias:

$$MSE(\hat{\theta}) = Var(\hat{\theta}) + [bias(\hat{\theta})]^2.$$

The standard error of an estimator $\hat{\theta}$ is the square root of the variance: $se(\hat{\theta}) = \sqrt{Var(\hat{\theta})}$. An important example is the standard error of the mean

$$se(\overline{X}) = \sqrt{Var(\overline{X})} = \sqrt{\frac{Var(X)}{n}} = \frac{\sigma_x}{\sqrt{n}}.$$

A sample proportion $\hat{p}$ is an unbiased estimator of the population proportion p. The standard error of a sample proportion is $\sqrt{p(1-p)/n}$. Note that $se(\hat{p}) \leq 0.5/\sqrt{n}$.

For each fixed $x \in \mathbb{R}$, the ecdf $F_n(x)$ is an unbiased estimator of the cdf $F(x)$. The standard error of $F_n(x)$ is $\sqrt{F(x)(1 - F(x))/n} \leq 0.5/\sqrt{n}$.

The variance of the q sample quantile [63, 2.7] is

$$Var(\hat{x}_q) = \frac{q(1-q)}{nf(x_q)^2},\qquad (2.14)$$

where f is the density of the sampled distribution. When quantiles are estimated, the density f is usually unknown, but (2.14) shows that larger samples are needed for estimates of quantiles in the part of the support set where the density is close to zero.

Method of Moments

The rth sample moment $m'_r = \frac{1}{n}\sum_{i=1}^{n} X_i^r$, $r = 1, 2, \ldots$ is an unbiased estimator of the rth population moment $E[X^r]$, provided that the rth moment exists. If X has density $f(x; \theta_1, \ldots, \theta_k)$, then the method of moments estimator of $\theta = (\theta_1, \ldots, \theta_k)$ is given by the simultaneous solution $\hat{\theta} = (\hat{\theta}_1, \ldots, \hat{\theta}_k)$ of the equations

$$E[X^r] = m'_r(x_1, \ldots, x_n) = \frac{1}{n}\sum_{i=1}^{n} x_i^r, \qquad r = 1, \ldots, k.$$

The Likelihood Function

Suppose that the sample observations are iid from a distribution with density function $f(X|\theta)$, where θ is a parameter. The likelihood function is the conditional probability of observing the sample, given θ, which is given by

$$L(\theta) = \prod_{i=1}^{n} f(x_i|\theta).\qquad (2.15)$$

The parameter θ could be a vector of parameters, $\theta = (\theta_1, \ldots, \theta_p)$. The likelihood function regards the data as a function of the parameter(s) θ. As $L(\theta)$ is a product, it is usually easier to work with the logarithm of $L(\theta)$, called the log likelihood function,

$$l(\theta) = \log(L(\theta)) = \sum_{i=1}^{n} \log f(x_i|\theta).\qquad (2.16)$$

Maximum Likelihood Estimation

The method of maximum likelihood was introduced by R. A. Fisher. By maximizing the likelihood function $L(\theta)$ with respect to θ, we are looking for the most likely value of θ given the information available, namely the sample data. Suppose that Θ is the parameter space of possible values of θ. If the maximum of $L(\theta)$ exists and it occurs at a unique point $\hat{\theta} \in \Theta$, then $\hat{\theta}$ is called the maximum likelihood estimator of $L(\theta)$. If the maximum exists but is not unique, then any of the points where the maximum is attained is an MLE of θ. For many problems, the MLE can be determined analytically. However, it is often the case that the optimization cannot be solved analytically, and in that case numerical optimization or other computational approaches can be applied.

Maximum likelihood estimators have an invariance property. This property states that if $\hat{\theta}$ is an MLE of θ and τ is a function of θ, then $\tau(\hat{\theta})$ is an MLE of $\tau(\theta)$.

Note that the maximum likelihood principle can also be applied in problems where the observed variables are not independent or identically distributed (the likelihood function (2.15) given above is for the iid case).

Example 2.5 (Maximum likelihood estimation of two parameters)

Find the maximum likelihood estimator of $\theta = (\lambda, \eta)$ for the two-parameter exponential distribution (see Example 2.3). Suppose that $x_1, \ldots, x_n$ is a random sample from the $\text{Exp}(\lambda, \eta)$ distribution. The likelihood function is

$$L(\theta) = L(\lambda, \eta) = \prod_{i=1}^{n} \lambda e^{-\lambda(x_i - \eta)} I(x_i \geq \eta),$$

where $I(\cdot)$ is the indicator variable ($I(A) = 1$ on set A and $I(A) = 0$ on the complement of A). Then if $x_{(1)} = \min\{x_1, \ldots, x_n\}$, we have

$$L(\theta) = L(\lambda, \eta) = \lambda^n \exp\{-\lambda \sum_{i=1}^{n}(x_i - \eta)\}, \qquad x_{(1)} \geq \eta,$$

and the log-likelihood is given by

$$l(\theta) = l(\lambda, \eta) = n \log \lambda - \lambda \sum_{i=1}^{n}(x_i - \eta), \qquad x_{(1)} \geq \eta.$$

Then $l(\theta)$ is an increasing function of η for every fixed λ, and $\eta \leq x_{(1)}$, so $\hat{\eta} = x_{(1)}$. To find the maximum of $l(\theta)$ with respect to λ, solve

$$\frac{\partial l(\lambda, \eta)}{\partial \lambda} = \frac{n}{\lambda} - \sum_{i=1}^{n}(x_i - \eta) = 0,$$

to find the critical point $\lambda = 1/(\bar{x} - \eta)$. The MLE of $\theta = (\lambda, \eta)$ is

$$(\hat{\lambda}, \hat{\eta}) = \left(\frac{1}{\bar{x} - x_{(1)}}, \ x_{(1)} \right).$$

◇

Example 2.6 (Invariance property of MLE)

Find the maximum likelihood estimator of the α-quantile of the $\text{Exp}(\lambda, \eta)$ distribution in Examples 2.3 and 2.5. From Example 2.3 we have

$$F(x) = 1 - e^{-\lambda(x-\eta)}, \qquad x \geq \eta.$$

Therefore $F(x_\alpha) = \alpha$ implies that

$$x_\alpha = -\frac{1}{\lambda} \log(1 - \alpha) + \eta,$$

and by the invariance property of maximum likelihood, the MLE of x_α is

$$\hat{x}_\alpha = -(\bar{x} - x_{(1)}) \log(1 - \alpha) + x_{(1)}.$$

$\diamond$

2.7   Bayes' Theorem and Bayesian Statistics

The Law of Total Probability

If events $A_1, \ldots, A_k$ partition a sample space S into mutually exclusive and exhaustive nonempty events, then the *Law of Total Probability* states that the total probability of an event B is given by

$$
\begin{aligned}
P(B) &= P(A_1 B) + P(A_2 B) + \cdots + P(A_k B) \\
&= P(B|A_1)P(A_1) + P(B|A_2)P(A_2) + \cdots + P(B|A_k)P(A_k) \\
&= \sum_{j=1}^{k} P(B|A_j)P(A_j).
\end{aligned}
$$

For continuous random variables X and Y we have the distributional form of the Law of Total Probability

$$f_Y(y) = \int_{-\infty}^{\infty} f_{Y|X=x}(y) f_X(x)\, dx.$$

For discrete random variables X and Y we can write the distributional form of the Law of Total Probability as

$$f_Y(y) = P(Y = y) = \sum_{x} P(Y = y | X = x) P(X = x).$$

Bayes' Theorem

Bayes' Theorem provides a method for inverting conditional probabilities. In its simplest form, if A and B are events and $P(B) > 0$, then

$$P(A|B) = \frac{P(B|A)P(A)}{P(B)}.$$

Often the Law of Total Probability is applied to compute $P(B)$ in the denominator. These formulas follow from the definitions of conditional and joint probability.

For continuous random variables the distributional form of Bayes' Theorem is

$$f_{X|Y=y}(x) = \frac{f_{Y|X=x}(y)f_X(x)}{f_Y(y)} = \frac{f_{Y|X=x}(y)f_X(x)}{\int_{-\infty}^{\infty} f_{Y|X=x}(y)f_X(x)\,dx}.$$

For discrete random variables

$$f_{X|Y=y}(x) = P(X = x|Y = y) = \frac{P(Y = y|X = x)P(X = x)}{\sum_x \{P(Y = y|X = x)P(X = x)\}}.$$

These formulas follow from the definitions of conditional and joint probability.

Bayesian Statistics

In the frequentist approach to statistics, the parameters of a distribution are considered to be fixed but unknown constants. The Bayesian approach views the unknown parameters of a distribution as random variables. Thus, in Bayesian analysis, probabilities can be computed for parameters as well as the sample statistics.

Bayes' Theorem allows one to revise his/her prior belief about an unknown parameter based on observed data. The *prior* belief reflects the relative weights that one assigns to the possible values for the parameters. Suppose that X has the density $f(x|\theta)$. The conditional density of θ given the sample observations $x_1, \ldots, x_n$ is called the *posterior* density, defined by

$$f_{\theta|x}(\theta) = \frac{f(x_1, \ldots, x_n|\theta)f_\theta(\theta)}{\int f(x_1, \ldots, x_n|\theta)f_\theta(\theta)\,d\theta},$$

where $f_\theta(\theta)$ is the pdf of the prior distribution of θ. The posterior distribution summarizes our modified beliefs about the unknown parameters, taking into account the data that has been observed. Then one is interested in computing posterior quantities such as posterior means, posterior modes, posterior standard deviations, etc.

Note that any constant in the likelihood function cancels out of the posterior density. The basic relation is

$$posterior \propto prior \times likelihood,$$

which describes the shape of the posterior density up to a multiplicative constant. Often the evaluation of the constant is difficult and the integral cannot be obtained in closed form. However, Monte Carlo methods are available that do not require the evaluation of the constant in order to sample from the posterior distribution and estimate posterior quantities of interest. See e.g. [44, 103, 106, 120, 228] on development of Markov Chain Monte Carlo sampling.

Readers are referred to Lee [171] for an introductory presentation of Bayesian statistics. Albert [5] is a good introduction to computational Bayesian methods with R. A textbook covering probability and mathematical statistics from both a classical and Bayesian perspective at an advanced undergraduate level is DeGroot and Schervish [64].

2.8   Markov Chains

In this section we briefly review discrete time, discrete state space Markov chains. A basic understanding of Markov chains is necessary background for Chapter 9 on Markov Chain Monte Carlo methods. Readers are referred to Ross [234, Ch. 4] for an excellent introduction to Markov chains.

A Markov chain is a stochastic process $\{X_t\}$ indexed by time $t \geq 0$. Our goal is to generate a chain by simulation, so we consider discrete time Markov chains. The time index will be the nonnegative integers, so that the process starts in state X_0 and makes successive transitions to $X_1, X_2, \ldots, X_t, \ldots$. The set of possible values of X_t is the state space.

Suppose that the state space of a Markov chain is finite or countable. Without loss of generality, we can suppose that the states are $0, 1, 2, \ldots$. The sequence $\{X_t | t \geq 0\}$ is a Markov chain if

$$P(X_{t+1} = j | X_0 = i_0, X_1 = i_1, \ldots, X_{t-1} = i_{t-1}, X_t = i) =$$
$$P(X_{t+1} = j | X_t = i),$$

for all pairs of states (i, j), $t \geq 0$. In other words, the transition probability depends only on the current state, and not on the past.

If the state space is finite, the transition probabilities $P(X_{t+1} | X_t)$ can be represented by a transition matrix $\mathbb{P} = (p_{ij})$ where the entry p_{ij} is the probability that the chain makes a transition to state j in one step starting from state i. The probability that the chain moves from state i to state j in k steps is $p_{ij}^{(k)}$, and the Chapman-Kolmogorov equations (see e.g. [234, Ch. 4]) provide that the k-step transition probabilities are the entries of the matrix $\mathbb{P}^k$. That is, $\mathbb{P}^{(k)} = (p_{ij}^{(k)}) = \mathbb{P}^k$, the kth power of the transition matrix.

A Markov chain is *irreducible* if all states communicate with all other states: given that the chain is in state i, there is a positive probability that the chain can enter state j in finite time, for all pairs of states (i, j). A state i is *recurrent* if the chain returns to i with probability 1; otherwise state i is *transient*. If the expected time until the chain returns to i is finite, then i is *nonnull* or *positive recurrent*. The *period* of a state i is the greatest common divisor of the lengths of paths starting and ending at i. In an irreducible chain, the periods of all states are equal, and the chain is *aperiodic* if the states all have period 1. Positive recurrent, aperiodic states are ergodic. In a finite-state Markov chain all recurrent states are positive recurrent.

In an irreducible, ergodic Markov chain the transition probabilities converge to a stationary distribution π on the state space, independent of the initial state of the chain.

In a finite-state Markov chain, irreducibility and aperiodicity imply that for all states j

$$\pi_j = \lim_{n \to \infty} p_{ij}^{(n)}$$

exists and is independent of the initial state i. The probability distribution $\pi = \{\pi_j\}$ is called the *stationary distribution*, and π is the unique nonnegative solution to the system of equations

$$\pi_j = \sum_{i=0}^{\infty} \pi_i p_{ij}, \quad j \geq 0; \qquad \sum_{j=0}^{\infty} \pi_j = 1. \tag{2.17}$$

We can interpret π_j as the (limiting) proportion of time that the chain is in state j.

Example 2.7 (Finite state Markov chain)

Ross [234] gives the following example of a Markov chain model for mutations of DNA. A DNA nucleotide has four possible values. For each unit of time the model specifies that the nucleotide changes with probability 3α, for some $0 < \alpha < 1/3$. If it does change, then it is equally likely to change to any of the other three values. Thus $p_{ii} = 1 - 3\alpha$ and $p_{ij} = 3\alpha/3 = \alpha, i \neq j$. If we number the states 1 to 4, the transition matrix is

$$\mathbb{P} = \begin{bmatrix} 1 - 3\alpha & \alpha & \alpha & \alpha \\ \alpha & 1 - 3\alpha & \alpha & \alpha \\ \alpha & \alpha & 1 - 3\alpha & \alpha \\ \alpha & \alpha & \alpha & 1 - 3\alpha \end{bmatrix} \tag{2.18}$$

where $p_{ij} = \mathbb{P}_{i,j}$ is the probability of a mutation from state i to state j. The ith row of a transition matrix is the conditional probability distribution $P(X_{n+1} = j | X_n = i), j = 1, 2, 3, 4$ of a transition to state j given that the process is currently in state i. Thus each row must sum to 1 (the matrix is row stochastic). This matrix happens to be doubly stochastic because the columns also sum to 1, but in general a transition matrix need only be row stochastic.

Suppose that $\alpha = 0.1$. Then the two-step and the 16-step transition matrices are

$$\mathbb{P}^2 = \begin{bmatrix} 0.52 & 0.16 & 0.16 & 0.16 \\ 0.16 & 0.52 & 0.16 & 0.16 \\ 0.16 & 0.16 & 0.52 & 0.16 \\ 0.16 & 0.16 & 0.16 & 0.52 \end{bmatrix}, \quad \mathbb{P}^{16} \doteq \begin{bmatrix} 0.2626 & 0.2458 & 0.2458 & 0.2458 \\ 0.2458 & 0.2626 & 0.2458 & 0.2458 \\ 0.2458 & 0.2458 & 0.2626 & 0.2458 \\ 0.2458 & 0.2458 & 0.2458 & 0.2626 \end{bmatrix}.$$

The three-step transition matrix is $\mathbb{P}^2 \mathbb{P} = \mathbb{P}^3$, etc. The probability $p_{14}^{(2)}$ of transition from state 1 to state 4 in two steps is $\mathbb{P}_{1,4}^2 = 0.16$, and the probability that the process returns to state 2 from state 2 in 16 steps is $p_{22}^{(16)} = \mathbb{P}_{2,2}^{16} = 0.2626$.

All entries of $\mathbb{P}$ are positive, hence all states communicate; the chain is irreducible and ergodic. The transition probabilities in every row are converging to the same stationary distribution π on the four states. The stationary distribution is the solution of equations (2.17); in this case $\pi(i) = \frac{1}{4}, i = 1, 2, 3, 4$. (In this example, it can be shown that the limiting probabilities do not depend on α: $\mathbb{P}_{ii}^n = \frac{1}{4} + \frac{3}{4}(1 - 4\alpha)^n \to \frac{1}{4}$ as $n \to \infty$.)   ◇

Example 2.8 (Random walk)

An example of a discrete-time Markov chain with an infinite state space is the random walk. The state space is the set of all integers, and the transition probabilities are

$$\begin{aligned} p_{i,i+1} &= p, & i &= 0, \pm 1, \pm 2, \ldots, \\ p_{i,i-1} &= 1 - p, & i &= 0, \pm 1, \pm 2, \ldots, \\ p_{i,j} &= 0, & j &\notin \{i - 1, i + 1\}. \end{aligned}$$

In the random walk model, at each transition a step of unit length is taken at random to the right with probability p or left with probability $1 - p$. The state of the process at time n is the current location of the walker at time n. Another interpretation considers the gambler who bets \$1 on a sequence of Bernoulli(p) trials and wins or loses \$1 at each transition; if $X_0 = 0$, the state of the process at time n is his gain or loss after n trials.

In the random walk model all states communicate, so the chain is irreducible. All states have period 2. For example, it is impossible to return to state 0 starting from 0 in an odd number of steps. The probability that the first return to 0 from state 0 occurs in exactly $2n$ steps is

$$p_{00}^{(2n)} = \binom{2n}{n} p^n (1 - p)^n = \frac{(2n)!}{n!n!} (p(1 - p))^n.$$

It can be shown that $\sum_{n=1}^{\infty} p_{00}^{(2n)} < \infty$ if and only if $p \neq 1/2$. Thus, the expected number of visits to 0 is finite if and only if $p \neq 1/2$. Recurrence and transience are class properties, hence the chain is recurrent if and only if $p = 1/2$ and otherwise all states are transient. When $p = 1/2$ the process is called a symmetric random walk. The symmetric random walk is discussed in Example 3.26. ◇

Chapter 3

Methods for Generating Random Variables

3.1 Introduction

One of the fundamental tools required in computational statistics is the ability to simulate random variables from specified probability distributions. On this topic many excellent references are available. On the general subject of methods for generating random variates from specified probability distributions, readers are referred to [69, 94, 112, 114, 154, 228, 223, 233, 238]. On specific topics, also see [3, 4, 31, 43, 68, 98, 155, 159, 190].

In the simplest case, to simulate drawing an observation at random from a finite population, a method of generating random observations from the discrete uniform distribution is required. Therefore a suitable generator of uniform pseudo random numbers is essential. Methods for generating random variates from other probability distributions all depend on the uniform random number generator.

In this text we assume that a suitable uniform pseudo random number generator is available. Refer to the help topic for .Random.seed or RNGkind for details about the default random number generator in R. For reference about different types of random number generators and their properties see Gentle [112] and Knuth [164].

The uniform pseudo random number generator in R is runif. To generate a vector of n (pseudo) random numbers between 0 and 1 use runif(n). Throughout this text, whenever computer generated random numbers are mentioned, it is understood that these are pseudo random numbers. To generate n random Uniform(a, b) numbers use runif(n, a, b). To generate an n by m matrix of random numbers between 0 and 1 use matrix(runif(n*m), nrow=n, ncol=m) or matrix(runif(n*m), n, m).

In the examples of this chapter, several functions are given for generating random variates from continuous and discrete probability distributions. Generators for many of these distributions are available in R (e.g. rbeta, rgeom, rchisq, etc.), but the methods presented below are general and apply to many other types of distributions. These methods are also applicable for external libraries, stand alone programs, or nonstandard simulation problems.

Most of the examples include a comparison of the generated sample with the theoretical distribution of the sampled population. In some examples, histograms, density curves, or QQ plots are constructed. In other examples summary statistics such as sample moments, sample percentiles, or the empirical distribution are compared with the corresponding theoretical values. These are informal approaches to check the implementation of an algorithm for simulating a random variable.

Example 3.1 (Sampling from a finite population)

The sample function can be used to sample from a finite population, with or without replacement.

```
> #toss some coins
> sample(0:1, size = 10, replace = TRUE)
[1] 0 1 1 1 0 1 1 1 1 0

> #choose some lottery numbers
> sample(1:100, size = 6, replace = FALSE)
[1] 51 89 26 99 74 73

> #permuation of letters a-z
> sample(letters)
[1] "d" "n" "k" "x" "s" "p" "j" "t" "e" "b" "g"
    "a" "m" "y" "i" "v" "l" "r" "w" "q" "z"
[22] "u" "h" "c" "f" "o"

> #sample from a multinomial distribution
> x <- sample(1:3, size = 100, replace = TRUE,
              prob = c(.2, .3, .5))
> table(x)
x
 1  2  3
17 35 48
```

◇

Random Generators of Common Probability Distributions in R

In the sections that follow, various methods of generating random variates from specified probability distributions are presented. Before discussing those methods, however, it is useful to summarize some of the probability functions available in R. The probability mass function (pmf) or density (pdf), cumulative distribution function (cdf), quantile function, and random generator of many commonly used probability distributions are available. For example, four functions are documented in the help topic `Binomial`:

```
dbinom(x, size, prob, log = FALSE)
pbinom(q, size, prob, lower.tail = TRUE, log.p = FALSE)
qbinom(p, size, prob, lower.tail = TRUE, log.p = FALSE)
rbinom(n, size, prob)
```

The same pattern is applied to other probability distributions. In each case, the abbreviation for the name of the distribution is combined with first letter d for density or pmf, p for cdf, q for quantile, or r for random generation from the distribution.

A partial list of available probability distributions and parameters is given in Table 3.1. For a complete list, refer to the R documentation [279, Ch. 8]. In addition to the parameters listed, some of the functions take optional log, lower.tail, or log.p arguments, and some take an optional ncp (noncentrality) parameter.

TABLE 3.1: Selected Univariate Probability Functions Available in R

Distribution	cdf	Generator	Parameters
beta	pbeta	rbeta	shape1, shape2
binomial	pbinom	rbinom	size, prob
chi-squared	pchisq	rchisq	df
exponential	pexp	rexp	rate
F	pf	rf	df1, df2
gamma	pgamma	rgamma	shape, rate or scale
geometric	pgeom	rgeom	prob
lognormal	plnorm	rlnorm	meanlog, sdlog
negative binomial	pnbinom	rnbinom	size, prob
normal	pnorm	rnorm	mean, sd
Poisson	ppois	rpois	lambda
Student's t	pt	rt	df
uniform	punif	runif	min, max

3.2 The Inverse Transform Method

The inverse transform method of generating random variables is based on the following well known result (see e.g. [16, p. 201] or [231, p. 203]).

THEOREM 3.1 (Probability Integral Transformation) *If X is a continuous random variable with cdf $F_X(x)$, then $U = F_X(X) \sim Uniform(0, 1)$.*

The inverse transform method of generating random variables applies the probability integral transformation. Define the inverse transformation

$$F_X^{-1}(u) = \inf\{x : F_X(x) = u\}, \qquad 0 < u < 1.$$

If $U \sim \text{Uniform}(0, 1)$, then for all $x \in \mathbb{R}$

$$P(F_X^{-1}(U) \le x) = P(\inf\{t : F_X(t) = U\} \le x)$$
$$= P(U \le F_X(x))$$
$$= F_U(F_X(x)) = F_X(x),$$

and therefore $F_X^{-1}(U)$ has the same distribution as X. Thus, to generate a random observation X, first generate a Uniform(0,1) variate u and deliver the inverse value $F_X^{-1}(u)$. The method is easy to apply, provided that F_X^{-1} is easy to compute. The method can be applied for generating continuous or discrete random variables. The method can be summarized as follows.

1. Derive the inverse function $F_X^{-1}(u)$.

2. Write a command or function to compute $F_X^{-1}(u)$.

3. For each random variate required:

   (a) Generate a random u from Uniform(0,1).
   (b) Deliver $x = F_X^{-1}(u)$

3.2.1   Inverse Transform Method, Continuous Case

Example 3.2 (Inverse transform method, continuous case)

This example uses the inverse transform method to simulate a random sample from the distribution with density $f_X(x) = 3x^2$, $0 < x < 1$.

Here $F_X(x) = x^3$ for $0 < x < 1$, and $F_X^{-1}(u) = u^{1/3}$. Generate all n required random uniform numbers as vector u. Then u^(1/3) is a vector of length n containing the sample $x_1, \ldots, x_n$.

```
n <- 1000
u <- runif(n)
x <- u^(1/3)
hist(x, prob = TRUE) #density histogram of sample
y <- seq(0, 1, .01)
lines(y, 3*y^2)     #density curve f(x)
```

The histogram and density plot in Figure 3.1 suggests that the empirical and theoretical distributions approximately agree.   ◇

R note 3.1 *In Figure 3.1, the title includes a math expression. This title is obtained by specifying the main title using the **expression** function as follows:*

```
hist(x, prob = TRUE, main = expression(f(x)==3*x^2))
```

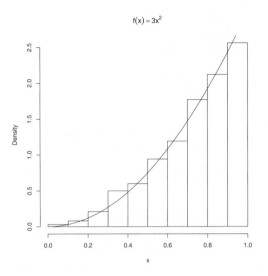

FIGURE 3.1: Probability density histogram of a random sample generated by the inverse transform method in Example 3.2, with the theoretical density $f(x) = 3x^2$ superimposed.

Alternately, `main = bquote(f(x)==3*x^2)` *produces the same title. Math annotation is covered in the help topic for* ***plotmath.*** *Also see the help topics for* ***text*** *and* ***axis.***

Example 3.3 (Exponential distribution)

This example applies the inverse transform method to generate a random sample from the exponential distribution with mean $1/\lambda$.

If $X \sim \text{Exp}(\lambda)$, then for $x > 0$ the cdf of X is $F_X(x) = 1 - e^{-\lambda x}$. The inverse transformation is $F_X^{-1}(u) = -\frac{1}{\lambda}\log(1 - u)$. Note that U and $1 - U$ have the same distribution and it is simpler to set $x = -\frac{1}{\lambda}\log(u)$. To generate a random sample of size n with parameter `lambda`:

```
-log(runif(n)) / lambda
```

A generator `rexp` is available in R. However, this algorithm is very useful for implementation in other situations, such as a C program. ◇

3.2.2 Inverse Transform Method, Discrete Case

The inverse transform method can also be applied to discrete distributions. If X is a discrete random variable and

$$\ldots < x_{i-1} < x_i < x_{i+1} < \ldots$$

are the points of discontinuity of $F_X(x)$, then the inverse transformation is $F_X^{-1}(u) = x_i$, where $F_X(x_{i-1}) < u \le F_X(x_i)$.

For each random variate required:

1. Generate a random u from Uniform(0,1).

2. Deliver x_i where $F(x_{i-1}) < u \le F(x_i)$.

The solution of $F(x_{i-1}) < u \le F(x_i)$ in Step (2) may be difficult for some distributions. See Devroye [69, Ch. III] for several different methods of implementing the inverse transform method in the discrete case.

Example 3.4 (Two point distribution)

This example applies the inverse transform to generate a random sample of Bernoulli($p = 0.4$) variates. Although there are simpler methods to generate a two point distribution in R, this example illustrates computing the inverse cdf of a discrete random variable in the simplest case.

In this example, $F_X(0) = f_X(0) = 1 - p$ and $F_X(1) = 1$. Thus, $F_X^{-1}(u) = 1$ if $u > 0.6$ and $F_X^{-1}(u) = 0$ if $u \le 0.6$. The generator should therefore deliver the numerical value of the logical expression $u > 0.6$.

```
n <- 1000
p <- 0.4
u <- runif(n)
x <- as.integer(u > 0.6)     #(u > 0.6) is a logical vector

> mean(x)
[1] 0.41
> var(x)
[1] 0.2421421
```

Compare the sample statistics with the theoretical moments. The sample mean of a generated sample should be approximately $p = 0.4$ and the sample variance should be approximately $p(1 - p) = 0.24$. Our sample statistics are $\bar{x} = 0.41$ ($se = \sqrt{0.24/1000} \doteq 0.0155$) and $s^2 \doteq 0.242$.   ◇

R note 3.2 *In R one can use the* `rbinom` *(random binomial) function with* `size=1` *to generate a Bernoulli sample. Another method is to sample from the vector (0,1) with probabilities $(1 - p, p)$.*

```
rbinom(n, size = 1, prob = p)
sample(c(0,1), size = n, replace = TRUE, prob = c(.6,.4))
```

Also see Example 3.1.

Example 3.5 (Geometric distribution)

Use the inverse transform method to generate a random geometric sample with parameter $p = 1/4$.

The pmf is $f(x) = pq^x$, $x = 0, 1, 2, \ldots$, where $q = 1 - p$. At the points of discontinuity $x = 0, 1, 2, \ldots$, the cdf is $F(x) = 1 - q^{x+1}$. For each sample element we need to generate a random uniform u and solve

$$1 - q^x < u \leq 1 - q^{x+1}.$$

This inequality simplifies to $x < \log(1 - u)/\log(q) \leq x + 1$. The solution is $x + 1 = \lceil \log(1 - u)/\log(q) \rceil$, where $\lceil t \rceil$ denotes the ceiling function (the smallest integer not less than t).

```
n <- 1000
p <- 0.25
u <- runif(n)
k <- ceiling(log(1-u) / log(1-p)) - 1
```

Here again there is a simplification, because U and $1 - U$ have the same distribution. Also, the probability that $\log(1-u)/\log(1-p)$ equals an integer is zero. The last step can therefore be simplified to

```
k <- floor(log(u) / log(1-p))
```

$\diamond$

The geometric distribution was particularly easy to simulate by the inverse transform method because it was easy to solve the inequality

$$F(x - 1) < u \leq F(x)$$

rather than compare each u to all the possible values $F(x)$. The same method applied to the Poisson distribution is more complicated because we do not have an explicit formula for the value of x such that $F(x - 1) < u \leq F(x)$.

The R function `rpois` generates random Poisson samples. The basic method to generate a Poisson(λ) variate (see e.g. [233]) is to generate and store the cdf via the recursive formula

$$f(x + 1) = \frac{\lambda f(x)}{x + 1}; \qquad F(x + 1) = F(x) + f(x + 1).$$

For each Poisson variate required, a random uniform u is generated, and the cdf vector is searched for the solution to $F(x - 1) < u \leq F(x)$.

To illustrate the main idea of the inverse transform method for generating Poisson variates, here is a similar example for which there is no R generator available: the logarithmic distribution. The logarithmic distribution is a one parameter discrete distribution supported on the positive integers.

Example 3.6 (Logarithmic distribution)

This example implements a function to simulate a Logarithmic(θ) random sample by the inverse transform method. A random variable X has the logarithmic distribution (see [158], Ch. 7) if

$$f(x) = P(X = x) = \frac{a\,\theta^x}{x}, \qquad x = 1, 2, \ldots \tag{3.1}$$

where $0 < \theta < 1$ and $a = (-\log(1 - \theta))^{-1}$. A recursive formula for $f(x)$ is

$$f(x+1) = \frac{\theta^x}{x+1} f(x), \qquad x = 1, 2, \ldots. \tag{3.2}$$

Theoretically, the pmf can be evaluated recursively using (3.2), but the calculation is not sufficiently accurate for large values of x and ultimately produces $f(x) = 0$ with $F(x) < 1$. Instead we compute the pmf from (3.1) as $\exp(\log a + x \log \theta - \log x)$. In generating a large sample, there will be many repetitive calculations of the same values $F(x)$. It is more efficient to store the cdf values. Initially choose a length N for the cdf vector, and compute $F(x)$, $x = 1, 2, \ldots, N$. If necessary, N will be increased.

To solve $F(x-1) < u \le F(x)$ for a particular u, it is necessary to count the number of values x such that $F(x-1) < u$. If F is a vector and u_i is a scalar, then the expression $F < u_i$ produces a logical vector; that is, a vector the same length as F containing logical values TRUE or FALSE. In an arithmetic expression, TRUE has value 1 and FALSE has value 0. Notice that the sum of the logical vector $(u_i > F)$ is exactly $x - 1$.

The code for `logarithmic` is on the next page. Generate random samples from a Logarithmic(0.5) distribution.

```
n <- 1000
theta <- 0.5
x <- rlogarithmic(n, theta)
#compute density of logarithmic(theta) for comparison
k <- sort(unique(x))
p <- -1 / log(1 - theta) * theta^k / k
se <- sqrt(p*(1-p)/n)    #standard error
```

In the following results, the relative frequencies of the sample (first line) match the theoretical distribution (second line) of the Logarithmic(0.5) distribution within two standard errors.

```
> round(rbind(table(x)/n, p, se),3)
       1     2     3     4     5     6     7
   0.741 0.169 0.049 0.026 0.008 0.003 0.004
p  0.721 0.180 0.060 0.023 0.009 0.004 0.002
se 0.014 0.012 0.008 0.005 0.003 0.002 0.001
```

◇

```
rlogarithmic <- function(n, theta) {
    #returns a random logarithmic(theta) sample size n
    u <- runif(n)
    #set the initial length of cdf vector
    N <- ceiling(-16 / log10(theta))
    k <- 1:N
    a <- -1/log(1-theta)
    fk <- exp(log(a) + k * log(theta) - log(k))
    Fk <- cumsum(fk)
    x <- integer(n)
    for (i in 1:n) {
        x[i] <- as.integer(sum(u[i] > Fk))  #F^{-1}(u)-1
        while (x[i] == N) {
            #if x==N we need to extend the cdf
            #very unlikely because N is large
            logf <- log(a) + (N+1)*log(theta) - log(N+1)
            fk <- c(fk, exp(logf))
            Fk <- c(Fk, Fk[N] + fk[N+1])
            N <- N + 1
            x[i] <- as.integer(sum(u[i] > Fk))
        }
    }
    x + 1
}
```

Remark 3.1 *A more efficient generator for the Logarithmic(θ) distribution is implemented in Example 3.9 of Section 3.4.*

3.3   The Acceptance-Rejection Method

Suppose that X and Y are random variables with density or pmf f and g respectively, and there exists a constant c such that

$$\frac{f(t)}{g(t)} \leq c$$

for all t such that $f(t) > 0$. Then the acceptance-rejection method (or rejection method) can be applied to generate the random variable X.

The Acceptance-Rejection Method

1. Find a random variable Y with density g satisfying $f(t)/g(t) \leq c$, for all t such that $f(t) > 0$. Provide a method to generate random Y.

2. For each random variate required:

    (a) Generate a random y from the distribution with density g.

    (b) Generate a random u from the Uniform$(0, 1)$ distribution.

    (c) If $u < f(y)/(cg(y))$ accept y and deliver $x = y$; otherwise reject y and repeat from step (2a).

Note that in step (2c),

$$P(\text{accept}|Y) = P\left(U < \frac{f(Y)}{cg(Y)} \,\middle|\, Y\right) = \frac{f(Y)}{cg(Y)}.$$

The last equality is simply evaluating the cdf of U. The total probability of acceptance for any iteration is therefore

$$\sum_y P(\text{accept}|y)P(Y = y) = \sum_y \frac{f(y)}{cg(y)}g(y) = \frac{1}{c},$$

and the number of iterations until acceptance has the geometric distribution with mean c. Hence, on average each sample value of X requires c iterations. For efficiency, Y should be easy to simulate and c small.

To see that the accepted sample has the same distribution as X, apply Bayes' Theorem. In the discrete case, for each k such that $f(k) > 0$,

$$P(k\,|\text{accepted}) = \frac{P(\text{accepted }|k)g(k)}{P(\text{accepted})} = \frac{[f(k)/(cg(k))]\,g(k)}{1/c} = f(k).$$

The continuous case is similar.

Example 3.7 (Acceptance-rejection method)

This example illustrates the acceptance-rejection method for the beta distribution. On average, how many random numbers must be simulated to generate 1000 variates from the Beta($\alpha = 2$, $\beta = 2$) distribution by this method? It depends on the upper bound c of $f(x)/g(x)$, which depends on the choice of the function $g(x)$.

The Beta(2,2) density is $f(x) = 6x(1 - x)$, $0 < x < 1$. Let $g(x)$ be the Uniform(0,1) density. Then $f(x)/g(x) \leq 6$ for all $0 < x < 1$, so $c = 6$. A random x from $g(x)$ is accepted if

$$\frac{f(x)}{cg(x)} = \frac{6x(1 - x)}{6(1)} = x(1 - x) > u.$$

On average, $cn = 6000$ iterations (12000 random numbers) will be required for a sample size 1000. In the following simulation, the counter j for iterations is not necessary, but included to record how many iterations were actually needed to generate the 1000 beta variates.

```
n <- 1000
k <- 0        #counter for accepted
j <- 0        #iterations
y <- numeric(n)

while (k < n) {
    u <- runif(1)
    j <- j + 1
    x <- runif(1)  #random variate from g
    if (x * (1-x) > u) {
        #we accept x
        k <- k + 1
        y[k] <- x
    }
}


> j
[1] 5873
```

In this simulation, 5873 iterations (11746 random numbers) were required to generate the 1000 beta variates. Compare the empirical and theoretical percentiles.

```
#compare empirical and theoretical percentiles
p <- seq(.1, .9, .1)
Qhat <- quantile(y, p)     #quantiles of sample
Q <- qbeta(p, 2, 2)        #theoretical quantiles
se <- sqrt(p * (1-p) / (n * dbeta(Q, 2, 2))) #see Ch. 1
```

The sample percentiles (first line) approximately match the Beta(2,2) percentiles computed by qbeta (second line), most closely near the center of the distribution. Larger numbers of replicates are required for estimation of percentiles where the density is close to zero.

```
> round(rbind(Qhat, Q, se), 3)
        10%    20%    30%    40%    50%    60%    70%    80%    90%
Qhat 0.189 0.293 0.365 0.449 0.519 0.589 0.665 0.741 0.830
Q     0.196 0.287 0.363 0.433 0.500 0.567 0.637 0.713 0.804
se    0.010 0.011 0.012 0.013 0.013 0.013 0.012 0.011 0.010
```

Repeating the simulation with $n = 10000$ produces more precise estimates.

```
>  round(rbind(Qhat, Q, se), 3)
        10%    20%    30%    40%    50%    60%    70%    80%    90%
Qhat 0.194 0.292 0.368 0.436 0.504 0.572 0.643 0.716 0.804
Q     0.196 0.287 0.363 0.433 0.500 0.567 0.637 0.713 0.804
se    0.003 0.004 0.004 0.004 0.004 0.004 0.004 0.004 0.003
```

◇

Remark 3.2 *See Example 3.8 for a more efficient beta generator based on the ratio of gammas method.*

3.4   Transformation Methods

Many types of transformations other than the probability inverse transformation can be applied to simulate random variables. Some examples are

1. If $Z \sim N(0,1)$, then $V = Z^2 \sim \chi^2(1)$.

2. If $U \sim \chi^2(m)$ and $V \sim \chi^2(n)$ are independent, then $F = \frac{U/m}{V/n}$ has the F distribution with (m, n) degrees of freedom.

3. If $Z \sim N(0,1)$ and $V \sim \chi^2(n)$ are independent, then $T = \frac{Z}{\sqrt{V/n}}$ has the Student t distribution with n degrees of freedom.

4. If $U, V \sim \text{Unif}(0,1)$ are independent, then

$$Z_1 = \sqrt{-2 \log U} \, \cos(2\pi V),$$
$$Z_2 = \sqrt{-2 \log V} \, \sin(2\pi U)$$

   are independent standard normal variables (see e.g. [238, p. 86]).

5. If $U \sim \text{Gamma}(r, \lambda)$ and $V \sim \text{Gamma}(s, \lambda)$ are independent, then $X = \frac{U}{U+V}$ has the Beta(r, s) distribution.

6. If $U, V \sim \text{Unif}(0,1)$ are independent, then

$$X = \left\lfloor 1 + \frac{\log(V)}{\log(1 - (1 - \theta)^U)} \right\rfloor$$

   has the Logarithmic(θ) distribution, where $\lfloor x \rfloor$ denotes the integer part of x.

Generators based on transformations (5) and (6) are implemented in Examples 3.8 and 3.9. Sums and mixtures are special types of transformations that are discussed in Section 3.5. Example 3.21 uses a multivariate transformation to generate points uniformly distributed on the unit sphere.

Example 3.8 (Beta distribution)

The following relation between beta and gamma distributions provides another beta generator.

If $U \sim \text{Gamma}(r, \lambda)$ and $V \sim \text{Gamma}(s, \lambda)$ are independent, then

$$X = \frac{U}{U + V}$$

has the Beta(r, s) distribution [238, p.64]. This transformation determines an algorithm for generating random Beta(a, b) variates.

1. Generate a random u from Gamma$(a, 1)$.

2. Generate a random v from Gamma$(b, 1)$.

3. Deliver $x = \frac{u}{u+v}$.

This method is applied below to generate a random Beta$(3, 2)$ sample.

```
n <- 1000
a <- 3
b <- 2
u <- rgamma(n, shape=a, rate=1)
v <- rgamma(n, shape=b, rate=1)
x <- u / (u + v)
```

The sample data can be compared with the Beta$(3, 2)$ distribution using a quantile-quantile (QQ) plot. If the sampled distribution is Beta$(3, 2)$, the QQ plot should be nearly linear.

```
q <- qbeta(ppoints(n), a, b)
qqplot(q, x, cex=0.25, xlab="Beta(3, 2)", ylab="Sample")
abline(0, 1)
```

The line $x = q$ is added for reference. The QQ plot of the ordered sample vs the Beta$(3, 2)$ quantiles in Figure 3.2 is very nearly linear, as it should be if the generated sample is in fact a Beta$(3, 2)$ sample.   ◇

Example 3.9 (Logarithmic distribution, version 2)

This example provides another, more efficient generator for the logarithmic distribution (see Example 3.6). If U, V are independent Uniform(0,1) random variables, then

$$X = \left\lfloor 1 + \frac{\log(V)}{\log(1 - (1 - \theta)^U)} \right\rfloor \tag{3.3}$$

has the Logarithmic(θ) distribution ([69, pp. 546-8], [159]). This transformation provides a simple and efficient generator for the logarithmic distribution.

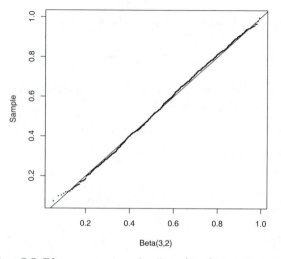

FIGURE 3.2:   QQ Plot comparing the Beta(3, 2) distribution with a simulated random sample generated by the ratio of gammas method in Example
3.8.

1. Generate u from Unif(0,1).
2. Generate v from Unif(0,1).
3. Deliver $x = \lfloor 1 + \log(v)/\log(1 - (1 - \theta)^u) \rfloor$.

Below is a comparison of the Logarithmic(0.5) distribution with a sample
generated using transformation (3.3). The empirical probabilities `p.hat` are
within two standard errors of the theoretical probabilities `p`.

```
n <- 1000
theta <- 0.5
u <- runif(n)    #generate logarithmic sample
v <- runif(n)
x <- floor(1 + log(v) / log(1 - (1 - theta)^u))
k <- 1:max(x)    #calc. logarithmic probs.
p <- -1 / log(1 - theta) * theta^k / k
se <- sqrt(p*(1-p)/n)
p.hat <- tabulate(x)/n

> print(round(rbind(p.hat, p, se), 3))
       [,1]  [,2]  [,3]  [,4]  [,5]  [,6]  [,7]
p.hat 0.740 0.171 0.052 0.018 0.010 0.006 0.003
p     0.721 0.180 0.060 0.023 0.009 0.004 0.002
se    0.014 0.012 0.008 0.005 0.003 0.002 0.001
```

The following function is a simple replacement for `rlogarithmic` in Example 3.6 on page 54.

```
rlogarithmic <- function(n, theta) {
    stopifnot(all(theta > 0 & theta < 1))
    th <- rep(theta, length=n)
    u <- runif(n)
    v <- runif(n)
    x <- floor(1 + log(v) / log(1 - (1 - th)^u))
    return(x)
}
```

◇

R note 3.3 *The & operator performs an elementwise AND comparison. The && operator evaluates from left to right until a logical result is obtained. For example*

```
x <- 1:5
> 1 < x & x < 5
[1] FALSE  TRUE   TRUE   TRUE FALSE
> 1 < x && x < 5
[1] FALSE
> any( 1 < x & x < 5 )
[1] TRUE
> any( 1 < x && x < 5 )
[1] FALSE
> any(1 < x) && any(x < 5)
[1] TRUE
> all(1 < x) && all(x < 5)
[1] FALSE
```

Similarly, | performs elementwise an OR comparison and || evaluates from left to right.

R note 3.4 *The* `tabulate` *function bins positive integers, so it can be used on the logarithmic sample. For other types of data, recode the data to positive integers or use* `table`. *If the data are not positive integers,* `tabulate` *will truncate real numbers and ignore without warning integers less than 1.*

3.5 Sums and Mixtures

Sums and mixtures of random variables are special types of transformations. In this section we focus on sums of independent random variables (convolutions) and several examples of discrete and continuous mixtures.

Convolutions

Let $X_1, \ldots, X_n$ be independent and identically distributed with distribution $X_j \sim X$, and let $S = X_1 + \cdots + X_n$. The distribution function of the sum S is called the n-fold convolution of X and denoted $F_X^{*(n)}$. It is straightforward to simulate a convolution by directly generating $X_1, \ldots, X_n$ and computing the sum.

Several distributions are related by convolution. If $\nu > 0$ is an integer, the chisquare distribution with ν degrees of freedom is the convolution of ν iid squared standard normal variables. The negative binomial distribution NegBin(r, p) is the convolution of r iid Geom(p) random variables. The convolution of r independent Exp(λ) random variables has the Gamma(r, λ) distribution. See Bean [23] for an introductory level presentation of these and many other interesting relationships between families of distributions.

In R it is of course easier to use the functions `rchisq`, `rgeom` and `rnbinom` to generate chisquare, geometric and negative binomial random samples. The following example is presented to illustrate a general method that can be applied whenever distributions are related by convolutions.

Example 3.10 (Chisquare)

This example generates a chisquare $\chi^2(\nu)$ random variable as the convolution of ν squared normals. If $Z_1, \ldots, Z_\nu$ are iid N(0,1) random variables, then $V = Z_1^2 + \cdots + Z_\nu^2$ has the $\chi^2(\nu)$ distribution. Steps to generate a random sample of size n from $\chi^2(\nu)$ are as follows.

1. Fill an $n \times \nu$ matrix with $n\nu$ random N(0,1) variates.

2. Square each entry in the matrix (1).

3. Compute the row sums of the squared normals. Each row sum is one random observation from the $\chi^2(\nu)$ distribution.

4. Deliver the vector of row sums.

An example with $n = 1000$ and $\nu = 2$ is shown below.

```
n <- 1000
nu <- 2
X <- matrix(rnorm(n*nu), n, nu)^2 #matrix of sq. normals
#sum the squared normals across each row: method 1
y <- rowSums(X)
#method 2
y <- apply(X, MARGIN=1, FUN=sum)   #a vector length n
> mean(y)
[1] 2.027334
> mean(y^2)
[1] 7.835872
```

A $\chi^2(\nu)$ random variable has mean ν and variance 2ν. Our sample statistics below agree very closely with the theoretical moments $E[Y] = \nu = 2$ and $E[Y^2] = 2\nu + \nu^2 = 8$. Here the standard errors of the sample moments are 0.063 and 0.089 respectively. $\diamond$

R note 3.5 *This example introduces the* `apply` *function. The* `apply` *function applies a function to the margins of an array. To sum across the rows of matrix* X, *the function* `(FUN=sum)` *is applied to the rows* `(MARGIN=1)`. *Notice that a loop is not used to compute the row sums. In general for efficient programming in R, avoid unnecessary loops. (For row and column sums it is easier to use* `rowSums` *and* `colSums`.)

Mixtures

A random variable X is a discrete mixture if the distribution of X is a weighted sum $F_X(x) = \sum \theta_i F_{X_i}(x)$ for some sequence of random variables $X_1, X_2, \ldots$ and $\theta_i > 0$ such that $\sum_i \theta_i = 1$. The constants θ_i are called the mixing weights or mixing probabilities. Although the notation is similar for sums and mixtures, the distributions represented are different.

A random variable X is a continuous mixture if the distribution of X is $F_X(x) = \int_{-\infty}^{\infty} F_{X|Y=y}(x) f_Y(y)\, dy$ for a family $X|Y = y$ indexed by the real numbers y and weighting function f_Y such that $\int_{-\infty}^{\infty} f_Y(y)\, dy = 1$.

Compare the methods for simulation of a convolution and a mixture of normal variables. Suppose $X_1 \sim N(0,1)$ and $X_2 \sim N(3,1)$ are independent. The notation $S = X_1 + X_2$ denotes the *convolution* of X_1 and X_2. The distribution of S is normal with mean $\mu_1 + \mu_2 = 3$ and variance $\sigma_1^2 + \sigma_2^2 = 2$.

To simulate the *convolution*:

1. Generate $x_1 \sim N(0, 1)$.

2. Generate $x_2 \sim N(3, 1)$.

3. Deliver $s = x_1 + x_2$.

We can also define a 50% normal *mixture* X, denoted $F_X(x) = 0.5F_{X_1}(x) + 0.5F_{X_2}(x)$. Unlike the convolution above, the distribution of the mixture X is distinctly non-normal; it is bimodal.

To simulate the *mixture*:

1. Generate an integer $k \in \{1, 2\}$, where $P(1) = P(2) = 0.5$.

2. If $k = 1$ deliver random x from $N(0, 1)$;
   if $k = 2$ deliver random x from $N(3, 1)$.

In the following example we will compare simulated distributions of a convolution and a mixture of gamma random variables.

Example 3.11 (Convolutions and mixtures)

Let $X_1 \sim \text{Gamma}(2, 2)$ and $X_2 \sim \text{Gamma}(2, 4)$ be independent. Compare the histograms of the samples generated by the convolution $S = X_1 + X_2$ and the mixture $F_X = 0.5F_{X_1} + 0.5F_{X_2}$.

```
n <- 1000
x1 <- rgamma(n, 2, 2)
x2 <- rgamma(n, 2, 4)
s <- x1 + x2                 #the convolution
u <- runif(n)
k <- as.integer(u > 0.5)     #vector of 0's and 1's
x <- k * x1 + (1-k) * x2     #the mixture

par(mfcol=c(1,2))            #two graphs per page
hist(s, prob=TRUE)
hist(x, prob=TRUE)
par(mfcol=c(1,1))            #restore display
```

The histograms shown in Figure 3.3, of the convolution S and mixture X, are clearly different.   ◇

R note 3.6 *The* par *function can be used to set (or query) certain graphical parameters. A list of all graphical parameters is returned by* par()*. The command* par(mfcol=c(n,m)) *configures the graphical device to display nm graphs per screen, in n rows and m columns.*

The method of generating the mixture in this example is simple for a mixture of two distributions, but not for arbitrary mixtures. The next example illustrates how to generate a mixture of several distributions with arbitrary mixing probabilities.

Example 3.12 (Mixture of several gamma distributions)

This example is similar to the previous one, but there are several components to the mixture and the mixing weights are not uniform. The mixture is

$$F_X = \sum_{i=1}^{5} \theta_j F_{X_j},$$

where $X_j \sim \text{Gamma}(r = 3, \lambda_j = 1/j)$ are independent and the mixing probabilities are $\theta_j = j/15$, $j = 1, \ldots, 5$.

To simulate one random variate from the mixture F_X:

1. Generate an integer $k \in \{1, 2, 3, 4, 5\}$, where $P(k) = \theta_k$, $k = 1, \ldots, 5$.
2. Deliver a random $\text{Gamma}(r, \lambda_k)$ variate.

Histogram of s   Histogram of x

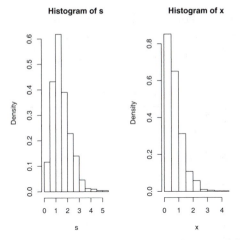

FIGURE 3.3:   Histogram of a simulated convolution of Gamma(2, 2) and Gamma(2, 4) random variables (left), and a 50% mixture of the same variables (right), from Example 3.11.

To generate a sample size n, steps (1) and (2) are repeated n times. Notice that the algorithm stated above suggests using a `for` loop, but `for` loops are really inefficient in R. The algorithm can be translated into a vectorized approach.

1. Generate a random sample $k_1, \ldots, k_n$ of integers in a vector k, where $P(k) = \theta_k$, $k = 1, \ldots, 5$. Then k[i] indicates which of the five gamma distributions will be sampled to get the ith element of the sample (use `sample`).

2. Set `rate` equal to the length n vector $\lambda = (\lambda_k)$.

3. Generate a gamma sample size n, with shape parameter r and rate vector `rate` (use `rgamma`).

Then an efficient way to implement this in R is shown by the following example.

```
n <- 5000
k <- sample(1:5, size=n, replace=TRUE, prob=(1:5)/15)
rate <- 1/k
x <- rgamma(n, shape=3, rate=rate)

#plot the density of the mixture
#with the densities of the components
plot(density(x), xlim=c(0,40), ylim=c(0,.3),
    lwd=3, xlab="x", main="")
for (i in 1:5)
    lines(density(rgamma(n, 3, 1/i)))
```

The plot in Figure 3.4 shows the density of each X_j and the density of the mixture (thick line). The density curves in Figure 3.4 are actually density estimates, which will be discussed in Chapter 10.   ◇

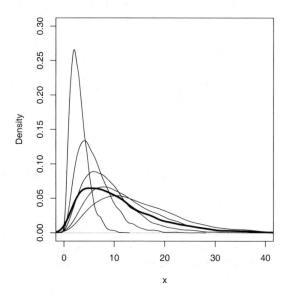

FIGURE 3.4:   Density estimates from Example 3.12: A mixture (thick line) of several gamma densities (thin lines).

Example 3.13 (Mixture of several gamma distributions)

Let

$$F_X = \sum_{j=1}^{5} \theta_j F_{X_j}$$

where $X_j \sim \text{Gamma}(3, \lambda_j)$ are independent, with rates $\lambda = (1, 1.5, 2, 2.5, 3)$, and mixing probabilities $\theta = (0.1, 0.2, 0.2, 0.3, 0.2)$.

  This example is similar to the previous one. Sample from 1:5 with probability weights θ to get a vector length n. The ith position in this vector indicates which of the five gamma distributions is sampled to get the ith element of the sample. This vector is used to select the correct rate parameter from the vector λ.

```
n <- 5000
p <- c(.1,.2,.2,.3,.2)
```

```
lambda <- c(1,1.5,2,2.5,3)
k <- sample(1:5, size=n, replace=TRUE, prob=p)
rate <- lambda[k]
x <- rgamma(n, shape=3, rate=rate)
```

Note that `lambda[k]` is a vector the same length as `k`, containing the elements of `lambda` indexed by the vector `k`. In mathematical notation, `lambda[k]` is equal to $(\lambda_{k_1}, \lambda_{k_2}, \dots, \lambda_{k_n})$.

Compare the first few entries of `k` and the corresponding values of `rate` with λ.

```
> k[1:8]
[1] 5 1 4 2 1 3 2 3
> rate[1:8]
[1] 3.0 1.0 2.5 1.5 1.0 2.0 1.5 2.0
```

◇

Example 3.14 (Plot density of mixture)

Plot the densities (not density estimates) of the gamma distributions and the mixture in Example 3.13. (This example is a programming exercise that involves vectors of parameters and repeated use of the `apply` function.)

The density of the mixture is

$$f(x) = \sum_{j=1}^{5} \theta_j f_j(x), \quad x > 0, \tag{3.4}$$

where f_j is the Gamma(3, λ_j) density. To produce the plot, we need a function to compute the density $f(x)$ of the mixture.

```
f <- function(x, lambda, theta) {
    #density of the mixture at the point x
    sum(dgamma(x, 3, lambda) * theta)
}
```

The function `f` computes the density of the mixture (3.4) for a single value of `x`. If `x` has length 1, `dgamma(x, 3, lambda)` is a vector the same length as `lambda`; in this case $(f_1(x), \dots, f_5(x))$. Then `dgamma(x, 3, lambda)*theta` is the vector $(\theta_1 f_1(x), \dots, \theta_5 f_5(x))$. The sum of this vector is the density of the mixture (3.3) evaluated at the point `x`.

```
x <- seq(0, 8, length=200)
dim(x) <- length(x)   #need for apply

#compute density of the mixture f(x) along x
y <- apply(x, 1, f, lambda=lambda, theta=p)
```

The density of the mixture is computed by function `f` applied to the vector `x`. The function `f` takes several arguments, so the additional arguments `lambda=lambda, theta=prob` are supplied after the name of the function, `f`.

A plot of the five densities with the mixture is shown in Figure 3.5. The code to produce the plot is listed below. The densities f_k can be computed by the `dgamma` function. A sequence of points `x` is defined and each of the densities are computed along `x`.

```
#plot the density of the mixture
plot(x, y, type="l", ylim=c(0,.85), lwd=3, ylab="Density")

for (j in 1:5) {
    #add the j-th gamma density to the plot
    y <- apply(x, 1, dgamma, shape=3, rate=lambda[j])
    lines(x, y)
}
```

◇

R note 3.7 *The* `apply` *function requires a dimension attribute for* x. *Since* x *is a vector, it does not have a dimension attribute by default. The dimension of* x *is assigned by* `dim(x) <- length(x)`. *Alternately,* x `<- as.matrix(x)` *converts* x *to a matrix (a column vector), which has a dimension attribute.*

Example 3.15 (Poisson-Gamma mixture)

This is an example of a continuous mixture. The negative binomial distribution is a mixture of Poisson(Λ) distributions, where Λ has a gamma distribution. Specifically, if $(X|\Lambda = \lambda) \sim \text{Poisson}(\lambda)$ and $\Lambda \sim \text{Gamma}(r, \beta)$, then X has the negative binomial distribution with parameters r and $p = \beta/(1 + \beta)$ (see e.g. [23]). This example illustrates a method of sampling from a Poisson-Gamma mixture and compares the sample with the negative binomial distribution.

```
#generate a Poisson-Gamma mixture
n <- 1000
r <- 4
beta <- 3
lambda <- rgamma(n, r, beta) #lambda is random

#now supply the sample of lambda's as the Poisson mean
x <- rpois(n, lambda)        #the mixture

#compare with negative binomial
mix <- tabulate(x+1) / n
```

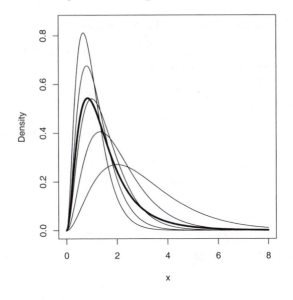

FIGURE 3.5: Densities from Example 3.14: A mixture (thick line) of several gamma densities (thin lines).

```
negbin <- round(dnbinom(0:max(x), r, beta/(1+beta)), 3)
se <- sqrt(negbin * (1 - negbin) / n)
```

The empirical distribution (first line below) of the mixture agrees very closely with the pmf of NegBin$(4, 3/4)$ (second line).

```
> round(rbind(mix, negbin, se), 3)
        [,1]  [,2]  [,3]  [,4]  [,5]  [,6]  [,7]  [,8]  [,9]
mix    0.334 0.305 0.201 0.091 0.042 0.018 0.005 0.003 0.001
negbin 0.316 0.316 0.198 0.099 0.043 0.017 0.006 0.002 0.001
se     0.015 0.015 0.013 0.009 0.006 0.004 0.002 0.001 0.001
```

◇

3.6 Multivariate Distributions

Generators for the multivariate normal distribution, multivariate normal mixtures, Wishart distribution, and uniform distribution on the sphere in $\mathbb{R}^d$ are presented in this section.

3.6.1   Multivariate Normal Distribution

A random vector $X = (X_1, \ldots, X_d)$ has a d-dimensional mutivariate normal (MVN) distribution denoted $N_d(\mu, \Sigma)$ if the density of X is

$$f(x) = \frac{1}{(2\pi)^{d/2} |\Sigma|^{1/2}} \exp\{-(1/2)(x - \mu)^T \Sigma^{-1}(x - \mu)\}, \quad x \in \mathbb{R}^d, \quad (3.5)$$

where $\mu = (\mu_1, \ldots, \mu_d)^T$ is the mean vector and Σ is a $d \times d$ symmetric positive definite matrix

$$\Sigma = \begin{bmatrix} \sigma_{11} & \sigma_{12} & \cdots & \sigma_{1d} \\ \sigma_{21} & \sigma_{22} & \cdots & \sigma_{2d} \\ \vdots & \vdots & & \vdots \\ \sigma_{d1} & \sigma_{d2} & \cdots & \sigma_{dd} \end{bmatrix}$$

with entries $\sigma_{ij} = Cov(X_i, X_j)$. Here Σ^{-1} is the inverse of Σ, and $|\Sigma|$ is the determinant of Σ. The bivariate normal distribution is the special case $N_2(\mu, \Sigma)$.

A random $N_d(\mu, \Sigma)$ variate can be generated in two steps. First generate $Z = (Z_1, \ldots, Z_d)$, where $Z_1, \ldots, Z_d$ are iid standard normal variates. Then transform the random vector Z so that it has the desired mean vector μ and covariance structure Σ. The transformation requires factoring the covariance matrix Σ.

Recall that if $Z \sim N_d(\mu, \Sigma)$, then the linear transformation $CZ + b$ is multivariate normal with mean $C\mu + b$ and covariance $C\Sigma C^T$. If Z is $N_d(0, I_d)$, then

$$CZ + b \sim N_d(b, CC^T).$$

Suppose that Σ can be factored so that $\Sigma = CC^T$ for some matrix C. Then

$$CZ + \mu \sim N_d(\mu, \Sigma),$$

and $CZ + \mu$ is the required transformation.

The required factorization of Σ can be obtained by the spectral decomposition method (eigenvector decomposition), Choleski factorization, or singular value decomposition (svd). The corresponding R functions are `eigen`, `chol`, and `svd`.

Usually, one does not apply a linear transformation to the random vectors of a sample one at a time. Typically, one applies the transformation to a data matrix and transforms the entire sample. Suppose that $Z = (Z_{ij})$ is an $n \times d$ matrix where Z_{ij} are iid N(0,1). Then the rows of Z are n random observations from the d-dimensional standard MVN distribution. The required transformation applied to the data matrix is

$$X = ZQ + J\mu^T, \quad (3.6)$$

where $Q^T Q = \Sigma$ and J is a column vector of ones. The rows of X are n random observations from the d-dimensional MVN distribution with mean vector μ and covariance matrix Σ.

Method for generating multivariate normal samples

To generate a random sample of size n from the $N_d(\mu, \Sigma)$ distribution:

1. Generate an $n \times d$ matrix Z containing nd random $N(0,1)$ variates (n random vectors in $\mathbb{R}^d$).
2. Compute a factorization $\Sigma = Q^T Q$.
3. Apply the transformation $X = ZQ + J\mu^T$.
4. Deliver the $n \times d$ matrix X.
   Each row of X is a random variate from the $N_d(\mu, \Sigma)$ distribution.

The $X = ZQ + J\mu^T$ transformation can be coded in R as follows. Recall that the matrix multiplication operator is `%*%`.

```
Z <- matrix(rnorm(n*d), nrow = n, ncol = d)
X <- Z %*% Q + matrix(mu, n, d, byrow = TRUE)
```

The matrix product $J\mu^T$ is equal to `matrix(mu, n, d, byrow = TRUE)`. This saves a matrix multiplication. The argument `byrow = TRUE` is necessary here; the default is `byrow = FALSE`. The matrix is filled row by row with the entries of the mean vector `mu`.

In this section each method of generating MVN random samples is illustrated with examples. Also note that there are functions provided in R packages for generating multivariate normal samples. See the `mvrnorm` function in the `MASS` package [278], and `rmvnorm` in the `mvtnorm` package [115]. In all of the examples below, the `rnorm` function is used to generate standard normal random variates.

Spectral decomposition method for generating $N_d(\mu, \Sigma)$ samples

The square root of the covariance is $\Sigma^{1/2} = P\Lambda^{1/2}P^{-1}$, where Λ is the diagonal matrix with the eigenvalues of Σ along the diagonal and P is the matrix whose columns are the eigenvectors of Σ corresponding to the eigenvalues in Λ. This method can also be called the eigen-decomposition method. In the eigen-decomposition we have $P^{-1} = P^T$ and therefore $\Sigma^{1/2} = P\Lambda^{1/2}P^T$. The matrix $Q = \Sigma^{1/2}$ is a factorization of Σ such that $Q^T Q = \Sigma$.

Example 3.16 (Spectral decomposition method)

This example provides a function `rmvn.eigen` to generate a multivariate normal random sample. It is applied to generate a bivariate normal sample with zero mean vector and

$$\Sigma = \begin{bmatrix} 1.0 & 0.9 \\ 0.9 & 1.0 \end{bmatrix}.$$

```
# mean and covariance parameters
mu <- c(0, 0)
Sigma <- matrix(c(1, .9, .9, 1), nrow = 2, ncol = 2)
```

The `eigen` function returns the eigenvalues and eigenvectors of a matrix.

```
rmvn.eigen <-
function(n, mu, Sigma) {
    # generate n random vectors from MVN(mu, Sigma)
    # dimension is inferred from mu and Sigma
    d <- length(mu)
    ev <- eigen(Sigma, symmetric = TRUE)
    lambda <- ev$values
    V <- ev$vectors
    R <- V %*% diag(sqrt(lambda)) %*% t(V)
    Z <- matrix(rnorm(n*d), nrow = n, ncol = d)
    X <- Z %*% R + matrix(mu, n, d, byrow = TRUE)
    X
}
```

Print summary statistics and display a scatterplot as a check on the results of the simulation.

```
# generate the sample
X <- rmvn.eigen(1000, mu, Sigma)

plot(X, xlab = "x", ylab = "y", pch = 20)

> print(colMeans(X))
[1] -0.001628189  0.023474775

> print(cor(X))
          [,1]      [,2]
[1,] 1.0000000 0.8931007
[2,] 0.8931007 1.0000000
```

Output from Example 3.16 shows the sample mean vector is $(-0.002, 0.023)$ and sample correlation is 0.893, which agree closely with the specified parameters. The scatter plot of the sample data shown in Figure 3.6 exhibits the elliptical symmetry of multivariate normal distributions.   ◇

SVD Method of generating $N_d(\mu, \Sigma)$ samples

The singular value decomposition (svd) generalizes the idea of eigenvectors to rectangular matrices. The svd of a matrix X is $X = UDV^T$, where D is a vector containing the singular values of X, U is a matrix whose columns contain the left singular vectors of X, and V is a matrix whose columns contain the right singular vectors of X. The matrix X in this case is the population covariance matrix Σ, and $UV^T = I$. The svd of a symmetric positive definite

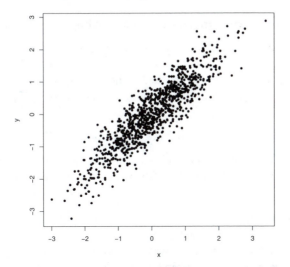

FIGURE 3.6: Scatterplot of a random bivariate normal sample with mean vector zero, variances $\sigma_1^2 = \sigma_2^2 = 1$ and correlation $\rho = 0.9$, from Example 3.16.

matrix Σ gives $U = V = P$ and $\Sigma^{1/2} = UD^{1/2}V^T$. Thus the svd method for this application is equivalent to the spectral decomposition method, but is less efficient because the svd method does not take advantage of the fact that the matrix Σ is square symmetric.

Example 3.17 (SVD method)

This example provides a function rmvn.svd to generate a multivariate normal sample, using the svd method to factor Σ.

```
rmvn.svd <-
function(n, mu, Sigma) {
    # generate n random vectors from MVN(mu, Sigma)
    # dimension is inferred from mu and Sigma
    d <- length(mu)
    S <- svd(Sigma)
    R <- S$u %*% diag(sqrt(S$d)) %*% t(S$v) #sq. root Sigma
    Z <- matrix(rnorm(n*d), nrow=n, ncol=d)
    X <- Z %*% R + matrix(mu, n, d, byrow=TRUE)
    X
}
```

This function is applied in Example 3.19 on page 76.   ⋄

Choleski factorization method of generating $N_d(\mu, \Sigma)$ samples

The Choleski factorization of a real symmetric positive-definite matrix is $X = Q^T Q$, where Q is an upper triangular matrix. The Choleski factorization is implemented in the R function `chol`. The basic syntax is `chol(X)` and the return value is an upper triangular matrix R such that $R^T R = X$.

Example 3.18 (Choleski factorization method)

The Choleski factorization method is applied to generate 200 random observations from a four-dimensional multivariate normal distribution.

```
rmvn.Choleski <-
function(n, mu, Sigma) {
    # generate n random vectors from MVN(mu, Sigma)
    # dimension is inferred from mu and Sigma
    d <- length(mu)
    Q <- chol(Sigma) # Choleski factorization of Sigma
    Z <- matrix(rnorm(n*d), nrow=n, ncol=d)
    X <- Z %*% Q + matrix(mu, n, d, byrow=TRUE)
    X
}
```

In this example, we will generate the samples according to the same mean and covariance structure as the four-dimensional iris virginica data.

```
y <- subset(x=iris, Species=="virginica")[, 1:4]
mu <- colMeans(y)
Sigma <- cov(y)
> mu
Sepal.Length  Sepal.Width Petal.Length  Petal.Width
       6.588        2.974        5.552        2.026
> Sigma
             Sepal.Length Sepal.Width Petal.Length Petal.Width
Sepal.Length   0.40434286  0.09376327   0.30328980  0.04909388
Sepal.Width    0.09376327  0.10400408   0.07137959  0.04762857
Petal.Length   0.30328980  0.07137959   0.30458776  0.04882449
Petal.Width    0.04909388  0.04762857   0.04882449  0.07543265
```

```
#now generate MVN data with this mean and covariance
X <- rmvn.Choleski(200, mu, Sigma)
pairs(X)
```

The pairs plot of the data in Figure 3.7 gives a 2-D view of the bivariate distribution of each pair of marginal distributions. The joint distribution of each pair of marginal distributions is theoretically bivariate normal. The plot can be compared with Figure 4.1, which displays the iris virginica data. (The iris virginica data are not multivariate normal, but means and correlation for each pair of variables should be similar to the simulated data.) ◇

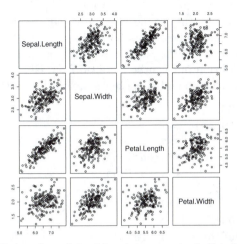

FIGURE 3.7: Pairs plot of the bivariate marginal distributions of a simulated multivariate normal random sample in Example 3.18. The parameters match the mean and covariance of the *iris virginica* data.

Remark 3.3 *To standardize a multivariate normal sample, we invert the procedure above, substituting the sample mean vector and sample covariance matrix if the parameters are unknown. The transformed d-dimensional sample then has zero mean vector and covariance I_d. This is not the same as scaling the columns of the data matrix.* ◇

Comparing Performance of Generators

We have discussed several methods for generating random samples from specified probability distributions. When several methods are available, which method is preferred? One consideration may be the computational time required (the time complexity). Another important consideration, if the purpose of the simulation is to estimate one or more parameters, is the variance of the estimator. The latter topic is considered in Chapter 5. To compare the empirical performance with respect to computing time, we can time each procedure.

R provides the `system.time` function, which times the evaluation of its argument. This function can be used as a rough benchmark to compare the performance of different algorithms. In the next example, the `system.time` function is used to compare the CPU time required for several different methods of generating multivariate normal samples.

Example 3.19 (Comparing performance of MVN generators)

This example generates multivariate normal samples in a higher dimension $(d = 30)$ and compares the timing of each of the methods presented in Section 3.6.1 and two generators available in R packages. This example uses a function rmvnorm in the package mvtnorm [115]. This package is not part of the standard R distribution but can be installed from CRAN. The MASS package [278] is one of the recommended packages included with the R distribution.

```
library(MASS)
library(mvtnorm)
n <- 100            #sample size
d <- 30             #dimension
N <- 2000           #iterations
mu <- numeric(d)

set.seed(100)
system.time(for (i in 1:N)
    rmvn.eigen(n, mu, cov(matrix(rnorm(n*d), n, d))))
set.seed(100)
system.time(for (i in 1:N)
    rmvn.svd(n, mu, cov(matrix(rnorm(n*d), n, d))))
set.seed(100)
system.time(for (i in 1:N)
    rmvn.Choleski(n, mu, cov(matrix(rnorm(n*d), n, d))))
set.seed(100)
system.time(for (i in 1:N)
    mvrnorm(n, mu, cov(matrix(rnorm(n*d), n, d))))
set.seed(100)
system.time(for (i in 1:N)
    rmvnorm(n, mu, cov(matrix(rnorm(n*d), n, d))))
set.seed(100)
system.time(for (i in 1:N)
    cov(matrix(rnorm(n*d), n, d)))
```

Most of the work involved in generating a multivariate normal sample is the factorization of the covariance matrix. The covariances used for this example are actually the sample covariances of standard multivariate normal samples. Thus, the randomly generated Σ varies with each iteration, but Σ is close to an identity matrix. In order to time each method on the same covariance matrices, the random number seed is restored before each run. The last run simply generates the covariances, for comparison with the total time.

The results below (summarized from the console output) suggest that there are differences in performance among these five methods when the covariance matrix is close to identity. The Choleski method is somewhat faster, while `rmvn.eigen` and `mvrnorm` (`MASS`) [278] appear to perform about equally well. The similar performance of `rmvn.eigen` and `mvrnorm` is not surprising, because according to the documentation for `mvrnorm`, the method of matrix decomposition is the eigendecomposition. Documentation for `mvrnorm` states that "although a Choleski decomposition might be faster, the eigendecomposition is stabler."

```
Timings of MVN generators

                 user  system elapsed
rmvn.eigen       7.36    0.00    7.37
rmvn.svd         9.93    0.00    9.94
rmvn.choleski    5.32    0.00    5.35
mvrnorm          7.95    0.00    7.96
rmvnorm         11.91    0.00   11.93
generate Sigma   2.78    0.00    2.78
```

◇

The `system.time` function was also used to compare the methods in Examples 3.22 and 3.23. The code (not shown) is similar to the examples above.

3.6.2   Mixtures of Multivariate Normals

A multivariate normal mixture is denoted

$$pN_d(\mu_1, \Sigma_1) + (1-p)N_d(\mu_2, \Sigma_2) \tag{3.7}$$

where the sampled population is $N_d(\mu_1, \Sigma_1)$ with probability p, and $N_d(\mu_2, \Sigma_2)$ with probability $1 - p$. As the mixing parameter p and other parameters are varied, the multivariate normal mixtures have a wide variety of types of departures from normality. For example, a 50% normal location mixture is symmetric with light tails, and a 90% normal location mixture is skewed with heavy tails. A normal location mixture with $p = 1 - \frac{1}{2}(1 - \frac{\sqrt{3}}{3}) \doteq 0.7887$, provides an example of a skewed distribution with normal kurtosis [140]. Parameters can be varied to generate a wide variety of distributional shapes. Johnson [154] gives many examples for the bivariate normal mixtures. Many commonly applied statistical procedures do not perform well under this type of departure from normality, so normal mixtures are often chosen to compare the properties of competing robust methods of analysis.

If X has the distribution (3.7) then a random observation from the distribution of X can be generated as follows.

To generate a random sample from $pN_d(\mu_1, \Sigma_1) + (1-p)N_d(\mu_2, \Sigma_2)$

1. Generate $U \sim \text{Uniform}(0,1)$.

2. If $U \leq p$ generate X from $N_d(\mu_1, \Sigma_1)$;
   otherwise generate X from $N_d(\mu_2, \Sigma_2)$.

The following procedure is equivalent.

1. Generate $N \sim \text{Bernoulli}(p)$.

2. If $N = 1$ generate X from $N_d(\mu_1, \Sigma_1)$;
   otherwise generate X from $N_d(\mu_2, \Sigma_2)$.

Example 3.20 (Multivariate normal mixture)

Write a function to generate a multivariate normal mixture with two components. The components of a location mixture differ in location only. Use the mvrnorm (MASS) function [278] to generate the multivariate normal observations.

First we write this generator in an inefficient loop to clearly illustrate the steps outlined above. (We will eliminate the loop later.)

```
library(MASS)  #for mvrnorm
#ineffecient version loc.mix.0 with loops

loc.mix.0 <- function(n, p, mu1, mu2, Sigma) {
    #generate sample from BVN location mixture
    X <- matrix(0, n, 2)

    for (i in 1:n) {
        k <- rbinom(1, size = 1, prob = p)
        if (k)
            X[i,] <- mvrnorm(1, mu = mu1, Sigma) else
            X[i,] <- mvrnorm(1, mu = mu2, Sigma)
    }
    return(X)
}
```

Although the code above will generate the required mixture, the loop is rather inefficient. Generate n_1, the number of observations realized from the first component, from Binomial(n, p). Generate n_1 variates from component 1 and $n_2 = n - n_1$ from component 2 of the mixture. Generate a random permutation of the indices 1:n to indicate the order in which the sample observations appear in the data matrix. See Appendix B.1 for details about permutations of rows of a matrix.

```
#more efficient version

loc.mix <- function(n, p, mu1, mu2, Sigma) {
    #generate sample from BVN location mixture
    n1 <- rbinom(1, size = n, prob = p)
    n2 <- n - n1
    x1 <- mvrnorm(n1, mu = mu1, Sigma)
    x2 <- mvrnorm(n2, mu = mu2, Sigma)
    X <- rbind(x1, x2)              #combine the samples
    return(X[sample(1:n), ])       #mix them
}
```

To illustrate the normal mixture generator, we apply `loc.mix` to generate a random sample of $n = 1000$ observations from a 50% 4-dimensional normal location mixture with $\mu_1 = (0,0,0,0)$ and $\mu_2 = (2,3,4,5)$ and covariance I_4.

```
x <- loc.mix(1000, .5, rep(0, 4), 2:5, Sigma = diag(4))
r <- range(x) * 1.2
par(mfrow = c(2, 2))
for (i in 1:4)
    hist(x[ , i], xlim = r, ylim = c(0, .3), freq = FALSE,
        main = "", breaks = seq(-5, 10, .5))
par(mfrow = c(1, 1))
```

It is difficult to visualize data in $\mathbb{R}^4$, so we display only the histograms of the marginal distributions in Figure 3.8. All of the one dimensional marginal distributions are univariate normal location mixtures. Methods for visualization of multivariate data are covered in Chapter 4. Also, an interesting view of a bivariate normal mixture with three components is shown in Figure 10.13 on page 313.   ◇

3.6.3   Wishart Distribution

Suppose $M = X^T X$, where X is an $n \times d$ data matrix of a random sample from a $N_d(\mu, \Sigma)$ distribution. Then M has a Wishart distribution with scale matrix Σ and n degrees of freedom, denoted $M \sim W_d(\Sigma, n)$ (see e.g. [8, 188]). Note that when $d = 1$, the elements of X are a univariate random sample from $N(\mu, \sigma^2)$ so $W_1(\sigma^2, n) \overset{D}{=} \sigma^2 \chi^2(n)$.

An obvious, but inefficient approach to generating random variates from a Wishart distribution, is to generate multivariate normal random samples and compute the matrix product $X^T X$. This method is computationally expensive because nd random normal variates must be generated to determine the $d(d+1)/2$ distinct entries in M.

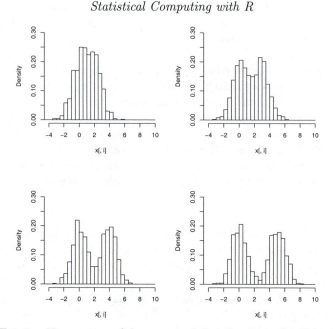

FIGURE 3.8: Histograms of the marginal distributions of multivariate normal location mixture data generated in Example 3.20.

A more efficient method based on Bartlett's decomposition [21] is summarized by Johnson [154, p. 204] as follows. Let $T = (T_{ij})$ be a lower triangular $d \times d$ random matrix with independent entries satisfying

1. $T_{ij} \overset{iid}{\sim} N(0,1)$, $i > j$.

2. $T_{ii} \sim \sqrt{\chi^2(n-i+1)}$, $i = 1, \ldots, d$.

Then the matrix $A = TT^T$ has a $W_d(I_d, n)$ distribution. To generate $W_d(\Sigma, n)$ random variates, obtain the Choleski factorization $\Sigma = LL^T$, where L is lower triangular. Then $LAL^T \sim W_d(\Sigma, n)$ [21, 133, 207]. Implementation is left as an exercise.

3.6.4   Uniform Distribution on the d-Sphere

The d-sphere is the set of all points $x \in \mathbb{R}^d$ such that $\|x\| = (x^T x)^{1/2} = 1$. Random vectors uniformly distributed on the d-sphere have equally likely directions. A method of generating this distribution uses a property of the multivariate normal distribution (see e.g. [94, 154]). If $X_1, \ldots, X_d$ are iid $N(0,1)$, then $U = (U_1, \ldots, U_d)$ is uniformly distributed on the unit sphere in $\mathbb{R}^d$, where

$$U_j = \frac{X_j}{(X_1^2 + \cdots + X_d^2)^{1/2}}, \quad j = 1, \ldots, d. \tag{3.8}$$

Algorithm to generate uniform variates on the d-Sphere

1. For each variate u_i, $i = 1, \ldots, n$ repeat

    (a) Generate a random sample $x_{i1}, \ldots, x_{id}$ from $N(0, 1)$.

    (b) Compute the Euclidean norm $\|x_i\| = (x_{i1}^2 + \cdots + x_{id}^2)^{1/2}$.

    (c) Set $u_{ij} = x_{ij}/\|x_i\|$, $j = 1, \ldots, d$.

    (d) Deliver $u_i = (u_{i1}, \ldots, u_{id})$.

To implement these steps efficiently in R for a sample size n,

1. Generate nd univariate normals in $n \times d$ matrix M. The ith row of M corresponds to to the ith random vector u_i.

2. Compute the denominator of (3.8) for each row, storing the n norms in vector L.

3. Divide each number M[i,j] by the norm L[i], to get the matrix U, where U[i,] $= u_i = (u_{i1}, \ldots, u_{id})$.

4. Deliver matrix U containing n random observations in rows.

Example 3.21 (Generating variates on a sphere)

This example provides a function to generate random variates uniformly distributed on the unit d-sphere.

```
runif.sphere <- function(n, d) {
    # return a random sample uniformly distributed
    # on the unit sphere in R ^d
    M <- matrix(rnorm(n*d), nrow = n, ncol = d)
    L <- apply(M, MARGIN = 1,
               FUN = function(x){sqrt(sum(x*x))})
    D <- diag(1 / L)
    U <- D %*% M
    U
}
```

The function `runif.sphere` is used to generate a sample of 200 points uniformly distributed on the circle.

```
#generate a sample in d=2 and plot
X <- runif.sphere(200, 2)
par(pty = "s")
plot(X, xlab = bquote(x[1]), ylab = bquote(x[2]))
par(pty = "m")
```

The circle of points is shown in Figure 3.9. ◇

R note 3.8 *The `apply` function in `runif.sphere` returns a vector contain-ing the n norms $\|x_1\|, \|x_2\|, \ldots, \|x_n\|$ of the sample vectors in matrix M.*

R note 3.9 *The command `par(pty = "s")` sets the square plot type so the circle is round rather than elliptical; `par(pty = "m")` restores the type to maximal plotting region. See the help topic `?par` for other plot parameters.*

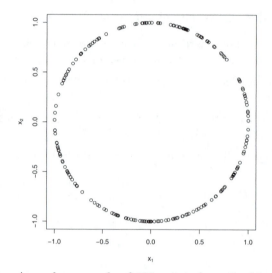

FIGURE 3.9:   A random sample of 200 points from the bivariate distribu-tion (X_1, X_2) that is uniformly distributed on the unit circle in Example 3.21.

Uniformly distributed points on a hyperellipsoid can be generated by ap-plying a suitable linear transformation to a Uniform sample on the d-sphere. Fishman [94, 3.28] gives an algorithm for generating points in and on a sim-plex.

3.7   Stochastic Processes

A stochastic process is a collection $\{X(t) : t \in T\}$ of random variables indexed by the set T, which usually represents time. The *index set T* could be discrete or continuous. The set of possible values $X(t)$ can take is the *state space*, which also can be discrete or continuous. Ross [234] is an excellent introduction to stochastic processes, and includes a chapter on simulation.

A *counting process* records the number of events or arrivals that occur by time t. A counting process has *independent increments* if the number of arrivals in disjoint time intervals are independent. A counting process has *stationary increments* if the number of events occurring in an interval depends only on the length of the interval. An example of a counting process is a Poisson process.

To study a counting process through simulation, we can generate a realization of the process that records events for a finite period of time. The set of times of consecutive arrivals records the outcome and determines the state $X(t)$ at any time t. In a simulation, the sequence of arrival times must be finite. One method of simulation for a counting process is to choose a sufficiently long time interval and generate the arrival times or the interarrival times in this interval.

Poisson Processes

A *homogeneous Poisson process* $\{N(t), t \geq 0\}$ with rate λ is a counting process, with independent increments, such that $N(0) = 0$ and

$$P(N(s+t) - N(s) = n) = \frac{e^{\lambda t}(\lambda t)^n}{n!}, \qquad n \geq 0, \, t, s > 0. \qquad (3.9)$$

Thus, a homogeneous Poisson process has stationary increments and the number of events $N(t)$ in $[0, t]$ has the Poisson(λt) distribution. If T_1 is the time until the first arrival,

$$P(T_1 > t) = P(N(t) = 0) = e^{-\lambda t}, \quad t \geq 0,$$

so T_1 is exponentially distributed with rate λ. The *interarrival times* $T_1, T_2, \ldots$ are the times between successive arrivals. The interarrival times are iid exponentials with rate λ, which follows from (3.9) and the memoryless property of the exponential distribution.

One method of simulating a Poisson process is to generate the interarrival times. Then the time of the nth arrival is the sum $S_n = T_1 + \cdots + T_n$ (the waiting time until n^{th} arrival). A sequence of interarrival times $\{T_n\}_{n=1}^{\infty}$ or sequence of arrival times $\{S_n\}_{n=1}^{\infty}$ are a realization of the process. Thus, a realization is an infinite sequence, rather than a single number. In a simulation, the finite sequence of interarrival times $\{T_n\}_{n=1}^{N}$ or arrival times $\{S_n\}_{n=1}^{N}$ are a simulated realization of the process on the interval $[0, S_N)$.

Another method of simulating a Poisson process is to use the fact that the conditional distribution of the (unordered) arrival times given $N(t) = n$ is the same as that of a random sample of size n from a Uniform$(0, t)$ distribution.

The state of the process at a given time t is equal to the number of arrivals in $[0, t]$, which is the number $\min(k : S_k > t) - 1$. That is, $N(t) = n - 1$, where S_n is the smallest arrival time exceeding t.

Algorithm for simulating a homogeneous Poisson process on an interval $[0, t_0]$ by generating interarrival times.

1. Set $S_1 = 0$.

2. For $j = 1, 2, \ldots$ while $S_j \leq t_0$:

   (a) Generate $T_j \sim \text{Exp}(\lambda)$.
   (b) Set $S_j = T_1 + \cdots + T_j$.

3. $N(t_0) = \min_j (S_j > t_0) - 1$.

It is inefficient to implement this algorithm in R using a `for` loop. It should be translated into vectorized operations, as shown in the next example.

Example 3.22 (Poisson process)

This example illustrates a simple approach to simulation of a Poisson process with rate λ. Suppose we need $N(3)$, the number of arrivals in $[0, 3]$. Generate iid exponential times T_i with rate λ and find the index n where the cumulative sum $S_n = T_1 + \cdots + T_n$ first exceeds 3. It follows that the number of arrivals in $[0, 3]$ is $n - 1$. On average this number is $E[N(3)] = 3\lambda$.

```
lambda <- 2
t0 <- 3
Tn <- rexp(100, lambda)          #interarrival times
Sn <- cumsum(Tn)                 #arrival times
n <- min(which(Sn > t0))         #arrivals+1 in [0, t0]
```

Results from two runs are shown below.

```
> n-1
[1] 8
> round(Sn[1:n], 4)
[1] 1.2217 1.3307 1.3479 1.4639 1.9631 2.0971
      2.3249 2.3409 3.9814

> n-1
[1] 5
> round(Sn[1:n], 4)
[1] 0.4206 0.8620 1.0055 1.6187 2.6418 3.4739
```

For this example, the average of simulated values $N(3) = n - 1$ for a large number of runs should be close to $E[N(3)] = 3\lambda = 6$.   ◇

An alternate method of generating the arrival times of a Poisson process is based on the fact that *given* the number of arrivals in an interval $(0, t)$, the *conditional* distribution of the unordered arrival times are uniformly distributed on $(0, t)$. That is, *given that the number of arrivals in $(0, t)$ is n,* the arrival times $S_1, \ldots, S_n$ are jointly distributed as an ordered random sample of size n from a Uniform$(0, t)$ distribution.

Applying the conditional distribution of the arrival times, it is possible to simulate a Poisson(λ) process on an interval $(0, t)$ by first generating a random observation n from the Poisson(λt) distribution, then generating a random sample of n Uniform$(0, t)$ observations and ordering the uniform sample to obtain the arrival times.

Example 3.23 (Poisson process, cont.)

Returning to Example 3.22, simulate a Poisson(λ) process and find $N(3)$, using the conditional distribution of the arrival times. As a check, we estimate the mean and variance of $N(3)$ from 10000 replications.

```
lambda <- 2
t0 <- 3
upper <- 100
pp <- numeric(10000)
for (i in 1:10000) {
    N <- rpois(1, lambda * upper)
    Un <- runif(N, 0, upper)        #unordered arrival times
    Sn <- sort(Un)                  #arrival times
    n <- min(which(Sn > t0))        #arrivals+1 in [0, t0]
    pp[i] <- n - 1                  #arrivals in [0, t0]
    }
```

Alternately, the loop can be replaced by `replicate`, as shown.

```
pp <- replicate(10000, expr = {
    N <- rpois(1, lambda * upper)
    Un <- runif(N, 0, upper)        #unordered arrival times
    Sn <- sort(Un)                  #arrival times
    n <- min(which(Sn > t0))        #arrivals+1 in [0, t0]
    n - 1 })                        #arrivals in [0, t0]
```

The mean and variance should both be equal to $\lambda t = 6$ in this example. Here the sample mean and sample variance of the generated values $N(3)$ are indeed very close to 6.

```
> c(mean(pp), var(pp))
[1] 5.977100 5.819558
```

Actually, it is possible that none of the generated arrival times exceed the time $t_0 = 3$. In this case, the process needs to be simulated for a longer time than the value in `upper`. Therefore, in practice, one should choose `upper` according to the parameters of the process, and do some error checking. For example, if we need $N(t_0)$, one approach is to wrap the min(which()) step with try and check that the result of try is an integer using is.integer. See the corresponding help topics for details.

Ross [234] discusses the computational efficiency of the two methods applied in Examples 3.22 and 3.23. Actually, the second method is considerably slower (by a factor of 4 or 5) than the previous method of Example 3.22 when coded in R. The rexp generator is almost as fast as runif, while the sort operation adds $O(n \log(n))$ time. Some performance improvement might be gained if this algorithm is coded in C and a faster sorting algorithm designed for uniform numbers is used.   ◇

Nonhomogeneous Poisson Processes

A counting process is a Poisson process with intensity function $\lambda(t)$, $t \geq 0$ if $N(t) = 0$, $N(t)$ has independent increments, and for $h > 0$,

$$P(N(t+h) - N(t) \geq 2) = o(h), \text{ and}$$
$$P(N(t+h) - N(t) = 1) = \lambda(t)h + o(h).$$

The Poisson process $N(t)$ is *nonhomogeneous* if the intensity function $\lambda(t)$ is not constant. A nonhomogeneous Poisson process has independent increments but does not have stationary increments. The distribution of

$$N(s + t) - N(s)$$

is Poisson with mean $\int_s^{s+t} \lambda(y)dy$. The function $m(t) = E[N(t)] = \int_0^t \lambda(y)dy$ is called the *mean value function* of the process. Note that $m(t) = \lambda$ in the case of the homogeneous Poisson process, where the intensity function is a constant.

Every nonhomogeneous Poisson process with a bounded intensity function can be obtained by time sampling a homogeneous Poisson process. Suppose that $\lambda(t) \leq \lambda < \infty$ for all $t \geq 0$. Then sampling a Poisson(λ) process such that an event happening at time t is accepted or counted with probability $\lambda(t)/\lambda$ generates the nonhomogeneous process with intensity function $\lambda(t)$. To see this, let $N(t)$ be the number of accepted events in $[0, t]$. Then $N(t)$ has the Poisson distribution with mean

$$E[N(t)] = \lambda \int_0^t \frac{\lambda(y)}{\lambda} \, dy = \int_0^t \lambda(y)dy.$$

To simulate a nonhomogeneous Poisson process on an interval $[0, t_0]$, find $\lambda_0 < \infty$ such that $\lambda(t) <= \lambda_0$, $0 \leq t \leq t_0$. Then generate from the homogeneous Poisson(λ_0) process the arrival times $\{S_j\}$, and accept each arrival with probability $\lambda(S_j)/\lambda_0$. The steps to simulate the process on an interval $[0, t_0)$ are as follows.

Algorithm for simulating a nonhomogeneous Poisson process on an interval $[0, t_0]$ by sampling from a homogeneous Poisson process.

1. Set $S_1 = 0$.
2. For $j = 1, 2, \ldots$ while $S_j \leq t_0$:

    (a) Generate $T_j \sim \text{Exp}(\lambda_0)$ and set $S_j = T_1 + \cdots + T_j$.
    (b) Generate $U_j \sim \text{Uniform}(0,1)$.
    (c) If $U_j \leq \lambda(S_j)/\lambda_0$ accept (count) this arrival and set $I_j = 1$; otherwise $I_j = 0$.

3. Deliver the arrival times $\{S_j : I_j = 1\}$.

Although this algorithm is quite simple, for implementation in R it is more efficient if translated into vectorized operations. This is shown in the next example.

Example 3.24 (Nonhomogeneous Poisson process)

Simulate a realization from a nonhomogeneous Poisson process with intensity function $\lambda(t) = 3\cos^2(t)$. Here the intensity function is bounded above by $\lambda = 3$, so the jth arrival is accepted if $U_j \leq 3\cos^2(S_j)/3 = \cos^2(S_j)$.

```
lambda <- 3
upper <- 100
N <- rpois(1, lambda * upper)
Tn <- rexp(N, lambda)
Sn <- cumsum(Tn)
Un <- runif(N)
keep <- (Un <= cos(Sn)^2)      #indicator, as logical vector
Sn[keep]
```

Now, the values in `Sn[keep]` are the ordered arrival times of the nonhomogeneous Poisson process.

```
> round(Sn[keep], 4)
 [1]   0.0237  0.5774  0.5841  0.6885  2.3262
         2.4403  2.9984  3.4317  3.7588  3.9297
[11]   4.2962  6.2602  6.2862  6.7590  6.8354
         7.0150  7.3517  8.3844  9.4499  9.4646    . . .
```

To determine the state of the process at time $t = 2\pi$, for example, refer to the entries of `Sn` indexed by `keep`.

```
> sum(Sn[keep] <= 2*pi)
[1] 12
```

```
> table(keep)/N
keep
      FALSE       TRUE
  0.4969325  0.5030675
```

Thus $N(2\pi) = 12$, and in this example approximately 50% of the arrivals were counted.   ◇

Renewal Processes

A renewal process is a generalization of the Poisson process. If $\{N(t), t \geq 0\}$ is a counting process, such that the sequence of nonnegative interarrival times $T_1, T_2, \ldots$ are iid (not necessarily exponential distribution), then $\{N(t), t \geq 0\}$ is a *renewal process*. The function $m(t) = E[N(t)]$ is called the mean value function of the process, which uniquely determines the distribution of the interarrival times.

If the distribution $F_T(t)$ of the iid interarrival times is specified, then a renewal process can be simulated by generating the sequence of interarrival times, by a method similar to Example 3.22.

Example 3.25 (Renewal process)

Suppose the interarrival times of a renewal process have the geometric distribution with success probability p. (This example is discussed in [234, Sec. 7.2].) Then the interarrival times are nonnegative integers, and $S_j = T_1 + \cdots + T_j$ have the negative binomial distribution with size parameter $r = j$ and probability p. The process can be simulated by generating geometric interarrival times and computing the consecutive arrival times by the cumulative sum of interarrival times.

```
t0 <- 5
Tn <- rgeom(100, prob = .2)    #interarrival times
Sn <- cumsum(Tn)               #arrival times
n <- min(which(Sn > t0))       #arrivals+1 in [0, t0]
```

The distribution of $N(t_0)$ can be estimated by replicating the simulation above.

```
Nt0 <- replicate(1000, expr = {
    Sn <- cumsum(rgeom(100, prob = .2))
    min(which(Sn > t0)) - 1
    })
table(Nt0)/1000
Nt0
      0     1     2     3     4     5     6     7
  0.273 0.316 0.219 0.108 0.053 0.022 0.007 0.002
```

To estimate the means $E[N(t)]$, vary the time t_0.

```
t0 <- seq(0.1, 30, .1)
mt <- numeric(length(t0))

for (i in 1:length(t0)) {
    mt[i] <- mean(replicate(1000,
    {
    Sn <- cumsum(rgeom(100, prob = .2))
    min(which(Sn > t0[i])) - 1
    }))
}
plot(t0, mt, type = "l", xlab = "t", ylab = "mean")
```

Let us compare with the homogeneous Poisson process, where the interarrival times have a constant mean. Here we have $p = 0.2$ so the average interarrival time is $0.8/0.2 = 4$. The Poisson process that has mean interarrival time 4 has Poisson parameter $\lambda t = t/4$. We added a reference line to the plot corresponding to the Poisson process mean $\lambda t = t/4$ using `abline(0, .25)`.

The plot is shown in Figure 3.10. It should not be surprising that the mean of the renewal process is very close to λt, because the geometric distribution is the discrete analog of exponential; it has the memoryless property. That is, if $X \sim \text{Geometric}(p)$, then for all $j, k = 0, 1, 2, \ldots$

$$P(X > j + k \mid X > j) = \frac{(1-p)^{j+k}}{(1-p)^j} = (1-p)^k = P(X > k).$$

$\diamond$

Symmetric Random Walk

Let $X_1, X_2, \ldots$ be a sequence of iid random variables with probability distribution $P(X_i = 1) = P(X_i = -1) = 1/2$. Define the partial sum $S_n = \sum_{i=1}^n X_i$. The process $\{S_n, n \geq 0\}$ is called a *symmetric random walk*. For example, if a gambler bets \$1 on repeated trials of coin flipping, then S_n represents the gain/loss after n tosses.

Example 3.26 (Plot a partial realization of a random walk)

It is very simple to generate a symmetric random walk process over a short time span.

```
n <- 400
incr <- sample(c(-1, 1), size = n, replace = TRUE)
S <- as.integer(c(0, cumsum(incr)))
plot(0:n, S, type = "l", main = "", xlab = "i")
```

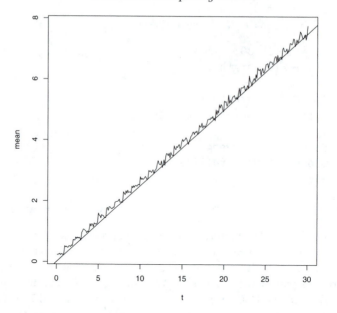

FIGURE 3.10:   Sequence of sample means of a simulated renewal process in Example 3.25. The reference line corresponds to the mean $\lambda t = t/4$ of a homogeneous Poisson process.

A partial realization of the symmetric random walk process starting at $S_0 = 0$ is shown in Figure 3.11. The process has returned to 0 several times within time $[1, 400]$.

```
> which(S == 0)
 [1]   1   3  27  29  31  37  41  95 225 229 233 237 239 241
```

The value of S_n can be determined by the partial random walk starting at the most recent time the process returned to 0.   ◇

If the state of the symmetric random walk S_n at time n is required, but not the history up to time n, then for large n it may be more efficient to generate S_n as follows.

Assume that $S_0 = 0$ is the initial state of the process. If the process has returned to the origin before time n, then to generate S_n we can ignore the past history up until the time the process most recently hit 0. Let T be the time until the first return to the origin. Then to generate S_n, one can simplify the problem by first generating the waiting times T until the total time first exceeds n. Then starting from the last return to the origin before time n, generate the increments X_i and sum them.

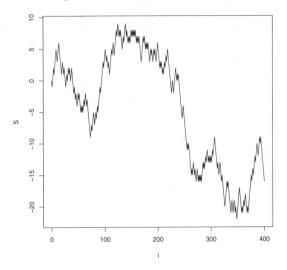

FIGURE 3.11: Partial realization of a symmetric random walk in Example 3.26.

Algorithm to simulate the state S_n of a symmetric random walk

The following algorithm is adapted from [69, XIV.6].
Let W_j be the waiting time until the jth return to the origin.

1. Set $W_1 = 0$.

2. For $j = 1, 2, \ldots$ while $W_j \leq n$:

   (a) Generate a random T_j from the distribution of the time until the first return to 0.

   (b) Set $W_j = T_1 + \cdots + T_j$.

3. Set $t_0 = W_j - T_j$ (time of last return to 0 in time n.)

4. Set $s_1 = 0$.

5. Generate the increments from time $t_0 + 1$ until time n:
   For $i = 1, 2, \ldots, n - t_0$

   (a) Generate a random increment $x_i \sim P(X = \pm 1) = 1/2$.

   (b) Set $s_i = x_1 + \cdots + x_i$.

   (c) If $s_i = 0$ reset the counter to $i = 1$ (another return to 0 is not accepted, so reject this partial random walk and generate a new sequence of increments starting again from time $t_0 + 1$.)

6. Deliver s_i.

To implement the algorithm, one needs to provide a generator for T, the time until the next return of the process to 0. The probability distribution of T [69, Thm. 6.1] is given by

$$P(T = 2n) = p_{2n} = \binom{2n-2}{n-1}\frac{1}{n\,2^{2n-1}} = \frac{\Gamma(2n-1)}{n\,2^{2n-1}\Gamma^2(n)}, \qquad n \geq 1,$$

$$P(T = 2n+1) = 0, \qquad n \geq 0.$$

Example 3.27 (Generator for the time until return to origin)

An efficient algorithm for generating from the distribution T is given by Devroye [69, p. 754]. Here we will apply an inefficient version that is easily implemented in R. Notice that p_{2n} equals $1/(2n)$ times the probability $P(X = n-1)$ where $X \sim$ Binomial $(2n - 2, p = 1/2)$.

The following methods are equivalent.

```
#compute the probabilities directly
n <- 1:10000
p2n <- exp(lgamma(2*n-1)
             - log(n) - (2*n-1)*log(2) - 2*lgamma(n))

#or compute using dbinom
P2n <- (.5/n) * dbinom(n-1, size = 2*n-2, prob = 0.5)
```

Recall that if X is a discrete random variable and

$$\ldots < x_{i-1} < x_i < x_{i+1} < \ldots$$

are the points of discontinuity of $F_X(x)$, then the inverse transformation is $F_X^{-1}(u) = x_i$, where $F_X(x_{i-1}) < u \leq F_X(x_i)$. Therefore, a generator can be written for values of T up to 20000 using the probability vector computed above.

```
pP2n <- cumsum(P2n)
#for example, to generate one T
u <- runif(1)
Tj <- 2 * (1 + sum(u > pP2n))
```

Here are two examples to illustrate the method of looking up the solution $F_X(x_{i-1}) < u \leq F_X(x_i)$ in the probability vector.

```
#first part of pP2n
[1] 0.5000000 0.6250000 0.6875000 0.7265625 0.7539062 0.7744141
```

In the first example $u = 0.6612458$ and the first return to the origin occurs at time $n = 6$, and in the second example $u = 0.5313384$ and the next return to the 0 occurs at time $n = 4$ after the first return to 0. Thus the second return to the origin occurs at time 10. (The case u > max(pP2n) must be handled separately.)

Suppose now that n is given and we need to compute the time of the last return to 0 in $(0, n]$.

```
n <- 200
sumT <- 0
while (sumT <= n) {
    u <- runif(1)
    s <- sum(u > pP2n)
    if (s == length(pP2n)) warning("T is truncated")
    Tj <- 2 * (1 + s)
    #print(c(Tj, sumT))
    sumT <- sumT + Tj
    }
sumT - Tj
```

In case the random uniform exceeds the maximal value in the cdf vector pP2n, a warning is issued. Here instead of issuing a warning, one could append to the vector and return a valid T. We leave that as an exercise. A better algorithm is suggested by Devroye [69, p. 754]. One run of the simulation above generates the times 110, 128, 162, 164, 166, 168, and 210 that the process visits 0 (uncomment the print statement to print the times). Therefore the last visit to 0 before $n = 200$ is at time 168.

Finally, S_{200} can be generated by simulating a symmetric random walk starting from $S_{168} = 0$ for $t = 169, \ldots, 200$ (rejecting the partial random walk if it hits 0). ◇

Packages and Further Reading

General references on discrete event simulation and simulation of stochastic processes include Banks et al. [18], Devroye [69], and Fishman [95]. Algorithms for generating random tours in general are discussed by Fishman [94, Ch. 5]. Also see Cornuejols and Tütüncü [53] on related optimization methods.

Ross [234, Ch. 10] has a nice introduction to Brownian Motion, starting with the interpretation of Brownian Motion as the limit of random walks. For a more theoretical treatment see Durrett [77, Ch. 7].

See Franklin [98] for simulation of Gaussian processes. Functions to simulate long memory time series processes, including fractional Brownian motion are available in the R package fSeries (see e.g. fbmSim) [299] and sde [149]. The FracSim package [65, 66] implements methods for simulation of multifractional Lévy motions. Also see Coeurjolly [51] for a bibliographical and comparative study on simulation and identification of fractional Brownian motion.

References on the general subject of methods for generating random variates from specified probability distributions have been given in Section 3.1.

Exercises

3.1   Write a function that will generate and return a random sample of size n from the two-parameter exponential distribution $\text{Exp}(\lambda, \eta)$ for arbitrary n, λ, and η. (See Examples 2.3 and 2.6.) Generate a large sample from $\text{Exp}(\lambda, \eta)$ and compare the sample quantiles with the theoretical quantiles.

3.2   The standard Laplace distribution has density $f(x) = \frac{1}{2}e^{-|x|}$, $x \in \mathbb{R}$. Use the inverse transform method to generate a random sample of size 1000 from this distribution. Use one of the methods shown in this chapter to compare the generated sample to the target distribution.

3.3   The $\text{Pareto}(a, b)$ distribution has cdf

$$F(x) = 1 - \left(\frac{b}{x}\right)^a, \qquad x \geq b > 0, a > 0.$$

Derive the probability inverse transformation $F^{-1}(U)$ and use the inverse transform method to simulate a random sample from the $\text{Pareto}(2, 2)$ distribution. Graph the density histogram of the sample with the $\text{Pareto}(2, 2)$ density superimposed for comparison.

3.4   The Rayleigh density [156, Ch. 18] is

$$f(x) = \frac{x}{\sigma^2} e^{-x^2/(2\sigma^2)}, \qquad x \geq 0, \sigma > 0.$$

Develop an algorithm to generate random samples from a $\text{Rayleigh}(\sigma)$ distribution. Generate $\text{Rayleigh}(\sigma)$ samples for several choices of $\sigma > 0$ and check that the mode of the generated samples is close to the theoretical mode σ (check the histogram).

3.5   A discrete random variable X has probability mass function

x	0	1	2	3	4
$p(x)$	0.1	0.2	0.2	0.2	0.3

Use the inverse transform method to generate a random sample of size 1000 from the distribution of X. Construct a relative frequency table and compare the empirical with the theoretical probabilities. Repeat using the R `sample` function.

3.6   Prove that the accepted variates generated by the acceptance-rejection sampling algorithm are a random sample from the target density f_X.

3.7   Write a function to generate a random sample of size n from the $\text{Beta}(a, b)$ distribution by the acceptance-rejection method. Generate a random sample of size 1000 from the $\text{Beta}(3,2)$ distribution. Graph the histogram of the sample with the theoretical $\text{Beta}(3,2)$ density superimposed.

3.8 Write a function to generate random variates from a Lognormal(μ, σ) distribution using a transformation method, and generate a random sample of size 1000. Compare the histogram with the lognormal density curve given by the dlnorm function in R.

3.9 The rescaled Epanechnikov kernel [85] is a symmetric density function

$$f_e(x) = \frac{3}{4}(1 - x^2), \qquad |x| \le 1. \tag{3.10}$$

Devroye and Györfi [71, p. 236] give the following algorithm for simulation from this distribution. Generate iid $U_1, U_2, U_3 \sim$ Uniform$(-1, 1)$. If $|U_3| \ge |U_2|$ and $|U_3| \ge |U_1|$, deliver U_2; otherwise deliver U_3. Write a function to generate random variates from f_e, and construct the histogram density estimate of a large simulated random sample.

3.10 Prove that the algorithm given in Exercise 3.9 generates variates from the density f_e (3.10).

3.11 Generate a random sample of size 1000 from a normal location mixture. The components of the mixture have $N(0, 1)$ and $N(3, 1)$ distributions with mixing probabilities p_1 and $p_2 = 1 - p_1$. Graph the histogram of the sample with density superimposed, for $p_1 = 0.75$. Repeat with different values for p_1 and observe whether the empirical distribution of the mixture appears to be bimodal. Make a conjecture about the values of p_1 that produce bimodal mixtures.

3.12 Simulate a continuous Exponential-Gamma mixture. Suppose that the rate parameter Λ has Gamma(r, β) distribution and Y has Exp(Λ) distribution. That is, $(Y|\Lambda = \lambda) \sim f_Y(y|\lambda) = \lambda e^{-\lambda y}$. Generate 1000 random observations from this mixture with $r = 4$ and $\beta = 2$.

3.13 It can be shown that the mixture in Exercise 3.12 has a Pareto distribution with cdf

$$F(y) = 1 - \left(\frac{\beta}{\beta + y}\right)^r, \qquad y \ge 0.$$

(This is an alternative parameterization of the Pareto cdf given in Exercise 3.3.) Generate 1000 random observations from the mixture with $r = 4$ and $\beta = 2$. Compare the empirical and theoretical (Pareto) distributions by graphing the density histogram of the sample and superimposing the Pareto density curve.

3.14 Generate 200 random observations from the 3-dimensional multivariate normal distribution having mean vector $\mu = (0, 1, 2)$ and covariance matrix

$$\Sigma = \begin{bmatrix} 1.0 & -0.5 & 0.5 \\ -0.5 & 1.0 & -0.5 \\ 0.5 & -0.5 & 1.0 \end{bmatrix}$$

using the Choleski factorization method. Use the R pairs plot to graph an array of scatter plots for each pair of variables. For each pair of variables, (visually) check that the location and correlation approximately agree with the theoretical parameters of the corresponding bivariate normal distribution.

3.15   Write a function that will standardize a multivariate normal sample for arbitrary n and d. That is, transform the sample so that the sample mean vector is zero and sample covariance is the identity matrix. To check your results, generate multivariate normal samples and print the sample mean vector and covariance matrix before and after standardization.

3.16   Efron and Tibshirani discuss the `scor (bootstrap)` test score data on 88 students who took examinations in five subjects [84, Table 7.1], [188, Table 1.2.1]. Each row of the data frame is a set of scores $(x_{i1}, \ldots, x_{i5})$ for the ith student. Standardize the scores by type of exam. That is, standardize the bivariate samples (X_1, X_2) (closed book) and the trivariate samples (X_3, X_4, X_5) (open book). Compute the covariance matrix of the transformed sample of test scores.

3.17   Compare the performance of the Beta generator of Exercise 3.7, Example 3.8 and the R generator `rbeta`. Fix the parameters $a = 2, b = 2$ and time each generator on 1000 iterations with sample size 5000. (See Example 3.19.) Are the results different for different choices of a and b?

3.18   Write a function to generate a random sample from a $W_d(\Sigma, n)$ (Wishart) distribution for $n > d + 1 \geq 1$, based on Bartlett's decomposition.

3.19   Suppose that A and B each start with a stake of \$10, and bet \$1 on consecutive coin flips. The game ends when either one of the players has all the money. Let S_n be the fortune of player A at time n. Then $\{S_n, n \geq 0\}$ is a symmetric random walk with absorbing barriers at 0 and 20. Simulate a realization of the process $\{S_n, n \geq 0\}$ and plot S_n vs the time index from time 0 until a barrier is reached.

3.20   A *compound Poisson process* is a stochastic process $\{X(t), t \geq 0\}$ that can be represented as the random sum $X(t) = \sum_{i=1}^{N(t)} Y_i$, $t \geq 0$, where $\{N(t), t \geq 0\}$ is a Poisson process and $Y_1, Y_2, \ldots$ are iid and independent of $\{N(t), t \geq 0\}$. Write a program to simulate a compound Poisson(λ)–Gamma process (Y has a Gamma distribution). Estimate the mean and the variance of $X(10)$ for several choices of the parameters and compare with the theoretical values. Hint: Show that $E[X(t)] = \lambda t E[Y_1]$ and $Var(X(t)) = \lambda t E[Y_1^2]$.

3.21   A nonhomogeneous Poisson process has mean value function $m(t) = t^2 + 2t$, $t \geq 0$. Determine the intensity function $\lambda(t)$ of the process, and write a program to simulate the process on the interval $[4, 5]$. Compute the probability distribution of $N(5) - N(4)$, and compare it to the empirical estimate obtained by replicating the simulation.

Chapter 4

Visualization of Multivariate Data

4.1 Introduction

The topic of visualization of multivariate data is related to more general subjects called exploratory data analysis (EDA) and statistical graphics. The term "exploratory" is in contrast to "confirmatory," which could describe hypothesis testing. Tukey [275] believed that it was important to do the exploratory work before hypothesis testing, to learn what are the appropriate questions to ask, and the most appropriate methods to answer them. With multivariate data, we may also be interested in dimension reduction or finding structure or groups in the data. Here we restrict attention to methods for visualizing multivariate data.

In this chapter several graphics functions are used. In addition to the R graphics package, which loads when R is started, other packages discussed in this chapter are lattice [239] and MASS (see [278]). Also see the rggobi [167] interface to GGobi and rgl [2] package for interactive 3D visualization. Table 1.4 lists some basic graphics functions in R (graphics) or other packages. Table 4.1 lists more 2D graphics functions and some of the 3D visualization methods.

Chapter 1 gives a brief summary of options for colors, plotting symbols, and line types.

4.2 Panel Displays

A panel display is an array of two-dimensional graphical summaries of pairs of variables in a multivariate dataset. For example, a scatterplot matrix displays the scatterplots for all pairs of variables in an array. The pairs function in the graphics package produces a scatterplot matrix, as shown in Figures 4.1 and 4.2 in Example 4.1, and Figure 3.7 on page 75. An example of a panel display of three-dimensional plots is Figure 4.5 on page 106.

TABLE 4.1: Graphics Functions for Multivariate Data in R (graphics) and Other Packages

Method	in (graphics)	in (package)
3D scatterplot		cloud (lattice)
Matrix of scatterplots	pairs	splom (lattice)
Bivariate density surface	persp	wireframe (lattice)
Contour plot	contour, image	contourplot (lattice)
	contourLines	contour (MASS)
	filled.contour	levelplot (lattice)
Parallel coord. plot		parallel (lattice)
		parcoord (MASS)
Star plot	stars	
Segment plot	stars	
Interactive 3D graphics		(rggobi), (rgl)

Example 4.1 (Scatterplot matrix)

We compare the four variables in the `iris` data for the species virginica, in a scatterplot matrix.

```
data(iris)
#virginica data in first 4 columns of the last 50 obs.
pairs(iris[101:150, 1:4])
```

In the plot produced by the `pairs` command above (not shown) the variable names will appear along the diagonal. The `pairs` function takes an optional argument `diag.panel`, which is a function that determines what is displayed along the diagonal. For example, to obtain a graph with estimated density curves along the diagonal, supply the name of a function to plot the densities. The function below called `panel.d` plots the densities.

```
panel.d <- function(x, ...) {
    usr <- par("usr")
    on.exit(par(usr))
    par(usr = c(usr[1:2], 0, .5))
    lines(density(x))
}
```

In `panel.d`, the graphics parameter `usr` specifies the extremes of the user coordinates of the plotting region. Before plotting, we apply the `scale` function to standardize each of the one-dimensional samples.

```
x <- scale(iris[101:150, 1:4])
r <- range(x)
pairs(x, diag.panel = panel.d, xlim = r, ylim = r)
```

The `pairs` plot is displayed in Figure 4.1. From the plot we can observe that the length variables are positively correlated, and the width variables appear to be positively correlated. Other structure could be present in the data that is not revealed by the bivariate marginal distributions.

The `lattice` package [239] provides functions to construct panel displays. Here we illustrate the scatterplot matrix function `splom` in `lattice`.

```
library(lattice)
splom(iris[101:150, 1:4])      #plot 1

#for all 3 at once, in color, plot 2
splom(iris[,1:4], groups = iris$Species)

#for all 3 at once, black and white, plot 3
splom(~iris[1:4], groups = Species, data = iris,
     col = 1, pch = c(1, 2, 3),   cex = c(.5,.5,.5))
```

The last plot (plot 3) is displayed in Figure 4.2. It is displayed here in black and white, but on screen the panel display is easier to interpret when displayed in color (plot 2). Also see the 3D scatterplot of the iris data in Figure 4.5. ◇

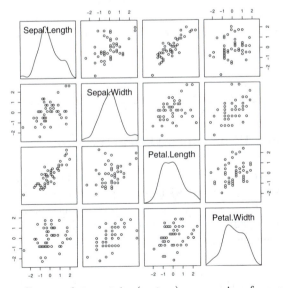

FIGURE 4.1: Scatterplot matrix (`pairs`) comparing four measurements of iris virginica species in Example 4.1.

For other types of panel displays, see the conditioning plots [42, 48, 49] implemented in `coplot`.

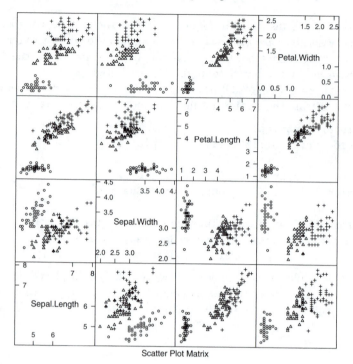

Scatter Plot Matrix

FIGURE 4.2: Scatterplot matrix comparing four measurements of iris data: setosa (circle), versicolor (triangle), virginica (cross) from Example 4.1.

4.3   Surface Plots and 3D Scatter Plots

Several packages provide surface and contour plots. The `persp` (graphics) function draws perspective plots of surfaces over the plane. Try running the demo examples for `persp`, to see many interesting graphs. The command is simply `demo(persp)`. We will also look at 3D methods in the `lattice` graphics package and the `rgl` package [239, 278, 2].

4.3.1   Surface plots

For certain graphs we need to mesh a grid of regularly spaced points in the plane. The command for this is `expand.grid`. If we do not need to save the x, y values, and only need the function values $\{z_{ij} = f(x_i, y_j)\}$, the `outer` function can be used.

Example 4.2 (Plot bivariate normal density)

Plot the standard bivariate normal density

$$f(x,y) = \frac{1}{2\pi} e^{-\frac{1}{2}(x^2+y^2)}, \quad (x,y) \in \mathbb{R}^2.$$

Code to plot the bivariate standard normal density surface using the `persp` function is below. Most of the parameters are optional; `x`, `y`, `z` are required. For this function we need the complete grid of z values, but only one vector of x and one vector of y values. In this example, $z_{ij} = f(x_i, y_j)$ are computed by the `outer` function.

```
#the standard BVN density
f <- function(x,y) {
    z <- (1/(2*pi)) * exp(-.5 * (x^2 + y^2))
    }

y <- x <- seq(-3, 3, length= 50)
z <- outer(x, y, f)    #compute density for all (x,y)

persp(x, y, z)         #the default plot

persp(x, y, z, theta = 45, phi = 30, expand = 0.6,
        ltheta = 120, shade = 0.75, ticktype = "detailed",
        xlab = "X", ylab = "Y", zlab = "f(x, y)")
```

The second version of the perspective plot is shown in Figure 4.3.   ◇

R note 4.1 *The `outer` function `outer(x, y, f)` in Example 4.2 applies the third argument, a bivariate function, to the grid of (x,y) values. The returned value is a matrix of function values for every point (x_i, y_j) in the grid. Storing the grid was not necessary.*

For a presentation, adding color (say, `col = "lightblue"`) produces a more attractive plot. The box can be suppressed by `box = FALSE`.

Adding elements to a perspective plot

The `persp` function returns the 'viewing transformation' in a 4×4 matrix. This transformation can be used to add elements to the plot.

Example 4.3 (Add elements to perspective plot)

This example uses the viewing transformation returned by the perspective plot of the standard bivariate normal density to add points, lines, and text to the plot.

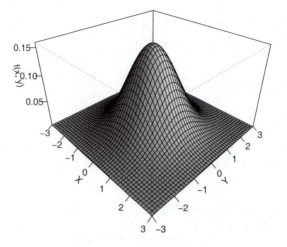

FIGURE 4.3:   Perspective plot of the standard bivariate normal density in Example 4.2.

```
#store viewing transformation in M
persp(x, y, z, theta = 45, phi = 30,
        expand = .4, box = FALSE) -> M
```

The transformation returned by the `persp` function call is

```
                [,1]         [,2]         [,3]          [,4]
[1,]    2.357023e-01 -0.1178511   0.2041241  -0.2041241
[2,]    2.357023e-01  0.1178511  -0.2041241   0.2041241
[3,]   -2.184757e-16  4.3700078   2.5230252  -2.5230252
[4,]    1.732284e-17 -0.3464960  -2.9321004   3.9321004
```

This transformation M is applied to (x, y, z, t) to project points onto the screen for display in the same coordinate system used to draw the perspective plot.

```
#add some points along a circle
a <- seq(-pi, pi, pi/16)
newpts <- cbind(cos(a), sin(a)) * 2
newpts <- cbind(newpts, 0, 1)   #z=0, t=1
N <- newpts %*% M
points(N[,1]/N[,4], N[,2]/N[,4], col=2)

#add lines
x2 <- seq(-3, 3, .1)
y2 <- -x2^2 / 3
z2 <- dnorm(x2) * dnorm(y2)
```

```
N <- cbind(x2, y2, z2, 1) %*% M
lines(N[,1]/N[,4], N[,2]/N[,4], col=4)

#add text
x3 <- c(0, 3.1)
y3 <- c(0, -3.1)
z3 <- dnorm(x3) * dnorm(y3) * 1.1
N <- cbind(x3, y3, z3, 1) %*% M
text(N[1,1]/N[1,4], N[1,2]/N[1,4], "f(x,y)")
text(N[2,1]/N[2,4], N[2,2]/N[2,4], bquote(y==-x^2/3))
```

The plot with added elements is shown in Figure 4.4 (Note: R provides a function `trans3d` to compute the coordinates above. Here we have shown the calculations.)   ◇

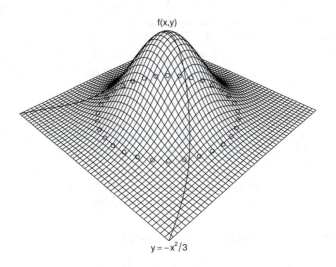

FIGURE 4.4: Perspective plot of the standard bivariate normal density with elements added using the viewing transformation returned by `persp` in Example 4.3.

Other functions for graphing surfaces

Surfaces can also be graphed using the `wireframe (lattice)` function [239]. Supply a formula z ~ x * y and a data frame or data matrix containing the points (x, y, z).

Example 4.4 (Surface plot using `wireframe(lattice)`)

The following code displays a surface plot of the bivariate normal density similar to Figure 4.3 using `wireframe(lattice)`. The wireframe function requires a formula $z \sim x * y$, where $z = f(x, y)$ is the surface to be plotted. The syntax for `wireframe` requires that x, y and z have the same number of rows. We can generate the matrix of (x, y) coordinates using `expand.grid`.

```
library(lattice)
x <- y <- seq(-3, 3, length= 50)

xy <- expand.grid(x, y)
z <- (1/(2*pi)) * exp(-.5 * (xy[,1]^2 + xy[,2]^2))
wireframe(z ~ xy[,1] * xy[,2])
```

The `wireframe` plot (not shown) looks very similar to the perspective plot of the bivariate normal density in Figure 4.3.   ◊

An interactive 3D display is provided by the graphics package `rgl` [2]. If the `rgl` package is installed, run the demo. One of the examples in the demo shows a bivariate normal density. (Actually, the data used to plot the surface in this demo is generated by smoothing simulated bivariate normal data.)

```
library(rgl)
demo(bivar)    #or demo(rgl) to see more
```

It may be helpful to enlarge the graph window. The graph can be rotated and tilted by the mouse to see the surface from different angles. For the source code of this demo, refer to the file *./demo/bivar.r* in the directory where `rgl` is installed.

Chapter 10 gives examples of methods to construct and plot density estimates for bivariate data. See e.g. Figures 10.11, 10.12(a), and 10.13.

4.3.2   Three-dimensional scatterplot

The `cloud (lattice)` [239] function produces 3D scatterplots. A possible application of this type of plot is to explore whether there are groups or clusters in the data. To apply the `cloud` function, provide a formula $z \sim x * y$, where $z = f(x, y)$ is the surface to be plotted. The first part of the following example is a simple application of `cloud` with groups identified by color. The second part of the example illustrates several options.

Example 4.5 (3D scatterplot)

This example uses the `cloud` function in the `lattice` package to display a 3D scatterplot of the `iris` data. There are three species of iris and each is measured on four variables. The following code produces a 3D scatterplot of sepal length, sepal width, and petal length. The plot produced is similar to (3) in Figure 4.5.

```
library(lattice)
attach(iris)
#basic 3 color plot with arrows along axes
print(cloud(Petal.Length ~ Sepal.Length * Sepal.Width,
        data=iris, groups=Species))
```

The iris data has four variables, so there are four subsets of three variables to graph. To see all four plots on the screen, use the more and split options. The split arguments determine the location of the plot within the panel display.

```
print(cloud(Sepal.Length ~ Petal.Length * Petal.Width,
        data = iris, groups = Species, main = "1", pch=1:3,
        scales = list(draw = FALSE), zlab = "SL",
        screen = list(z = 30, x = -75, y = 0)),
        split = c(1, 1, 2, 2), more = TRUE)

print(cloud(Sepal.Width ~ Petal.Length * Petal.Width,
        data = iris, groups = Species, main = "2", pch=1:3,
        scales = list(draw = FALSE), zlab = "SW",
        screen = list(z = 30, x = -75, y = 0)),
        split = c(2, 1, 2, 2), more = TRUE)

print(cloud(Petal.Length ~ Sepal.Length * Sepal.Width,
        data = iris, groups = Species, main = "3", pch=1:3,
        scales = list(draw = FALSE), zlab = "PL",
        screen = list(z = 30, x = -55, y = 0)),
        split = c(1, 2, 2, 2), more = TRUE)

print(cloud(Petal.Width ~ Sepal.Length * Sepal.Width,
        data = iris, groups = Species, main = "4", pch=1:3,
        scales = list(draw = FALSE), zlab = "PW",
        screen = list(z = 30, x = -55, y = 0)),
        split = c(2, 2, 2, 2))
detach(iris)
```

The four 3D scatterplots are shown in Figure 4.5. The plots show that the three species of iris are separated into groups or clusters in the three dimensional subspaces spanned by any three of the four variables. There is some structure evident in these plots. One might follow up with cluster analysis or principal components analysis to analyze the apparent structure in the data. ◇

R note 4.2 *Syntax for* `cloud`: *The* `screen` *option sets the orientation of the axes. Setting* `draw = FALSE` *suppresses arrows and tick marks on the axes.*

Syntax for `print(cloud)`: *To split the screen into n rows and m columns, and put the plot into position* (r, c), *set* `split` *equal to the vector* (r, c, n, m). *One unusual feature of* `cloud` *is that unlike most graphics functions in R,* `cloud` *does not plot a panel figure unless we* `print` *it. See* `print.trellis` *for documentation on the print method for* `cloud`.

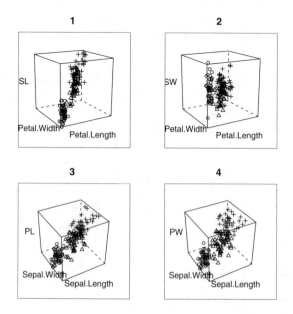

FIGURE 4.5: 3D scatterplots of iris data produced by `cloud` (`lattice`) in Example 4.5, with each species represented by a different plotting character.

4.4 Contour Plots

A contour plot represents a 3D surface $(x, y, f(x, y))$ in the plane by projecting the level curves $f(x, y) = c$ for selected constants c. The functions `contour` (`graphics`) and `contourplot` (`lattice`) [239] produce contour plots. The functions `filled.contour` in the graphics package and `levelplot` function in the `lattice` package produce filled contour plots. Both `contour` and `contourplot` label the contours by default. A variation of this type of plot is `image (graphics)`, which uses color to identify contour levels.

Example 4.6 (Contour plot)

A good example is provided in R using the `volcano` data. Information about this data is in the help file for `volcano`. The data is an 87 by 61 matrix containing topographic information for the Maunga Whau volcano.

```
#contour plot with labels
contour(volcano, asp = 1, labcex = 1)

#another version from lattice package
library(lattice)
contourplot(volcano) #similar to above
```

Figure 4.6(a) shows the contour plot of the `volcano` data produced by the `contour` function.

It may also be interesting to see the 3D surface of the volcano for comparison with the contour plots. A 3D view of the volcano surface is provided in the examples of the `persp` function. The R code for the example is in the `persp` help page. To run the example, type `example(persp)`.

If the `rgl` package is installed, an interactive 3D view of the volcano appears in the examples. When the volcano surface is displayed, use the mouse to rotate and tilt the surface, to view it from different angles.

```
library(rgl)
example(rgl)
```

Yet another 3D view of the `volcano` data, with shading to indicate contour levels, appears in the examples of the `wireframe` function in the `lattice` package. See the first example in the `wireframe` help file. ◇

Example 4.7 (Filled contour plots)

A contour plot with a 3D effect could be displayed in 2D by overlaying the contour lines on a color map corresponding to the height. The `image` function in the `graphics` package provides the color background for the plot. The plot produced below is similar to Figure 4.6(a), with the background of the plot in terrain colors.

```
image(volcano, col = terrain.colors(100), axes = FALSE)
contour(volcano, levels = seq(100,200,by = 10), add = TRUE)
```

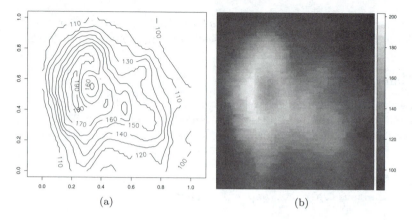

FIGURE 4.6: Contour plot and levelplot of volcano data in Examples 4.6 and 4.7.

Using `image` without `contour` produces essentially the same type of plot as `filled.contour` (graphics) and `levelplot` (lattice). The contours of `filled.contour` and `levelplot` are identified by a legend rather than superimposing the contour lines. Compare the plot produced by `image` with the following two plots.

```
filled.contour(volcano, color = terrain.colors, asp = 1)
levelplot(volcano, scales = list(draw = FALSE),
          xlab = "", ylab = "")
```

The plot produced by `levelplot` is shown in Figure 4.6(b). (The display on the screen will be in color.)   ◇

A limitation of 2D scatterplots is that for large data sets, there are often regions where data is very dense, and regions where data is quite sparse. In this case, the 2D scatterplot does not reveal much information about the bivariate density. Another approach is to produce a 2D or flat histogram, with the density estimate in each bin represented by an appropriate color.

Example 4.8 (2D histogram)

In this example, simulated bivariate normal data is displayed in a flat histogram with hexagonal bins. The `hexbin` function in package `hexbin` [38] (available from Bioconductor repository) produces a basic version of this plot in grayscale, shown in Figure 4.7.

```
library(hexbin)
x <- matrix(rnorm(4000), 2000, 2)
plot(hexbin(x[,1], x[,2]))
```

Compare the flat density histogram in Figure 4.7 with the bivariate histogram in Figure 10.11 on page 308. Note that the darker colors correspond to the regions where the density is highest, and colors are increasingly lighter along radial lines extending from the mode near the origin. The plot exhibits approximately circular symmetry, consistent with the standard bivariate normal density.

The bivariate histogram can also be displayed in 2D using a color palette, such as `heat.colors` or `terrain.colors`, to represent the density for each bin. A similar type of plot is implemented in the `gplots` package [290]. The plot (not shown) resulting from the following code is similar to Figure 4.7, but with color and square bins.

```
library(gplots)
hist2d(x, nbins = 30,
        col = c("white", rev(terrain.colors(30))))
```

◇

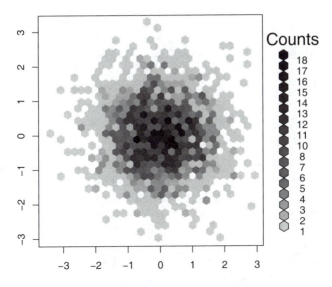

FIGURE 4.7: Flat density histogram of bivariate normal data with hexagonal bins produced by `hexbin` in Example 4.8.

4.5   Other 2D Representations of Data

In addition to contour plots and other projections of data into two dimensions, there are several other methods for representing multivariate data in two dimensions. These include, among others, Andrews curves, parallel coordinate plots, and various iconographic displays such as segment plots and star plots.

4.5.1   Andrews Curves

If $X_1, \ldots, X_n \in \mathbb{R}^d$, one approach to visualizing the data in two dimensions is to map each of the sample data vectors onto a real valued function. Andrews Curves [10] map each sample observation $x_i = x_{i1}, \ldots, x_{id}$ to the function

$$f_i(t) = \frac{x_{i1}}{\sqrt{2}} + x_{i2} \sin t + x_{i3} \cos t + x_{i4} \sin 2t + x_{i5} \cos 2t + \ldots$$

$$= \frac{x_{i1}}{\sqrt{2}} + \sum_{1 \le k \le d/2} x_{i,2k} \sin kt + \sum_{1 \le k < d/2} x_{i,2k+1} \cos kt, \qquad -\pi \le t \le \pi.$$

Thus, each observation is represented by its projection onto a set of orthogonal basis functions $\{2^{-1/2}, \{\sin kt\}_{k=1}^{\infty}, \{\cos kt\}_{k=1}^{\infty}\}$. Notice that differences between measurements are amplified more in the lower frequency terms, so that the representation depends on the order of the variables or features.

Example 4.9 (Andrews curves)

In this example, measurements of leaves taken at N. Queensland, Australia for two types of leaf architecture [162] are represented by Andrews curves. The data set is `leafshape17` in the `DAAG` package [184, 185]. Three measurements (leaf length, petiole, and leaf width) correspond to points in $\mathbb{R}^3$. It is easiest to interpret the plots if leaf architectures are identified by different colors, but here we use different line types. To plot the curves, define a function to compute $f_i(t)$ for arbitrary points x_i in $\mathbb{R}^3$ and $-\pi \le t \le \pi$. Evaluate the function along the interval $[-\pi, \pi]$ for each sample point x_i.

```
library(DAAG)
attach(leafshape17)

f <- function(a, v) {
    #Andrews curve f(a) for a data vector v in R^3
    v[1]/sqrt(2) + v[2]*sin(a) + v[3]*cos(a)
}
```

```
#scale data to range [-1, 1]
x <- cbind(bladelen, petiole, bladewid)
n <- nrow(x)
mins <- apply(x, 2, min)   #column minimums
maxs <- apply(x, 2, max)   #column maximums
r <- maxs - mins           #column ranges
y <- sweep(x, 2, mins)     #subtract column mins
y <- sweep(y, 2, r, "/")   #divide by range
x <- 2 * y - 1             #now has range [-1, 1]

#set up plot window, but plot nothing yet
plot(0, 0, xlim = c(-pi, pi), ylim = c(-3,3),
    xlab = "t", ylab = "Andrews Curves",
    main = "", type = "n")

#now add the Andrews curves for each observation
#line type corresponds to leaf architecture
#0=orthotropic, 1=plagiotropic
a <- seq(-pi, pi, len=101)
dim(a) <- length(a)
for (i in 1:n) {
    g <- arch[i] + 1
    y <- apply(a, MARGIN = 1, FUN = f, v = x[i,])
    lines(a, y, lty = g)
}
legend(3, c("Orthotropic", "Plagiotropic"), lty = 1:2)
detach(leafshape17)
```

The plot of Andrews curves for this example is shown in Figure 4.8. The plot reveals similarities within plagiotropic and orthotropic leaf architecture groups, and differences between these groups. In general, this type of plot may reveal possible clustering of data. ◇

R note 4.3 *In Example 4.9 the* sweep *operator is applied to subtract the column minimums above. The syntax is*

  sweep(x, MARGIN, STATS, FUN="-", ...)

By default, the statistic is subtracted but other operations are possible. Here

  y <- sweep(x, 2, mins)   #subtract column mins
  y <- sweep(y, 2, r, "/")   #divide by range

sweeps out (subtracts) the minimum of each columns (margin = 2). Then the ranges of each of the three columns (in r) are swept out; that is, each column is divided by its range. ◇

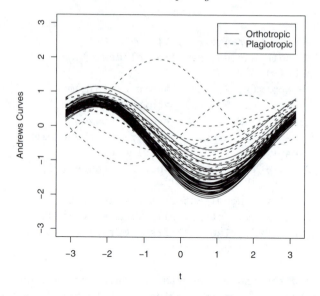

FIGURE 4.8: Andrews curves for `leafshape17` (DAAG) data at latitude 17.1: leaf length, width, and petiole measurements in Example 4.9. Curves are identified by leaf architecture.

R note 4.4 *In Figure 4.8 to identify the curves by color, replace* `lty` *with* `col` *parameters in the* `lines` *and* `legend` *statements.* ◇

4.5.2   Parallel Coordinate Plots

Parallel coordinate plots provide another approach to visualization of multivariate data. The representation of vectors by parallel coordinates was introduced by Inselberg [152] and applied for data analysis by Wegman [294].

Rather than represent axes as orthogonal, the parallel coordinate system represents axes as equidistant parallel lines. Usually these lines are horizontal with common origin, scale, and orientation. Then to represent vectors in $\mathbb{R}^d$, the parallel coordinates are simply the coordinates along the d copies of the real line. Each coordinate of a vector is then plotted along its corresponding axis, and the points are joined together with line segments.

Parallel coordinate plots are implemented by the `parcoord` function in the MASS package [278] and the `parallel` function in the `lattice` package [239]. The `parcoord` function displays the axes as vertical lines. The panel function `parallel` displays the axes as horizontal lines.

Example 4.10 (Parallel coordinates)

This example illustrates using the `parallel` (`lattice`) function to construct a panel display of parallel coordinate plots for the `crabs` (`MASS`) data [278]. The `crabs` data frame has 5 measurements on each of 200 crabs, from four groups of size 50. The groups are identified by species (blue or orange) and sex. The graph is best viewed in color. Here we use black and white, and for readability select only 1/5 of the data.

```
library(MASS)
library(lattice)
trellis.device(color = FALSE) #black and white display
x <- crabs[seq(5, 200, 5), ] #get every fifth obs.
parallel(~x[4:8] | sp*sex, x)
```

The resulting parallel coordinate plots are displayed in Figure 4.9(a). The labels along the vertical axis identify each axis corresponding to the five measurements (frontal lobe size, rear width, carapace length, carapace width, body depth). Much of the variability between groups is in overall size.

Adjusting the measurements of individual crabs for size may produce more interesting plots. Following the suggestion in Venables and Ripley [278] we adjust the measurements by the area of the carapace.

```
trellis.device(color = FALSE)    #black and white display
x <- crabs[seq(5, 200, 5), ]     #get every fifth obs.
a <- x$CW * x$CL                 #area of carapace
x[4:8] <- x[4:8] / sqrt(a)       #adjust for size
parallel(~x[4:8] | sp*sex, x)
```

In the resulting plot in Figure 4.9(b), differences in species and sex are much more evident after adjustment than in Figure 4.9(a).   ◇

4.5.3 Segments, stars, and other representations

Multivariate data can be represented by a two dimensional icon or glyph, such as a star. The Andrews curves in Example 4.9 are an example; the curves are the two-dimensional symbols. Andrews curves were displayed superimposed on the same coordinate system. Other representations as icons are best displayed in a table, so that features of observations can be compared. A tabular display does not have much practical value for high dimension or large data sets, but can be useful for some small data sets. Some examples include star plots and segment plots. This type of plot is easily obtained in R using the `stars` (`graphics`) function.

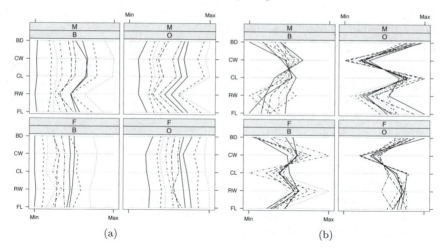

FIGURE 4.9:   Parallel coordinate plots in Example 4.10 for a subset of the crabs (MASS) data. (a) Differences between species (B=blue, O=orange) and sex (M, F) are largely obscured by large variation in overall size. (b) After adjusting the measurements for size of individual crabs, differences between groups are evident.

Example 4.11 (Segment plot)

This example uses the subset of crabs (MASS) data from Example 4.10. As in Example 4.10, individual measurements are adjusted for overall size by area of carapace.

```
#segment plot
library(MASS)  #for crabs data
attach(crabs)
x <- crabs[seq(5, 200, 5), ]          #get every fifth obs.
x <- subset(x, sex == "M")            #keep just the males
a <- x$CW * x$CL                      #area of carapace
x[4:8] <- x[4:8] / sqrt(a)            #adjust for size

#use default color palette or other colors
palette(gray(seq(.4, .95, len = 5))) #use gray scale
#palette(rainbow(6))                  #or use color
stars(x[4:8], draw.segments = TRUE,
      labels = x$sp, nrow = 4,
      ylim = c(-2,10), key.loc = c(3,-1))

#after viewing, restore the default colors
palette("default"); detach(crabs)
```

The plot is shown in Figure 4.10. The observations are labeled by species. The differences between the species (for males) in this sample are quite evident in the plot. The plot suggests, for example, that orange crabs have greater body depth relative to carapace width than blue crabs. ◇

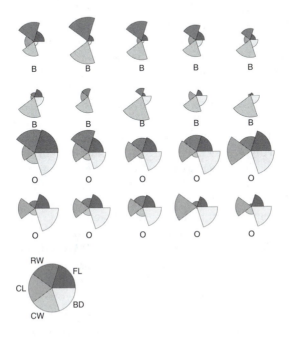

FIGURE 4.10: Segment plot of a subset of the males in the `crabs` (MASS) data set in Example 4.11. The measurements have been adjusted by overall size of the individual crab. The two species are blue (B) and orange (O).

4.6 Other Approaches to Data Visualization

Many other methods for data visualization are in the literature and we mention here only a few more. Asimov's grand tour [14] is an interactive graphical tool that projects data onto a plane, rotating through all angles to reveal any structure in the data. The grand tour is similar to projection pursuit exploratory data analysis (PPEDA) (Friedman and Tukey [100]). In both cases, structure might be defined as departure from normality. Once the structure is removed, the search can be repeated until no significant structure remains. Principal components analysis similarly uses projections (see e.g. [188, Ch. 8] and [278, Sec. 11.1]). When the data are projected onto the eigenvector corresponding to the maximal eigenvalue of the covariance matrix, this first principal component is in the direction that explains the most variation in the data. Dimension is reduced by projecting onto a small number of the principal components that collectively explain most of the variation. Pattern recognition and data mining are two broad areas of research that use some visualization methods. See Ripley [224] or Duda and Hart [75]. An interesting collection of topics on data mining and data visualization is found in Rao, Wegman, and Solka [222]. For an excellent resource on visualization of categorical data see Friendly [102] and `http://www.math.yorku.ca/SCS/vcd/`.

In addition to the R functions and packages mentioned in this chapter, several methods are available in other packages. Again, here we only name a few. Chernoff's faces [46] are implemented in `faces(aplpack)` [298] and in `faces(TeachingDemos)` [254]. Mosaic plots for visualization of categorical data are available in `mosaicplot`. Also see the package `vcd` [199] for visualization of categorical data. The functions `prcomp` and `princomp` provide principal components analysis. Many packages for R fall under the data mining or machine learning umbrella; for a start see `nnet` [278], `rpart` [268], and `randomForest` [176]. More packages are described on the Multivariate Task View and Machine Learning Task View on the CRAN web. Also see the graph gallery at `http://addictedtor.free.fr/graphiques/`.

The `rggobi` [167] package provides a command-line interface to GGobi, which is an open source visualization program for exploring high-dimensional data. GGobi has a graphical user interface, providing dynamic and interactive graphics. The GGobi software can be obtained from `http://www.ggobi.org/downloads/`. Readers are referred to documentation and examples at `http://www.ggobi.org/rggobi` and the book by Cook and Swayne [52] featuring examples using R and GGobi.

Exercises

4.1 Generate 200 random observations from the multivariate normal distribution having mean vector $\mu = (0, 1, 2)$ and covariance matrix

$$\Sigma = \begin{bmatrix} 1.0 & -0.5 & 0.5 \\ -0.5 & 1.0 & -0.5 \\ 0.5 & -0.5 & 1.0 \end{bmatrix}.$$

Construct a scatterplot matrix and verify that the location and correlation for each plot agrees with the parameters of the corresponding bivariate distributions.

4.2   Add a fitted smooth curve to each of the scatterplots in Figure 4.1 of Example 4.1. (`?panel.smooth`)

4.3   The random variables X and Y are independent and identically distributed with normal mixture distributions. The components of the mixture have $N(0, 1)$ and $N(3, 1)$ distributions with mixing probabilities p_1 and $p_2 = 1 - p_1$ respectively. Generate a bivariate random sample from the joint distribution of (X, Y) and construct a contour plot. Adjust the levels of the contours so that the the contours of the second mode are visible.

4.4   Construct a filled contour plot of the bivariate mixture in Exercise 4.3.

4.5   Construct a surface plot of the bivariate mixture in Exercise 4.3.

4.6   Repeat Exercise 4.3 for various different choices of the parameters of the mixture model, and compare the distributions through contour plots.

4.7   Create a parallel coordinates plot of the `crabs` (MASS) [278] data using all 200 observations. Compare the plots before and after adjusting the measurements by the size of the crab. Interpret the resulting plots.

4.8   Create a plot of Andrews curves for the `leafshape17` (DAAG) [185] data, using the logarithms of measurements (logwid, logpet, loglen). Set line type to identify leaf architecture as in Example 4.9. Compare with the plot in Figure 4.8.

4.9   Refer to the full `leafshape` (DAAG) data set. Produce Andrews curves for each of the six locations. Split the screen into six plotting areas, and display all six plots on one screen. Set line type or color to identify leaf architecture. Do the plots suggest differences in leaf shape by location?

4.10   Generalize the function in Example 4.9 to return the Andrews curve function for vectors in $\mathbb{R}^d$, where the dimension $d \geq 2$ is arbitrary. Test this function by producing Andrews curves for the `iris` data $(d = 4)$ and `crabs` (MASS) data $(d = 5)$.

4.11   Refer to the full `leafshape` (DAAG) data set. Display a segment style stars plot for leaf measurements at latitude 42 (Tasmania). Repeat using the logarithms of the measurements.

Chapter 5

Monte Carlo Integration and Variance Reduction

5.1 Introduction

Monte Carlo integration is a statistical method based on random sampling. Monte Carlo methods were developed in the late 1940's after World War II, but the idea of random sampling was not new. As early as 1777, Comte de Buffon used a random experiment to empirically check his probability calculation for the famous Buffon's needle experiment. Another well known example is that W. S. Gossett used random sampling to study the distribution of what are now called "Student t" statistics, publishing under the alias Student in 1908 [256]. The development of ENIAC, the first electronic computer, completed in 1946 at the University of Pennsylvania, and the seminal article by Metropolis and Ulam in 1949 [198] marked an important new era in the application of sampling methods. Teams of scientists at the Los Alamos National Laboratory and many other researchers contributed to the early development, including Ulam, Richtmyer, and von Neumann [276, 283]. For an interesting discussion of the history of the Monte Carlo method and scientific computing, see Eckhart [78] and Metropolis [195, 196].

5.2 Monte Carlo Integration

Let $g(x)$ be a function and suppose that we want to compute $\int_a^b g(x)dx$ (assuming that this integral exists). Recall that if X is a random variable with density $f(x)$, then the mathematical expectation of the random variable $Y = g(X)$ is

$$E[g(X)] = \int_{-\infty}^{\infty} g(x)f(x)dx.$$

If a random sample is available from the distribution of X, an unbiased estimator of $E[g(X)]$ is the sample mean.

5.2.1   Simple Monte Carlo estimator

Consider the problem of estimating $\theta = \int_0^1 g(x)dx$. If $X_1, \ldots, X_m$ is a random Uniform(0,1) sample then

$$\hat{\theta} = \overline{g_m(X)} = \frac{1}{m}\sum_{i=1}^{m} g(X_i)$$

converges to $E[g(X)] = \theta$ with probability 1, by the Strong Law of Large Numbers. The simple Monte Carlo estimator of $\int_0^1 g(x)dx$ is $\overline{g_m(X)}$.

Example 5.1 (Simple Monte Carlo integration)

Compute a Monte Carlo estimate of

$$\theta = \int_0^1 e^{-x}\, dx$$

and compare the estimate with the exact value.

```
m <- 10000
x <- runif(m)
theta.hat <- mean(exp(-x))
print(theta.hat)
print(1 - exp(-1))

[1] 0.6355289
[1] 0.6321206
```

The estimate is $\hat{\theta} \doteq 0.6355$ and $\theta = 1 - e^{-1} \doteq 0.6321$. ◇

To compute $\int_a^b g(t)dt$, make a change of variables so that the limits of integration are from 0 to 1. The linear transformation is $y = (t-a)/(b-a)$ and $dy = (1/(b-a))dt$. Substituting,

$$\int_a^b g(t)dt = \int_0^1 g(y(b-a)+a)(b-a)dy.$$

Alternately, we can replace the Uniform(0,1) density with any other density supported on the interval between the limits of integration. For example,

$$\int_a^b g(t)dt = (b-a)\int_a^b g(t)\frac{1}{b-a}dt$$

is $b - a$ times the expected value of $g(Y)$, where Y has the uniform density on (a, b). The integral is therefore $(b-a)$ times the average value of $g(\cdot)$ over (a, b).

Example 5.2 (Simple Monte Carlo integration, cont.)

Compute a Monte Carlo estimate of

$$\theta = \int_2^4 e^{-x}\,dx$$

and compare the estimate with the exact value of the integral.

```
m <- 10000
x <- runif(m, min=2, max=4)
theta.hat <- mean(exp(-x)) * 2
print(theta.hat)
print(exp(-2) - exp(-4))

[1] 0.1172158
[1] 0.1170196
```

The estimate is $\hat\theta \doteq 0.1172$ and $\theta = e^{-2} - e^{-4} \doteq 0.1170$. $\diamond$

To summarize, the simple Monte Carlo estimator of the integral $\theta = \int_a^b g(x)dx$ is computed as follows.

1. Generate $X_1, \ldots, X_m$, iid from Uniform(a, b).

2. Compute $\overline{g(X)} = \frac{1}{m}g(X_i)$.

3. $\hat\theta = (b - a)\overline{g(X)}$.

Example 5.3 (Monte Carlo integration, unbounded interval)

Use the Monte Carlo approach to estimate the standard normal cdf

$$\Phi(x) = \int_{-\infty}^x \frac{1}{\sqrt{2\pi}} e^{-t^2/2}\,dt.$$

First, notice that we cannot apply the algorithm above directly because the limits of integration cover an unbounded interval. However, we can break this problem into two cases: $x \geq 0$ and $x < 0$, and use the symmetry of the normal density to handle the second case. Then the problem is to estimate $\theta = \int_0^x e^{-t^2/2}dt$ for $x > 0$. This can be done by generating random Uniform$(0, x)$ numbers, but it would mean changing the parameters of the uniform distribution for each different value of the cdf required. Suppose that we prefer an algorithm that always samples from Uniform(0,1).

This can be accomplished by a change of variables. Making the substitution $y = t/x$, we have $dt = x\,dy$ and

$$\theta = \int_0^1 xe^{-(xy)^2/2}dy.$$

Thus, $\theta = E_Y[xe^{-(xY)^2/2}]$, where the random variable Y has the Uniform(0,1) distribution. Generate iid Uniform(0,1) random numbers $u_1, \ldots, u_m$, and compute

$$\hat{\theta} = \overline{g_m(u)} = \frac{1}{m} \sum_{i=1}^{m} x\, e^{-(u_i x)^2/2}.$$

The sample mean $\hat{\theta}$ converges to $E[\hat{\theta}] = \theta$ as $m \to \infty$. If $x > 0$, the estimate of $\Phi(x)$ is $0.5 + \hat{\theta}/\sqrt{2\pi}$. If $x < 0$ compute $\Phi(x) = 1 - \Phi(-x)$.

```
x <- seq(.1, 2.5, length = 10)
m <- 10000
u <- runif(m)
cdf <- numeric(length(x))
for (i in 1:length(x)) {
    g <- x[i] * exp(-(u * x[i])^2 / 2)
    cdf[i] <- mean(g) / sqrt(2 * pi) + 0.5
}
```

Now the estimates $\hat{\theta}$ for ten values of x are stored in the vector `cdf`. Compare the estimates with the value $\Phi(x)$ computed (numerically) by the `pnorm` function.

```
Phi <- pnorm(x)
print(round(rbind(x, cdf, Phi), 3))
```

Results for several values $x > 0$ are shown compared with the value of the normal cdf function `pnorm`. The Monte Carlo estimates appear to be very close to the `pnorm` values. (The estimates will be worse in the extreme upper tail of the distribution.)

```
      [,1]  [,2]  [,3]  [,4]  [,5]  [,6]  [,7]  [,8]  [,9] [,10]
x     0.10 0.367 0.633 0.900 1.167 1.433 1.700 1.967 2.233 2.500
cdf   0.54 0.643 0.737 0.816 0.879 0.925 0.957 0.978 0.990 0.997
Phi   0.54 0.643 0.737 0.816 0.878 0.924 0.955 0.975 0.987 0.994
```

Notice that it would have been simpler to generate random Uniform$(0, x)$ random variables and skip the transformation. This is left as an exercise. In fact, the integrand of the previous example is itself a density function, and we can generate random variables from this density. This provides a more direct approach to estimating the integral.   ◇

Example 5.4 (Example 5.3, cont.)

Let $I(\cdot)$ be the indicator function, and $Z \sim N(0, 1)$. Then for any constant x we have $E[I(Z \le x)] = P(Z \le x) = \Phi(x)$, the standard normal cdf evaluated at x.

Generate a random sample $z_1, \ldots, z_m$ from the standard normal distribution. Then the sample mean

$$\widehat{\Phi(x)} = \frac{1}{m} \sum_{i=1}^{m} I(z_i \leq x)$$

converges with probability one to its expected value $E[I(Z \leq x)] = P(Z \leq x) = \Phi(x)$.

```
x <- seq(.1, 2.5, length = 10)
m <- 10000
z <- rnorm(m)
dim(x) <- length(x)
p <- apply(x, MARGIN = 1,
           FUN = function(x, z) {mean(z < x)}, z = z)
```

Now the estimates in p for the sequence of x values can be compared to the result of the R normal cdf function pnorm.

```
Phi <- pnorm(x)
print(round(rbind(x, p, Phi), 3))
```

```
      [,1]   [,2]  [,3]  [,4]  [,5]  [,6]  [,7]  [,8]  [,9] [,10]
x     0.10  0.367 0.633 0.900 1.167 1.433 1.700 1.967 2.233 2.500
p     0.546 0.652 0.741 0.818 0.876 0.925 0.954 0.976 0.988 0.993
Phi   0.54  0.643 0.737 0.816 0.878 0.924 0.955 0.975 0.987 0.994
```

In this example, compared with the results in Example 5.3, it appears that we have better agreement with pnorm in the upper tail, but worse agreement near the center.   ◇

Summarizing, if $f(x)$ is a probability density function supported on a set A, (that is, $f(x) \geq 0$ for all $x \in \mathbb{R}$ and $\int_A f(x) = 1$), to estimate the integral

$$\theta = \int_A g(x)f(x)dx,$$

generate a random sample $x_1, \ldots, x_m$ from the distribution $f(x)$, and compute the sample mean

$$\hat{\theta} = \frac{1}{m} \sum_{i=1}^{m} g(x_i).$$

Then with probability one, $\hat{\theta}$ converges to $E[\hat{\theta}] = \theta$ as $m \to \infty$.

The standard error of $\hat{\theta} = \frac{1}{m} \sum_{i=1}^{m} g(x_i)$.

The variance of $\hat{\theta}$ is σ^2/m, where $\sigma^2 = Var_f(g(X))$. When the distribution of X is unknown we substitute for F_X the empirical distribution F_m of the sample $x_1, \ldots, x_m$. The variance of $\hat{\theta}$ can be estimated by

$$\frac{\hat{\sigma}^2}{m} = \frac{1}{m^2} \sum_{i=1}^{m} [g(x_i) - \overline{g(x)}]^2. \tag{5.1}$$

Note that

$$\frac{1}{m} \sum_{i=1}^{m} [g(x_i) - \overline{g(x)}]^2 \tag{5.2}$$

is the plug-in estimate of $Var(g(X))$. That is, (5.2) is the variance of U, where U is uniformly distributed on the set of replicates $\{g(x_i)\}$. The corresponding estimate of standard error of $\hat{\theta}$ is

$$\widehat{se}(\hat{\theta}) = \frac{\hat{\sigma}}{\sqrt{m}} = \frac{1}{m} \left\{ \sum_{i=1}^{m} [g(x_i) - \overline{g(x)}]^2 \right\}^{1/2}. \tag{5.3}$$

The Central Limit Theorem implies that

$$\frac{\hat{\theta} - E[\hat{\theta}]}{\sqrt{Var\,\hat{\theta}}}$$

converges in distribution to $N(0,1)$ as $m \to \infty$. Hence, if m is sufficiently large, $\hat{\theta}$ is approximately normal with mean θ. The large-sample, approximately normal distribution of $\hat{\theta}$ can be applied to put confidence limits or error bounds on the Monte Carlo estimate of the integral, and check for convergence.

Example 5.5 (Error bounds for MC integration)

Estimate the variance of the estimator in Example 5.4, and construct approximate 95% confidence intervals for the estimate of $\Phi(2)$ and $\Phi(2.5)$.

```
x <- 2
m <- 10000
z <- rnorm(m)
g <- (z < x)   #the indicator function
v <- mean((g - mean(g))^2) / m
cdf <- mean(g)
c(cdf, v)
c(cdf - 1.96 * sqrt(v), cdf + 1.96 * sqrt(v))

[1] 9.772000e-01 2.228016e-06
[1] 0.9742744 0.9801256
```

The probability $P(I(Z < x) = 1)$ is $\Phi(2) \approx 0.977$. Here $g(X)$ has the distribution of the sample proportion of 1's in $m = 10000$ Bernoulli trials with $p \doteq 0.977$, and the variance of $g(X)$ is therefore $(0.977)(1 - 0.977)/10000 = 2.223\text{e-}06$. The MC estimate 2.228e-06 of variance is quite close to this value.

For $x = 2.5$ the output is

```
[1] 9.94700e-01 5.27191e-07
[1] 0.9932769 0.9961231
```

The probability $P(I(Z < x) = 1)$ is $\Phi(2.5) \approx 0.995$. The Monte Carlo estimate 5.272e-07 of variance is approximately equal to the theoretical value $(0.995)(1 - 0.995)/10000 = 4.975\text{e-}07$. $\diamond$

5.2.2 Variance and Efficiency

We have seen that a Monte Carlo approach to estimating the integral $\int_a^b g(x)dx$ is to represent the integral as the expected value of a function of a uniform random variable. That is, if $X \sim \text{Uniform}(a, b)$, then $f(x) = \frac{1}{b-a}$, $a < x < b$, and

$$\theta = \int_a^b g(x)dx$$

$$= (b - a) \int_a^b g(x)\frac{1}{b - a}dx = (b - a)E[g(X)].$$

Recall that the sample-mean Monte Carlo estimator of the integral θ is computed as follows.

1. Generate $X_1, \ldots, X_m$, iid from $\text{Uniform}(a, b)$.
2. Compute $\overline{g(X)} = \frac{1}{m}g(X_i)$.
3. $\hat{\theta} = (b - a)\overline{g(X)}$.

The sample mean $\overline{g(X)}$ has expected value $g(X) = \theta/(b - a)$, and

$$Var(\overline{g(X)}) = (1/m)Var(g(X)).$$

Therefore $E[\hat{\theta}] = \theta$ and

$$Var(\hat{\theta}) = (b - a)^2 Var(\overline{g(X)}) = \frac{(b - a)^2}{m} Var(g(X)). \tag{5.4}$$

By the Central Limit Theorem, for large m, $\overline{g(X)}$ is approximately normally distributed, and therefore $\hat{\theta}$ is approximately normally distributed with mean θ and variance given by (5.4).

The "hit-or-miss" approach to Monte Carlo integration also uses a sample mean to estimate the integral, but the sample mean is taken over a different sample and therefore this estimator has a different variance than formula (5.4).

Suppose $f(x)$ is the density of a random variable X. The "hit-or-miss" approach to estimating $F(x) = \int_{-\infty}^x f(t)dt$ is as follows.

1. Generate a random sample $X_1, \ldots, X_m$ from the distribution of X.
2. For each observation X_i, compute

$$g(X_i) = I(X_i \leq x) = \begin{cases} 1, & X_i \leq x; \\ 0, & X_i > x. \end{cases}$$

3. Compute $\widehat{F(x)} = \overline{g(X)} = \frac{1}{m} \sum_{i=1}^{m} I(X_i \leq x)$.

Note that the random variable $Y = g(X)$ has the Binomial$(1, p)$ distribution, where the success probability is $p = P(X \leq x) = F(x)$. The transformed sample $Y_1, \ldots, Y_m$ are the outcomes of m independent, identically distributed Bernoulli trials. The estimator $\widehat{F(x)}$ is the sample proportion $\hat{p} = y/m$, where y is the total number of successes observed in m trials. Hence $E[\widehat{F(x)}] = p = F(x)$ and $Var(\widehat{F(x)}) = p(1-p)/m = F(x)(1 - F(x))/m$.

The variance of $\widehat{F(x)}$ can be estimated by $\hat{p}(1-\hat{p})/m = \widehat{F(x)}(1-\widehat{F(x)})/m$. The maximum variance occurs when $F(x) = 1/2$, so a conservative estimate of the variance of $\widehat{F(x)}$ is $1/(4m)$.

Efficiency

If $\hat{\theta}_1$ and $\hat{\theta}_2$ are two estimators for θ, then $\hat{\theta}_1$ is more efficient (in a statistical sense) than $\hat{\theta}_2$ if

$$\frac{Var(\hat{\theta}_1)}{Var(\hat{\theta}_2)} < 1.$$

If the variances of estimators $\hat{\theta}_i$ are unknown, we can estimate efficiency by substituting a sample estimate of the variance for each estimator.

Note that variance can alway be reduced by increasing the number of replicates, so computational efficiency is also relevant.

5.3   Variance Reduction

We have seen that Monte Carlo integration can be applied to estimate functions of the type $E[g(X)]$. In this section we consider several approaches to reducing the variance in the sample mean estimator of $\theta = E[g(X)]$.

If $\hat{\theta}_1$ and $\hat{\theta}_2$ are estimators of the parameter θ, and $Var(\hat{\theta}_2) < Var(\hat{\theta}_1)$, then the percent reduction in variance achieved by using $\hat{\theta}_2$ instead of $\hat{\theta}_1$ is

$$100 \left(\frac{Var(\hat{\theta}_1) - Var(\hat{\theta}_2)}{Var(\hat{\theta}_1)} \right).$$

The Monte Carlo approach to estimating $\theta = E[g(X)]$ is to compute the sample mean $\overline{g(X)}$ for a large number m of replicates from the distribution of $g(X)$. The function $g(\cdot)$ is often a statistic; that is, an n-variate function $g(X_1, \ldots, X_n)$ of a sample. When $g(X)$ is used in that context, we have $g(\mathcal{X}) = g(X_1, \ldots, X_n)$, where $\mathcal{X}$ denotes the sample elements. Unless it is not clear in context, however, for simplicity we use $g(X)$.

Let

$$X^{(j)} = \{X_1^{(j)}, \ldots, X_n^{(j)}\}, \qquad j = 1, \ldots, m$$

be iid from the distribution of X, and compute the corresponding replicates

$$Y_j = g(X_1^{(j)}, \ldots, X_n^{(j)}), \qquad j = 1, \ldots, m. \tag{5.5}$$

Then $Y_1, \ldots, Y_m$ are independent and identically distributed with distribution of $Y = g(\mathcal{X})$, and

$$E[\overline{Y}] = E\left[\frac{1}{m} \sum_{j=1}^{m} Y_j\right] = \theta.$$

Thus, the Monte Carlo estimator $\hat{\theta} = \overline{Y}$ is unbiased for $\theta = E[Y]$. The variance of the Monte Carlo estimator is

$$Var(\hat{\theta}) = Var\overline{Y} = \frac{Var_f g(X)}{m}.$$

Increasing the number of replicates m clearly reduces the variance of the Monte Carlo estimator. However, a large increase in m is needed to get even a small improvement in standard error. To reduce the standard error from 0.01 to 0.0001, we would need approximately 10000 times the number of replicates. In general, if standard error should be at most e and $Var_f(g(X)) = \sigma^2$, then $m \geq \lceil \sigma^2 / e^2 \rceil$ replicates are required.

Thus, although variance can always be reduced by increasing the number of Monte Carlo replicates, the computational cost is high. Other methods for reducing the variance can be applied that are less computationally expensive than simply increasing the number of replicates.

In the following sections some approaches to reducing the variance of this type of estimator are introduced. Several approaches have been covered in the literature. Readers are referred to [69, 112, 113, 121, 228, 233, 238] for reference and more examples.

5.4   Antithetic Variables

Consider the mean of two identically distributed random variables U_1 and U_2. If U_1 and U_2 are independent, then

$$Var\left(\frac{U_1 + U_2}{2}\right) = \frac{1}{4}(Var(U_1) + Var(U_2)),$$

but in general we have

$$Var\left(\frac{U_1 + U_2}{2}\right) = \frac{1}{4}(Var(U_1) + Var(U_2) + 2Cov(U_1, U_2)),$$

so the variance of $(U_1 + U_2)/2$ is smaller if U_1 and U_2 are negatively correlated than when the variables are independent. This fact leads us to consider negatively correlated variables as a possible method for reducing variance.

For example, suppose that $X_1, \dots, X_n$ are simulated via the inverse transform method. For each of the m replicates we have generated $U_j \sim$ Uniform$(0,1)$, and computed $X^{(j)} = F_X^{-1}(U_j)$, $j = 1, \dots, n$. Note that if U is uniformly distributed on $(0, 1)$ then $1 - U$ has the same distribution as U, but U and $1 - U$ are negatively correlated. Then in (5.5)

$$Y_j = g(F_X^{-1}(U_1^{(j)}), \dots, F_X^{-1}(U_n^{(j)}))$$

has the same distribution as

$$Y_j' = g(F_X^{-1}(1 - U_1^{(j)}), \dots, F_X^{-1}(1 - U_n^{(j)})).$$

Under what conditions are Y_j and Y_j' negatively correlated? Below it is shown that if the function g is *monotone*, the variables Y_j and Y_j' are negatively correlated.

Define $(x_1, \dots, x_n) \leq (y_1, \dots, y_n)$ if $x_j \leq y_j$, $j = 1, \dots, n$. An n-variate function $g = g(X_1, \dots, X_n)$ is *increasing* if it is increasing in its coordinates. That is, g is increasing if $g(x_1, \dots, x_n) \leq g(y_1, \dots, y_n)$ whenever $(x_1, \dots, x_n) \leq (y_1, \dots, y_n)$. Similarly g is *decreasing* if it is decreasing in its coordinates. Then g is *monotone* if it is increasing or decreasing.

PROPOSITION 5.1 *If $X_1, \dots, X_n$ are independent, and f and g are increasing functions, then*

$$E[f(X)g(X)] \geq E[f(X)]E[g(X)]. \tag{5.6}$$

Proof. Assume that f and g are increasing functions. The proof is by induction on n. Suppose $n = 1$. Then $(f(x) - f(y))(g(x) - g(y)) \geq 0$ for all $x, y \in \mathbb{R}$. Hence $E[(f(X) - f(Y))(g(X) - g(Y))] \geq 0$, and

$$E[f(X)g(X)] + E[f(Y)g(Y)] \geq E[f(X)g(Y)] + E[f(Y)g(X)].$$

Here X and Y are iid, so

$$2E[f(X)g(X)] = E[f(X)g(X)] + E[f(Y)g(Y)]$$
$$\geq E[f(X)g(Y)] + E[f(Y)g(X)] = 2E[f(X)]E[g(X)],$$

so the statement is true for $n = 1$. Suppose that the statement (5.6) is true for $X \in \mathbb{R}^{n-1}$. Condition on X_n and apply the induction hypothesis to obtain

$$E[f(X)g(X)|X_n = x_n] \geq E[f(X_1,\ldots,X_{n-1},x_n)]E[g(X_1,\ldots,X_{n-1},x_n)]$$
$$= E[f((X)|X_n = x_n]E[g((X)|X_n = x_n)],$$

or

$$E[f(X)g(X)|X_n] \geq E[f(X)|X_n]E[g(X)|X_n)].$$

Now $E[f(X)|X_n]$ and $E[g(X)|X_n)]$ are each increasing functions of X_n, so applying the result for $n = 1$ and taking the expected values of both sides

$$E[f(X)g(X)] \geq E[E[f(X)|X_n]\,E[g(X)|X_n)]] \geq E[f(X)]E[g(X)].$$

$\square$

COROLLARY 5.1 *If* $g = g(X_1,\ldots,X_n)$ *is monotone, then*

$$Y = g(F_X^{-1}(U_1),\ldots,F_X^{-1}(U_n))$$

and

$$Y' = g(F_X^{-1}(1 - U_1),\ldots,F_X^{-1}(1 - U_n)).$$

are negatively correlated.

Proof. Without loss of generality we can suppose that g is increasing. Then

$$Y = g(F_X^{-1}(U_1),\ldots,F_X^{-1}(U_n))$$

and

$$-Y' = f = -g(F_X^{-1}(1 - U_1),\ldots,F_X^{-1}(1 - U_n))$$

are both increasing functions. Therefore $E[g(U)f(U)] \geq E[g(U)]E[f(U)]$ and $E[YY'] \leq E[Y]E[Y']$, which implies that

$$Cov(Y,Y') = E[YY'] - E[Y]E[Y'] \leq 0,$$

so Y and Y' are negatively correlated. $\square$

The antithetic variable approach is easy to apply. If m Monte Carlo replicates are required, generate $m/2$ replicates

$$Y_j = g(F_X^{-1}(U_1^{(j)}),\ldots,F_X^{-1}(U_n^{(j)})) \tag{5.7}$$

and the remaining $m/2$ replicates

$$Y_j' = g(F_X^{-1}(1 - U_1^{(j)}), \ldots, F_X^{-1}(1 - U_n^{(j)})), \qquad (5.8)$$

where $U_i^{(j)}$ are iid Uniform(0,1) variables, $i = 1, \ldots, n$, $j = 1, \ldots, m/2$. Then the antithetic estimator is

$$\hat{\theta} = \frac{1}{m}\{Y_1 + Y_1' + Y_2 + Y_2' + \cdots + Y_{m/2} + Y_{m/2}'\}$$

$$= \frac{2}{m} \sum_{j=1}^{m/2} \left(\frac{Y_j + Y_j'}{2}\right).$$

Thus $nm/2$ rather than nm uniform variates are required, and the variance of the Monte Carlo estimator is reduced by using antithetic variables.

Example 5.6 (Antithetic variables)

Refer to Example 5.3, illustrating Monte Carlo integration applied to estimate the standard normal cdf

$$\Phi(x) = \int_{-\infty}^{x} \frac{1}{\sqrt{2\pi}} e^{-t^2/2} \, dt.$$

Repeat the estimation using antithetic variables, and find the approximate reduction in standard error. In this example (after change of variables) the target parameter is $\theta = E_U[xe^{-(xU)^2/2}]$, where U has the Uniform(0,1) distribution.

By restricting the simulation to the upper tail (see Example 5.3) the function $g(\cdot)$ is monotone, so the hypothesis of Corollary 5.1 is satisfied. Generate random numbers $u_1, \ldots, u_{m/2} \sim$ Uniform$(0, 1)$ and compute half of the replicates using

$$Y_j = g^{(j)}(u) = x \, e^{-(u_j x)^2/2}, \qquad j = 1, \ldots, m/2$$

as before, but compute the remaining half of the replicates using

$$Y_j' = x \, e^{-((1 - u_j)x)^2/2}, \qquad j = 1, \ldots, m/2.$$

The sample mean

$$\hat{\theta} = \overline{g_m(u)} = \frac{1}{m} \sum_{j=1}^{m/2} \left(x \, e^{-(u_j x)^2/2} + x \, e^{-((1 - u_j)x)^2/2}\right)$$

$$= \frac{1}{m/2} \sum_{j=1}^{m/2} \left(\frac{x \, e^{-(u_j x)^2/2} + x \, e^{-((1 - u_j)x)^2/2}}{2}\right)$$

converges to $E[\hat{\theta}] = \theta$ as $m \to \infty$. If $x > 0$, the estimate of $\Phi(x)$ is $0.5 + \hat{\theta}/\sqrt{2\pi}$. If $x < 0$ compute $\Phi(x) = 1 - \Phi(-x)$. The Monte Carlo estimation of the integral $\Phi(x)$ is implemented in the function MC.Phi below. Optionally MC.Phi will compute the estimate with or without antithetic sampling. The MC.Phi function could be made more general if an argument naming a function, the integrand, is added (see integrate for an example of this type of argument to a function).

```
MC.Phi <- function(x, R = 10000, antithetic = TRUE) {
    u <- runif(R/2)
    if (!antithetic) v <- runif(R/2) else
        v <- 1 - u
    u <- c(u, v)
    cdf <- numeric(length(x))
    for (i in 1:length(x)) {
        g <- x[i] * exp(-(u * x[i])^2 / 2)
        cdf[i] <- mean(g) / sqrt(2 * pi) + 0.5
    }
    cdf
}
```

A comparison of estimates obtained from a single Monte Carlo experiment is below.

```
x <- seq(.1, 2.5, length=5)
Phi <- pnorm(x)
set.seed(123)
MC1 <- MC.Phi(x, anti = FALSE)
set.seed(123)
MC2 <- MC.Phi(x)
print(round(rbind(x, MC1, MC2, Phi), 5))
```

```
        [,1]    [,2]    [,3]    [,4]    [,5]
x    0.10000 0.70000 1.30000 1.90000 2.50000
MC1  0.53983 0.75825 0.90418 0.97311 0.99594
MC2  0.53983 0.75805 0.90325 0.97132 0.99370
Phi  0.53983 0.75804 0.90320 0.97128 0.99379
```

The approximate reduction in variance can be estimated for given x by a simulation under both methods, the simple Monte Carlo integration approach and the antithetic variable approach.

```
m <- 1000
MC1 <- MC2 <- numeric(m)
x <- 1.95
for (i in 1:m) {
    MC1[i] <- MC.Phi(x, R = 1000, anti = FALSE)
    MC2[i] <- MC.Phi(x, R = 1000)
}

> print(sd(MC1))
[1] 0.007008661
> print(sd(MC2))
[1] 0.000470819
> print((var(MC1) - var(MC2))/var(MC1))
[1] 0.9954873
```

The antithetic variable approach achieved approximately 99.5% reduction in variance at $x = 1.95$.   ◇

5.5   Control Variates

Another approach to reduce the variance in a Monte Carlo estimator of $\theta = E[g(X)]$ is the use of control variates. Suppose that there is a function f, such that $\mu = E[f(X)]$ is known, and $f(X)$ is correlated with $g(X)$.

Then for any constant c, it is easy to check that $\hat{\theta}_c = g(X) + c(f(Y) - \mu)$ is an unbiased estimator of θ.

The variance

$$Var(\hat{\theta}_c) = Var(g(X)) + c^2 Var(f(X)) + 2c\, Cov(g(X), f(X)) \qquad (5.9)$$

is a quadratic function of c. It is minimized at $c = c^*$, where

$$c^* = -\frac{Cov(g(X), f(X))}{Var(f(X))}$$

and minimum variance is

$$Var(\hat{\theta}_{c^*}) = Var(g(X)) - \frac{[Cov(g(X), f(X))]^2}{Var(f(X))}. \qquad (5.10)$$

The random variable $f(X)$ is called a *control variate* for the estimator $g(X)$. In (5.10) we see that $Var(g(X))$ is reduced by

$$\frac{[Cov(g(X), f(X))]^2}{Var(f(X))},$$

hence the percent reduction in variance is

$$100\frac{[Cov(g(X), f(X))]^2}{Var(g(X))\, Var(f(X))} = 100[Cor(g(X), f(X))]^2.$$

Thus, it is advantageous if $f(X)$ and $g(X)$ are strongly correlated. No reduction of variance is possible in case $f(X)$ and $g(Y)$ are uncorrelated.

To compute the constant c^*, we need $Cov(g(X), f(X))$ and $Var(f(X))$, but these parameters can be estimated if necessary, from a preliminary Monte Carlo experiment.

Example 5.7 (Control variate)

Apply the control variate approach to compute

$$\theta = E[e^U] = \int_0^1 e^u du,$$

where $U \sim$ Uniform(0,1). In this example, we do not need simulation because $\theta = e - 1 = 1.718282$ by integration, but this provides an example where we can verify that the control variate approach is correctly implemented. If the simple Monte Carlo approach is applied with m replicates, the variance of the estimator is $Var(g(U))/m$, where

$$Var(g(U)) = Var(e^U) = E[e^{2U}] - \theta^2 = \frac{e^2 - 1}{2} - (e - 1)^2 \doteq 0.2420351.$$

A natural choice for a control variate is $U \sim$ Uniform(0,1). Then $E[U] = 1/2$, $Var(U) = 1/12$, and $Cov(e^U, U) = 1 - (1/2)(e - 1) \doteq 0.1408591$. Hence

$$c^* = \frac{-Cov(e^U, U)}{Var(U)} = -12 + 6(e - 1) \doteq -1.690309.$$

Our controlled estimator is $\hat{\theta}_{c^*} = e^U - 1.690309(U - 0.5)$. For m replicates, $mVar(\hat{\theta}_{c^*})$ is

$$Var(e^U) - \frac{[Cov(e^U, U)]^2}{Var(U)} = \frac{e^2 - 1}{2} - (e - 1)^2 - 12\left(1 - \frac{e - 1}{2}\right)$$

$$\doteq 0.2420356 - 12(0.1408591)^2$$

$$= 0.003940175.$$

The percent reduction in variance using the control variate compared with the simple Monte Carlo estimate is $100(1 - 0.003940175/0.2429355) = 98.3781\%$.

Now we implement the control variate method for this problem and compute empirically the percent reduction in variance achieved in the simulation. Comparing the simple Monte Carlo estimate with the control variate approach

```
m <- 10000
a <- - 12 + 6 * (exp(1) - 1)
U <- runif(m)
T1 <- exp(U)                    #simple MC
T2 <- exp(U) + a * (U - 1/2)    #controlled
```

gives the following results

```
> mean(T1)
[1] 1.717834
> mean(T2)
[1] 1.718229
> (var(T1) - var(T2)) / var(T1)
[1] 0.9838606
```

illustrating that the percent reduction 98.3781% in variance derived above is approximately achieved in this simulation.   ◇

Example 5.8 (MC integration using control variates)

Use the method of control variates to estimate

$$\int_0^1 \frac{e^{-x}}{1+x^2}\,dx.$$

(A version of this problem appears in [64, p. 734].) The parameter of interest is $\theta = E[g(X)]$ and $g(X) = e^{-x}/(1+x^2)$, where X is uniformly distributed on (0,1). We seek a function 'close' to $g(x)$ with known expected value, such that $g(X)$ and $f(X)$ are strongly correlated. For example, the function $f(x) = e^{-.5}(1+x^2)^{-1}$ is 'close' to $g(x)$ on (0,1) and we can compute its expectation. If U is uniformly distributed on (0,1), then

$$E[f(U)] = e^{-.5} \int_0^1 \frac{1}{1+u^2}\,du = e^{-.5}\arctan(1) = e^{-.5}\frac{\pi}{4}.$$

Setting up a preliminary simulation to obtain an estimate of the constant c^*, we also obtain an estimate of $Cor(g(U), f(U)) \approx 0.974$.

```
f <- function(u)
    exp(-.5)/(1+u^2)

g <- function(u)
    exp(-u)/(1+u^2)

set.seed(510) #needed later
u <- runif(10000)
B <- f(u)
A <- g(u)
```

Estimates of c^* and $Cor(f(U), g(U))$ are

```
> cor(A, B)
[1] 0.9740585
a <- -cov(A,B) / var(B)     #est of c*
> a
[1] -2.436228
```

Simulation results with and without the control variate follow.

```
m <- 100000
u <- runif(m)
T1 <- g(u)
T2 <- T1 + a * (f(u) - exp(-.5)*pi/4)

> c(mean(T1), mean(T2))
[1] 0.5253543 0.5250021
> c(var(T1), var(T2))
[1] 0.060231423 0.003124814
> (var(T1) - var(T2)) / var(T1)
[1] 0.9481199
```

Here the approximate reduction in variance of $g(X)$ compared with $g(X) + \hat{c}^*(f(X)-\mu)$ is 95%. We will return to this problem to apply another approach to variance reduction, the method of importance sampling. ◇

5.5.1 Antithetic variate as control variate.

The antithetic variate estimator of the previous section is actually a special case of the control variate estimator. First notice that the control variate estimator is a linear combination of unbiased estimators of θ. In general, if $\hat{\theta}_1$ and $\hat{\theta}_2$ are any two unbiased estimators of θ, then for every constant c,

$$\hat{\theta}_c = c\hat{\theta}_1 + (1 - c)\hat{\theta}_2$$

is also unbiased for θ. The variance of $c\hat{\theta}_1 + (1 - c)\hat{\theta}_2$ is

$$Var(\hat{\theta}_2) + c^2 Var(\hat{\theta}_1 - \hat{\theta}_2) + 2c\, Cov(\hat{\theta}_2, \hat{\theta}_1 - \hat{\theta}_2). \tag{5.11}$$

In the special case of antithetic variates in (5.7) and (5.8), $\hat{\theta}_1$ and $\hat{\theta}_2$ are identically distributed and $Cor(\hat{\theta}_1, \hat{\theta}_2) = -1$. Then $Cov(\hat{\theta}_1, \hat{\theta}_2) = -Var(\hat{\theta}_1)$, and the variance in (5.11) is

$$Var\hat{\theta}_c = 4c^2 Var(\hat{\theta}_1) - 4cVar(\hat{\theta}_1) + Var(\hat{\theta}_1) = (4c^2 - 4c + 1)Var(\hat{\theta}_1),$$

and the optimal constant is $c^* = 1/2$. The control variate estimator in this case is

$$\hat{\theta}_{c^*} = \frac{\hat{\theta}_1 + \hat{\theta}_2}{2},$$

which (for this particular choice of $\hat{\theta}_1$ and $\hat{\theta}_2$) is the antithetic variable estimator of θ.

5.5.2 Several control variates.

The idea of combining unbiased estimators of the target parameter θ to reduce variance can be extended to several control variables. In general, if $E[\hat{\theta}_i] = \theta$, $i = 1, 2, \ldots k$ and $c = (c_1, \ldots, c_k)$ such that $\sum_{i=1}^{k} c_i = 1$, then

$$\sum_{i=1}^{k} c_i \hat{\theta}_i$$

is also unbiased for θ. The corresponding control variate estimator is

$$\hat{\theta}_c = g(X) + \sum_{i=1}^{k} c_i^*(f_i(X) - \mu_i))$$

where $\mu_i = E[f_i(X)]$, $i = 1, \ldots, k$, and

$$E[\hat{\theta}_c] = E[g(X)] + \sum_{i=1}^{k} c_i^* E[f_i(X) - \mu_i] = \theta.$$

The controlled estimate $\hat{\theta}_{\hat{c}^*}$, and estimates for the optimal constants c_i^*, can be obtained by fitting a linear regression model. The details are discussed in section 5.5.3.

5.5.3 Control variates and regression.

In this section we will discuss the duality between the control variate approach and simple linear regression. This provides more insight into how the control variate reduces the variance in Monte Carlo integration. In addition, we have a convenient method for estimating the optimal constant c^*, the target parameter, the percent reduction in variance, and the standard error of the estimator, all by fitting a simple linear regression model.

Suppose that $(X_1, Y_1), \ldots, (X_n, Y_n)$ is a random sample from a bivariate distribution with mean (μ_X, μ_Y) and variances (σ_X^2, σ_Y^2). Let us compare the least squares estimators for regression of X on Y with the control variate estimator.

If there is a linear relation $X = \beta_1 Y + \beta_0 + \varepsilon$, and $E[\varepsilon] = 0$, then

$$E[X] = E[E[X|Y]] = E[\beta_0 + \beta_1 Y + \varepsilon] = \beta_0 + \beta_1 \mu_Y.$$

Here β_0 and β_1 are constant parameters and ε is a random error variable.

Let us consider the bivariate sample $(g(X_1), f(X_1)), \ldots, (g(X_n), f(X_n))$. Now if $g(X)$ replaces X and $f(X)$ replaces Y, we have $g(X) = \beta_0 + \beta_1 f(X) + \varepsilon$, and

$$E[g(X)] = \beta_0 + \beta_1 E[f(X)].$$

The least squares estimator of the slope is

$$\hat{\beta}_1 = \frac{\sum_{i=1}^{n}(X_i - \overline{X})(Y_i - \overline{Y})}{\sum_{i=1}^{n}(Y_i - \overline{Y})^2} = \frac{\widehat{Cov}(X, Y)}{\widehat{Var}(Y)} = \frac{\widehat{Cov}(g(X), f(X))}{\widehat{Var}(f(X))} = -\hat{c}^*.$$

This shows that a convenient way to estimate c^* is to use the estimated slope from the fitted simple linear regression model of $g(X)$ on $f(X)$:

```
L <- lm(gx ~ fx)
c.star <- -L$coeff[2]
```

The least squares estimator of the intercept is $\hat{\beta}_0 = \overline{g(X)} - (-\hat{c}^*)\overline{f(X)}$, so that the predicted response at $\mu = E[f(X)]$ is

$$\hat{\beta}_0 + \hat{\beta}_1\mu = \overline{g(X)} + \hat{c}^*(\overline{f(X)} - \hat{c}^*\mu)$$
$$= \overline{g(X)} + \hat{c}^*(\overline{f(X)} - \mu) = \hat{\theta}_{\hat{c}^*}.$$

Thus, the control variate estimate $\hat{\theta}_{\hat{c}^*}$ is the predicted value of the response variable $(g(X))$ at the point $\mu = E[f(X)]$.

The estimate of the error variance in the regression of X on Y is

$$\hat{\sigma}_{\varepsilon}^2 = \widehat{Var}(X - \hat{X}) = \widehat{Var}(X - (\hat{\beta}_0 + \hat{\beta}_1 Y))$$
$$= \widehat{Var}(X - \hat{\beta}_1 Y) = \widehat{Var}(X + \hat{c}^* Y),$$

the residual mean squared error (MSE). The estimate of variance of the control variate estimator is

$$\widehat{Var}(\overline{g(X)} + \hat{c}^*(\overline{f(X)} - \mu)) = \frac{\widehat{Var}(g(X) + \hat{c}^*(f(X) - \mu))}{n}$$
$$= \frac{\widehat{Var}(g(X) + \hat{c}^* f(X))}{n} = \frac{\hat{\sigma}_{\varepsilon}^2}{n}.$$

Thus, the estimated standard error of the control variate estimate is easily computed using R by applying the `summary` method to the `lm` object from the fitted regression model, for example using

```
se.hat <- summary(L)$sigma
```

to extract the value of $\hat{\sigma}_{\varepsilon} = \sqrt{MSE}$.

Finally, recall that the proportion of reduction in variance for the control variate is $[Cor(g(X), f(X))]^2$. In the simple linear regression model, the coefficient of determination is same number (R^2), which is the proportion of total variation in $g(X)$ about its mean explained by $f(X)$.

Example 5.9 (Control variate and regression)

Returning to Example 5.8, let us repeat the estimation by fitting a regression model. In this problem,

$$g(x) = \int_0^1 \frac{e^{-x}}{1 + x^2} dx$$

and the control variate is

$$f(x) = e^{-.5}(1 + x^2)^{-1}, \qquad 0 < x < 1,$$

with $\mu = E[f(X)] = e^{-.5}\pi/4$. To estimate the constant c^*,

```
set.seed(510)
u <- runif(10000)
f <- exp(-.5)/(1+u^2)
g <- exp(-u)/(1+u^2)
c.star <-  - lm(g ~ f)$coeff[2]    # beta[1]
mu <- exp(-.5)*pi/4
```

```
> c.star
       f
-2.436228
```

We used the same random number seed as in Example 5.8 and obtained the same estimate for c^*. Now $\hat{\theta}_{\hat{c}^*}$ is the predicted response at the point $\mu = 0.4763681$, so

```
u <- runif(10000)
f <- exp(-.5)/(1+u^2)
g <- exp(-u)/(1+u^2)
L <- lm(g ~ f)
theta.hat <- sum(L$coeff * c(1, mu))   #pred. value at mu
```

The estimate $\hat{\theta}$, residual mean squared error and the proportion of reduction in variance (R-squared) agree with the estimates obtained in Example 5.8.

```
> theta.hat
[1] 0.5253113
> summary(L)$sigma^2
[1] 0.003117644
> summary(L)$r.squared
[1] 0.9484514
```

$\diamond$

In case several control variates are used, similarly one can estimate a linear model

$$X = \beta_0 + \sum_{i=1}^{k} \beta_i Y_i + \varepsilon$$

to estimate the optimal constants $c^* = (c_1^*, \ldots, c_k^*)$. Then $-\hat{c}^* = (\hat{\beta}_1, \ldots, \hat{\beta}_k)$ and the estimate is the predicted response $\hat{X}$ at the point $\mu = (\mu_1, \ldots, \mu_k)$ (see section 5.5.2). The estimated variance of the controlled estimator is again $\hat{\sigma}_\varepsilon^2/n = MSE/n$, where n is the sample size (the number of replicates, in this case).

5.6 Importance Sampling

The average value of a function $g(x)$ over an interval (a, b) is usually defined (in calculus) by

$$\frac{1}{b-a} \int_a^b g(x)dx.$$

Here a uniform weight function is applied over the entire interval (a, b). If X is a random variable uniformly distributed on (a, b), then

$$E[g(X)] = \int_a^b g(x)\frac{1}{b-a}\,dx = \frac{1}{b-a} \int_a^b g(x)dx, \tag{5.12}$$

which is simply the average value of the function $g(x)$ over the interval (a, b) with respect to a uniform weight function. The simple Monte Carlo method generates a large number of replicates $X_1, \ldots, X_m$ uniformly distributed on $[a, b]$ and estimates $\int_a^b g(x)dx$ by the sample mean

$$\frac{b-a}{m} \sum_{i=1}^m g(X_i),$$

which converges to $\int_a^b g(x)dx$ with probability 1 by the strong law of large numbers. One limitation of this method is that it does not apply to unbounded intervals. Another drawback is that it can be inefficient to draw samples uniformly across the interval if the function $g(x)$ is not very uniform.

However, once we view the integration problem as an expected value problem (5.12), it seems reasonable to consider other weight functions (other densities) than uniform. This leads us to a general method called *importance sampling*.

Suppose X is a random variable with density function $f(x)$, such that $f(x) > 0$ on the set $\{x : g(x) > 0\}$. Let Y be the random variable $g(X)/f(X)$. Then

$$\int g(x)dx = \int \frac{g(x)}{f(x)} f(x)dx = E[Y].$$

Estimate $E[Y]$ by simple Monte Carlo integration. That is, compute the average

$$\frac{1}{m} \sum_{i=1}^m Y_i = \frac{1}{m} \sum_{i=1}^m \frac{g(X_i)}{f(X_i)},$$

where the random variables $X_1, \ldots, X_m$ are generated from the distribution with density $f(x)$. The density $f(x)$ is called the *importance function*.

In an importance sampling method, the variance of the estimator based on $Y = g(X)/f(X)$ is $Var(Y)/m$, so the variance of Y should be small. The variance of Y is small if Y is nearly constant, so the density $f(\cdot)$ should be 'close' to $g(x)$. Also, the variable with density $f(\cdot)$ should be reasonably easy to simulate.

In Example 5.5, random normals are generated to compute the Monte Carlo estimate of the standard normal cdf, $\Phi(2) = P(X \le 2)$. In the naive Monte Carlo approach, estimates in the tails of the distribution are less precise. Intuitively, we might expect a more precise estimate for a given sample size if the simulated distribution is not uniform. In this case, the average must be a weighted average rather than the unweighted sample mean, to correct for this bias. This method is called *importance sampling* (see e.g. Robert and Casella [228, Sec. 3.3]). The advantage of importance sampling is that the importance sampling distribution can be chosen so that variance of the Monte Carlo estimator is reduced.

Suppose that $f(x)$ is a density supported on a set A. If $\phi(x) > 0$ on A, then the the integral

$$\theta = \int_A g(x)f(x)dx,$$

can be written

$$\theta = \int_A g(x)\frac{f(x)}{\phi(x)}\phi(x)dx.$$

If $\phi(x)$ is a density on A, then an estimator of $\theta = E_\phi[g(x)f(x)/\phi(x)]$ is

$$\hat{\theta} = \frac{1}{n}\sum_{i=1}^{n} g(X_i)\frac{f(X_i)}{\phi(X_i)},$$

where $X_1, \ldots, X_n$ is a random sample from density $\phi(x)$. The function $\phi(\cdot)$ is called the *envelope* or the *importance sampling function*. There are many densities $\phi(x)$ that are convenient to simulate. Typically one should choose $\phi(x)$ so that $\phi(x) \approx |g(x)|f(x)$ on A (and $\phi(x)$ has finite variance).

Example 5.10 (Choice of the importance function)

In this example (from [64, p. 728]) several possible choices of importance functions to estimate

$$\int_0^1 \frac{e^{-x}}{1+x^2}\, dx$$

by importance sampling method are compared. The candidates for the importance functions are

$$
\begin{aligned}
f_0(x) &= 1, & 0 < x < 1, \\
f_1(x) &= e^{-x}, & 0 < x < \infty, \\
f_2(x) &= (1+x^2)^{-1}/\pi, & -\infty < x < \infty, \\
f_3(x) &= e^{-x}/(1-e^{-1}), & 0 < x < 1, \\
f_4(x) &= 4(1+x^2)^{-1}/\pi, & 0 < x < 1.
\end{aligned}
$$

The integrand is

$$g(x) = \begin{cases} e^{-x}/(1+x^2), & \text{if } (0 < x < 1); \\ 0, & \text{otherwise.} \end{cases}$$

While all five of the possible importance functions are positive on the set $0 < x < 1$ where $g(x) > 0$, f_1 and f_2 have larger ranges and many of the simulated values will contribute zeros to the sum, which is inefficient. All of these distributions are easy to simulate; f_2 is standard Cauchy or $t(\nu = 1)$. The densities are plotted on $(0,1)$ for comparison with $g(x)$ in Figure 5.1(a). The function that corresponds to the most nearly constant ratio $g(x)/f(x)$ appears to be f_3, which can be seen more clearly in Figure 5.1(b). From the graphs, we might prefer f_3 for the smallest variance.

```
m <- 10000
theta.hat <- se <- numeric(5)
g <- function(x) {
    exp(-x - log(1+x^2)) * (x > 0) * (x < 1)
    }

x <- runif(m)       #using f0
fg <- g(x)
theta.hat[1] <- mean(fg)
se[1] <- sd(fg)

x <- rexp(m, 1)     #using f1
fg <- g(x) / exp(-x)
theta.hat[2] <- mean(fg)
se[2] <- sd(fg)

x <- rcauchy(m)     #using f2
i <- c(which(x > 1), which(x < 0))
x[i] <- 2  #to catch overflow errors in g(x)
fg <- g(x) / dcauchy(x)
theta.hat[3] <- mean(fg)
se[3] <- sd(fg)

u <- runif(m)       #f3, inverse transform method
x <- - log(1 - u * (1 - exp(-1)))
fg <- g(x) / (exp(-x) / (1 - exp(-1)))
theta.hat[4] <- mean(fg)
se[4] <- sd(fg)

u <- runif(m)       #f4, inverse transform method
x <- tan(pi * u / 4)
fg <- g(x) / (4 / ((1 + x^2) * pi))
theta.hat[5] <- mean(fg)
se[5] <- sd(fg)
```

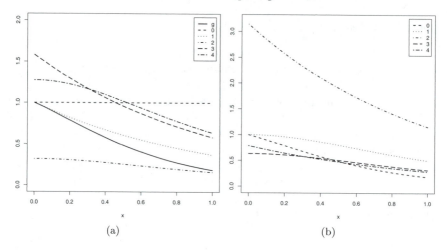

(a)   (b)

FIGURE 5.1:   Importance functions in Example 5.10: $f_0, \ldots, f_4$ (lines 0:4) with $g(x)$ in (a) and the ratios $g(x)/f(x)$ in (b).

Code to display Figures 5.1(a) and 5.1(b) is given on page 152.

The estimates (labeled `theta.hat`) of $\int_0^1 g(x)dx$ and the corresponding standard errors `se` for the simulation using each of the importance functions are

```
> rbind(theta.hat, se)
                [,1]        [,2]        [,3]         [,4]        [,5]
theta.hat 0.5241140  0.5313584  0.5461507  0.52506988  0.5260492
se         0.2436559  0.4181264  0.9661300  0.09658794  0.1427685
```

so the simulation indicates that f_3 and possibly f_4 produce smallest variance among these five importance functions, while f_2 produces the highest variance. The standard Monte Carlo estimate without importance sampling has $\widehat{se} \doteq 0.244$ ($f_0 = 1$). The importance functions f_1 and f_2 do not reduce error, but f_3 and f_4 each reduce the standard error in estimating θ.

The Cauchy density f_2 is supported on the entire real line, while the integrand $g(x)$ is evaluated on (0,1). There are a very large number of zeros (about 75%) produced in the ratio $g(x)/f(x)$ in this case, and all other values far from 0, resulting in a large variance. The following summary statistics for the ratio $g(x)/f_2(x)$ confirm this.

```
   Min. 1st Qu.  Median    Mean 3rd Qu.    Max.
 0.0000  0.0000  0.0000  0.5173  0.0000  3.1380
```

For f_1 there is a similar inefficiency, as f_1 is supported on $(0, \infty)$, which also generates many zeros in the sum of $g(x)/f(x)$ for the values outside of (0,1). The inefficiency for f_1 is not as bad as f_2 (about 37% zeros), however, because the tail of the distribution is lighter. The following summary statistics for the ratio $g(x)/f_1(x)$ also confirm this.

```
 Min. 1st Qu.  Median    Mean 3rd Qu.    Max.
0.0000  0.0000  0.6891  0.5314  0.9267  1.0000
```

$\diamond$

Example 5.10 illustrates that care must be taken to select an importance function that results in small variance of $Y = g(X)/f(X)$. The importance function should be an f that is supported on exactly the set where $g(x) > 0$, and such that the ratio $g(x)/f(x)$ is nearly constant.

Variance in Importance Sampling

If $\phi(x)$ is the importance sampling distribution (envelope), $f(x) = 1$ on A, and X has pdf $\phi(x)$ supported on A, then

$$\theta = \int_A g(x)dx = \int_A \frac{g(x)}{\phi(x)} \phi(x)dx = E\left[\frac{g(X)}{\phi(X)}\right].$$

If $X_1, \ldots, X_n$ is a random sample from the distribution of X, the estimator is again the sample-mean

$$\hat{\theta} = \overline{g(X)} = \frac{1}{n} \sum_{i=1}^{n} \frac{g(X_i)}{\phi(X_i)}.$$

Thus, the importance sampling method is a sample-mean method, and

$$Var(\hat{\theta}) = E[\hat{\theta}^2] - (E[\hat{\theta}])^2 = \int_A \frac{g^2(x)}{\phi(x)} \, ds - \theta^2.$$

The distribution of X can be chosen to reduce the variance of the sample-mean estimator. The minimum variance

$$\left(\int_A |g(x)|dx\right)^2 - \theta^2$$

is obtained when

$$\phi(x) = \frac{|g(x)|}{\int_A |g(x)|dx}.$$

Unfortunately, the problem is to estimate $\int_A g(x)dx$, so it is unlikely that the value of $\int_A |g(x)|dx$ in the denominator of $\phi(x)$ is available. Although it may be difficult to choose $\phi(x)$ to attain minimum variance, variance may be "close to" optimal if $\phi(x)$ is chosen so that the shape of the density $\phi(x)$ is "close to" $|g(x)|$ on A.

For general $f(x)$, choose $\phi(x)$ so that $\phi(x) \cong |g(x)|f(x)$ on A. If the ratio of the function being integrated to the importance function is bounded, then the importance sampling estimator will have finite variance. Considering the relative computational efficiency of estimators, one should also choose $\phi(x)$ so that the cost (time) to generate the Monte Carlo replicates is small.

5.7 Stratified Sampling

Another approach to variance reduction is stratified sampling, which aims to reduce the variance of the estimator by dividing the interval into strata and estimating the integral on each of the stratum with smaller variance. Linearity of the integral operator and the strong law of large numbers imply that the sum of these estimates converges to $\int g(x)dx$ with probability 1. In stratified sampling, the number of replicates m and number of replicates m_j to be drawn from each of k strata are fixed so that $m = m_1 + \cdots + m_k$, with the goal that

$$Var(\hat{\theta}_k(m_1, \ldots, m_k)) < Var(\hat{\theta}),$$

where $\hat{\theta}_k(m_1, \ldots, m_k)$ is the stratified estimator and $\hat{\theta}$ is the standard Monte Carlo estimator based on $m = m_1 + \cdots + m_k$ replicates.

To see how this might work, let us first see a numerical example.

Example 5.11 (Example 5.10, cont.)

In Figure 5.1(a) it is clear that our integrand $g(x)$ is not constant on $(0,1)$. Divide the interval into, say, four subintervals, and compute a Monte Carlo estimate of the integral on each subinterval using $1/4$ of the total number of replicates. Then combine these four estimates to obtain the estimate of $\int_0^1 e^{-x}(1 + x^2)^{-1} \ dx$. Does it appear that the variance of the estimator is reduced, compared with the variance of the standard Monte Carlo estimator?

The results are shown on the next page. Although 10 runs are not really enough to get good estimates of the standard errors, in this simulation it appears that stratification has improved variance by a factor of about 10. ◇

Intuitively, there can be more reduction in variance using stratification when the means of the strata are widely dispersed, as in Example 5.11, than if the means of the strata are approximately equal. For integrands that are monotone functions, stratification similar to Example 5.11 should be an effective way to reduce variance.

```
M <- 20    #number of replicates
T2 <- numeric(4)
estimates <- matrix(0, 10, 2)

g <- function(x) {
    exp(-x - log(1+x^2)) * (x > 0) * (x < 1) }

for (i in 1:10) {
    estimates[i, 1] <- mean(g(runif(M)))
    T2[1] <- mean(g(runif(M/4, 0, .25)))
    T2[2] <- mean(g(runif(M/4, .25, .5)))
    T2[3] <- mean(g(runif(M/4, .5, .75)))
    T2[4] <- mean(g(runif(M/4, .75, 1)))
    estimates[i, 2] <- mean(T2)
}
> estimates
          [,1]       [,2]
[1,]  0.6281555 0.5191537
[2,]  0.5105975 0.5265614
[3,]  0.4625555 0.5448566
[4,]  0.4999053 0.5151490
[5,]  0.4984972 0.5249923
[6,]  0.4886690 0.5179625
[7,]  0.5151231 0.5246307
[8,]  0.5503624 0.5171037
[9,]  0.5586109 0.5463568
[10,] 0.4831167 0.5548007

> apply(estimates, 2, mean)
[1] 0.5195593 0.5291568
> apply(estimates, 2, var)
[1] 0.0023031762 0.0002012629
```

PROPOSITION 5.2 *Denote the standard Monte Carlo estimator with M replicates by $\hat{\theta}^M$, and let*

$$\hat{\theta}^S = \frac{1}{k} \sum_{j=1}^{k} \hat{\theta}_j$$

denote the stratified estimator with equal size $m = M/k$ strata. Denote the mean and variance of $g(U)$ on stratum j by θ_j and σ_j^2, respectively. Then $Var(\hat{\theta}^M) \geq Var(\hat{\theta}^S)$.

Proof. By independence of $\hat{\theta}_j$'s,

$$Var(\hat{\theta}^S) = Var\left(\frac{1}{k} \sum_{j=1}^{k} \hat{\theta}_j\right) = \frac{1}{k^2} \sum_{j=1}^{k} \frac{\sigma_j^2}{m} = \frac{1}{Mk} \sum_{j=1}^{k} \sigma_j^2.$$

Now, if J is the randomly selected stratum, it is selected with uniform probability $1/k$, and applying the conditional variance formula

$$Var(\hat{\theta}^M) = \frac{Var(g(U))}{M} = \frac{1}{M}(Var(E[g(U|J)]) + E[Var(g(U|J)])$$

$$= \frac{1}{M}\left(Var(\theta_J) + E\left[\sigma_J^2\right]\right)$$

$$= \frac{1}{M}\left(Var(\theta_J) + \frac{1}{k}\sum_{j=1}^{k}\sigma_j^2\right)$$

$$= \frac{1}{M}Var(\theta_J) + Var(\hat{\theta}^S) \geq Var(\hat{\theta}^S).$$

The inequality is strict except in the case where all the strata have identical means.   □

From the above inequality it is clear that the reduction in variance is larger when the means of the strata are widely dispersed.

A similar proof can be applied in the general case when the strata have unequal probabilities. See Fishman [94, Sec. 4.3] for a proof of the general case.

Example 5.12 (Examples 5.10–5.11, cont., stratified sampling)

Stratified sampling is implemented in a more general way, for the Monte Carlo estimate of $\int_0^1 e^{-x}(1 + x^2)^{-1}dx$. The standard Monte Carlo estimate is also obtained for comparison.

```
M <- 10000  #number of replicates
k <- 10       #number of strata
r <- M / k  #replicates per stratum
N <- 50       #number of times to repeat the estimation
T2 <- numeric(k)
estimates <- matrix(0, N, 2)

g <- function(x) {
    exp(-x - log(1+x^2)) * (x > 0) * (x < 1)
    }

for (i in 1:N) {
    estimates[i, 1] <- mean(g(runif(M)))
    for (j in 1:k)
        T2[j] <- mean(g(runif(M/k, (j-1)/k, j/k)))
    estimates[i, 2] <- mean(T2)
}
```

The result of this simulation produces the following estimates.

```
> apply(estimates, 2, mean)
[1] 0.5251321 0.5247715
> apply(estimates, 2, var)
[1] 6.188117e-06 6.504485e-08
```

This represents a more than 98% reduction in variance. ◇

5.8 Stratified Importance Sampling

A modification to the importance sampling method of estimating $\theta = \int g(x)dx$ is stratified importance sampling.

Choose a suitable importance function f. Suppose that X is generated with density f and cdf F using the probability integral transformation. If M replicates are generated, the importance sampling estimate of θ has variance σ^2/M, where $\sigma^2 = Var(g(X)/f(X))$.

For the stratified importance sampling estimate, divide the real line into k intervals $I_j = \{x : a_{j-1} \leq x < a_j\}$ with endpoints $a_0 = -\infty$, $a_j = F^{-1}(j/k)$, $j = 1, \ldots, k-1$, and $a_k = \infty$. (The real line is divided into intervals corresponding to equal areas $1/k$ under the density $f(x)$. The interior endpoints are the percentiles or quantiles.) On each subinterval define $g_j(x) = g(x)$ if $x \in I_j$ and $g_j(x) = 0$ otherwise. We now have k parameters to estimate,

$$\theta_j = \int_{a_{j-1}}^{a_j} g_j(x)dx, \qquad j = 1, \ldots, k$$

and $\theta = \theta_1 + \cdots + \theta_k$. The conditional densities provide the importance functions on each subinterval. That is, on each subinterval I_j, the conditional density f_j of X is defined by

$$f_j(x) = f_{X|I_j}(x|I_j) = \frac{f(x, a_{j-1} \leq x < a_j)}{P(a_{j-1} \leq x < a_j)}$$

$$= \frac{f(x)}{1/k} = kf(x), \quad a_{j-1} \leq x < a_j.$$

Let $\sigma_j^2 = Var(g_j(X)/f_j(X))$. For each $j = 1, \ldots, k$ we simulate an importance sample size m, compute the importance sampling estimator $\hat{\theta}_j$ of θ_j on the jth subinterval, and compute $\hat{\theta}^{SI} = \frac{1}{k}\sum_{j=1}^{k} \hat{\theta}_j$. Then by independence of $\hat{\theta}_1, \ldots, \hat{\theta}_k$,

$$Var(\hat{\theta}^{SI}) = Var\left(\sum_{j=1}^{k} \hat{\theta}_j\right) = \sum_{j=1}^{k} \frac{\sigma_j^2}{m} = \frac{1}{m}\sum_{j=1}^{k} \sigma_j^2.$$

Denote the importance sampling estimator by $\hat{\theta}^I$. In order to determine whether $\hat{\theta}^{SI}$ is a better estimator of θ than $\hat{\theta}^I$, we need to check that $Var(\hat{\theta}^{SI})$ is smaller than the variance without stratification. The variance is reduced by stratification if

$$\frac{\sigma^2}{M} > \frac{1}{m}\sum_{j=1}^{k}\sigma_j^2 = \frac{k}{M}\sum_{j=1}^{k}\sigma_j^2 \Rightarrow \sigma^2 - k\sum_{j=1}^{k}\sigma_j^2 > 0.$$

Thus, we need to prove the following.

PROPOSITION 5.3 *Suppose $M = mk$ is the number of replicates for an importance sampling estimator $\hat{\theta}^I$, and $\hat{\theta}^{SI}$ is a stratified importance sampling estimator, with estimates $\hat{\theta}_j$ for θ_j on the individual strata, each with m replicates. If $Var(\hat{\theta}^I) = \sigma^2/M$ and $Var(\hat{\theta}_j) = \sigma_j^2/m$, $j = 1,\ldots,k$, then*

$$\sigma^2 - k\sum_{j=1}^{k}\sigma_j^2 \geq 0, \tag{5.13}$$

with equality if and only if $\theta_1 = \cdots = \theta_k$. Hence stratification never increases the variance, and there exists a stratification that reduces the variance except when $g(x)$ is constant.

Proof. To determine when the inequality (5.13) holds, we need to consider the relation between the random variables with densities f_j and the random variable X with density f.

Consider a two-stage experiment. First a number J is drawn at random from the integers 1 to k. After observing $J = j$, a random variable X^* is generated from the density f_j and

$$Y^* = \frac{g_j(X)}{f_j(X)} = \frac{g_j(X^*)}{kf(X^*)}.$$

To compute the variance of Y^* we apply the conditional variance formula

$$Var(Y^*) = E[Var(Y^*|J)] + Var(E[Y^*|J]). \tag{5.14}$$

Here

$$E[Var(Y^*|J)] = \sum_{j=1}^{k}\sigma_j^2 P(J=j) = \frac{1}{k}\sum_{j=1}^{k}\sigma_j^2$$

and $Var(E[Y^*|J]) = Var(\theta_J)$. Thus in (5.14) we have

$$Var(Y^*) = \frac{1}{k}\sum_{j=1}^{k}\sigma_j^2 + Var(\theta_J).$$

On the other hand,

$$k^2 Var(Y^*) = k^2 E[Var(Y^*|J)] + k^2 Var(E[Y^*|J]).$$

and

$$\sigma^2 = Var(Y) = Var(kY^*) = k^2 Var(Y^*)$$

which imply that

$$\sigma^2 = k^2 Var(Y^*) = k^2 \left(\frac{1}{k} \sum_{j=1}^{k} \sigma_j^2 + Var(\theta_J) \right) = k \sum_{j=1}^{k} \sigma_j^2 + k^2 Var(\theta_J).$$

Therefore

$$\sigma^2 - k \sum_{j=1}^{k} \sigma_j^2 = k^2 Var(\theta_J) \geq 0,$$

and equality holds if and only if $\theta_1 = \cdots = \theta_k$. $\qquad\square$

Example 5.13 (Example 5.10, cont.)

In Example 5.10 our best result was obtained with importance function $f_3(x) = e^{-x}/(1 - e^{-1})$, $0 < x < 1$. From 10000 replicates we obtained the estimate $\hat{\theta} = 0.5257801$ and an estimated standard error 0.0970314. Now divide the interval (0,1) into five subintervals, $(j/5, (j+1)/5)$, $j = 0, 1, \ldots, 4$.

Then on the jth subinterval variables are generated from the density

$$\frac{5e^{-x}}{1 - e^{-1}}, \qquad \frac{j-1}{5} < x < \frac{j}{5}.$$

The implementation is left as an exercise. $\qquad\diamond$

Exercises

5.1   Compute a Monte Carlo estimate of

$$\int_0^{\pi/3} \sin t \, dt$$

and compare your estimate with the exact value of the integral.

5.2   Refer to Example 5.3. Compute a Monte Carlo estimate of the standard normal cdf, by generating from the Uniform$(0,x)$ distribution. Compare your estimates with the normal cdf function `pnorm`. Compute an estimate of the variance of your Monte Carlo estimate of $\Phi(2)$, and a 95% confidence interval for $\Phi(2)$.

5.3   Compute a Monte Carlo estimate $\hat{\theta}$ of

$$\theta = \int_0^{0.5} e^{-x} \, dx$$

by sampling from Uniform(0, 0.5), and estimate the variance of $\hat{\theta}$. Find another Monte Carlo estimator θ^* by sampling from the exponential distribution. Which of the variances (of $\hat{\theta}$ and $\hat{\theta}^*$) is smaller, and why?

5.4   Write a function to compute a Monte Carlo estimate of the Beta(3, 3) cdf, and use the function to estimate $F(x)$ for $x = 0.1, 0.2, \ldots, 0.9$. Compare the estimates with the values returned by the pbeta function in R.

5.5   Compute (empirically) the efficiency of the sample mean Monte Carlo method of estimation of the definite integral in Example 5.3 relative to the "hit or miss" method in Example 5.4.

5.6   In Example 5.7 the control variate approach was illustrated for Monte Carlo integration of

$$\theta = \int_0^1 e^x \, dx.$$

Now consider the antithetic variate approach. Compute $Cov(e^U, e^{1-U})$ and $Var(e^U + e^{1-U})$, where $U \sim$ Uniform(0,1). What is the percent reduction in variance of $\hat{\theta}$ that can be achieved using antithetic variates (compared with simple MC)?

5.7   Refer to Exercise 5.6. Use a Monte Carlo simulation to estimate θ by the antithetic variate approach and by the simple Monte Carlo method. Compute an empirical estimate of the percent reduction in variance using the antithetic variate. Compare the result with the theoretical value from Exercise 5.6.

5.8   Let $U \sim$ Uniform(0,1), $X = aU$, and $X' = a(1 - U)$, where a is a constant. Show that $\rho(X, X') = -1$. Is $\rho(X, X') = -1$ if U is a symmetric beta random variable?

5.9   The Rayleigh density [156, (18.76)] is

$$f(x) = \frac{x}{\sigma^2} e^{-x^2/(2\sigma^2)}, \qquad x \geq 0, \sigma > 0.$$

Implement a function to generate samples from a Rayleigh(σ) distribution, using antithetic variables. What is the percent reduction in variance of $\frac{X+X'}{2}$ compared with $\frac{X_1+X_2}{2}$ for independent X_1, X_2?

5.10 Use Monte Carlo integration with antithetic variables to estimate

$$\int_0^1 \frac{e^{-x}}{1+x^2} dx,$$

and find the approximate reduction in variance as a percentage of the variance without variance reduction.

.11 If $\hat{\theta}_1$ and $\hat{\theta}_2$ are unbiased estimators of θ, and $\hat{\theta}_1$ and $\hat{\theta}_2$ are antithetic, we derived that $c^* = 1/2$ is the optimal constant that minimizes the variance of $\hat{\theta}_c = c\hat{\theta}_2 + (1 - c)\hat{\theta}_2$. Derive c^* for the general case. That is, if $\hat{\theta}_1$ and $\hat{\theta}_2$ are any two unbiased estimators of θ, find the value c^* that minimizes the variance of the estimator $\hat{\theta}_c = c\hat{\theta}_2 + (1 - c)\hat{\theta}_2$ in equation (5.11). (c^* will be a function of the variances and the covariance of the estimators.)

.12 Let $\hat{\theta}_f^{IS}$ be an importance sampling estimator of $\theta = \int g(x)dx$, where the importance function f is a density. Prove that if $g(x)/f(x)$ is bounded, then the variance of the importance sampling estimator $\hat{\theta}_f^{IS}$ is finite.

.13 Find two importance functions f_1 and f_2 that are supported on $(1, \infty)$ and are 'close' to

$$g(x) = \frac{x^2}{\sqrt{2\pi}} e^{-x^2/2}, \qquad x > 1.$$

Which of your two importance functions should produce the smaller variance in estimating

$$\int_1^\infty \frac{x^2}{\sqrt{2\pi}} e^{-x^2/2} \, dx$$

by importance sampling? Explain.

5.14 Obtain a Monte Carlo estimate of

$$\int_1^\infty \frac{x^2}{\sqrt{2\pi}} e^{-x^2/2} \, dx$$

by importance sampling.

5.15 Obtain the stratified importance sampling estimate in Example 5.13 and compare it with the result of Example 5.10.

R Code

Code to display the plot of importance functions in Figures 5.1(a) and 5.1(b) on page 142.

```
x <- seq(0, 1, .01)
w <- 2
f1 <- exp(-x)
f2 <- (1 / pi) / (1 + x^2)
f3 <- exp(-x) / (1 - exp(-1))
f4 <- 4 / ((1 + x^2) * pi)
g <- exp(-x) / (1 + x^2)

#figure (a)
plot(x, g, type = "l", main = "", ylab = "",
    ylim = c(0,2), lwd = w)
lines(x, g/g, lty = 2, lwd = w)
lines(x, f1, lty = 3, lwd = w)
lines(x, f2, lty = 4, lwd = w)
lines(x, f3, lty = 5, lwd = w)
lines(x, f4, lty = 6, lwd = w)
legend("topright", legend = c("g", 0:4),
        lty = 1:6, lwd = w, inset = 0.02)

#figure (b)
plot(x, g, type = "l", main = "", ylab = "",
    ylim = c(0,3.2), lwd = w, lty = 2)
lines(x, g/f1, lty = 3, lwd = w)
lines(x, g/f2, lty = 4, lwd = w)
lines(x, g/f3, lty = 5, lwd = w)
lines(x, g/f4, lty = 6, lwd = w)
legend("topright", legend = c(0:4),
        lty = 2:6, lwd = w, inset = 0.02)
```

Chapter 6

Monte Carlo Methods in Inference

6.1 Introduction

Monte Carlo methods encompass a vast set of computational tools in modern applied statistics. Monte Carlo integration was introduced in Chapter 5. Monte Carlo methods may refer to any method in statistical inference or numerical analysis where simulation is used. However, in this chapter only a subset of these methods are discussed. This chapter introduces some of the Monte Carlo methods for statistical inference. Monte Carlo methods can be applied to estimate parameters of the sampling distribution of a statistic, mean squared error (MSE), percentiles, or other quantities of interest. Monte Carlo studies can be designed to assess the coverage probability for confidence intervals, to find an empirical Type I error rate of a test procedure, to estimate the power of a test, and to compare the performance of different procedures for a given problem.

In statistical inference there is uncertainty in an estimate. The methods covered in this chapter use repeated sampling from a given probability model, sometimes called *parametric* bootstrap, to investigate this uncertainty. If we can simulate the stochastic process that generated our data, repeatedly drawing samples under identical conditions, then ultimately we hope to have a close replica of the process itself reflected in the samples. Other Monte Carlo methods, such as (nonparametric) bootstrap, are based on resampling from an observed sample. Resampling methods are covered in Chapters 7 and 8. Monte Carlo integration and Markov Chain Monte Carlo methods are covered in Chapters 5 and 9. Methods for generating random variates from specified probability distributions are covered in Chapter 3. See the references in Section 5.1 on some of the early history of Monte Carlo methods, and for general reference see e.g. [63, 84, 228].

6.2 Monte Carlo Methods for Estimation

Suppose $X_1, \ldots, X_n$ is a random sample from the distribution of X. An estimator $\hat{\theta}$ for a parameter θ is an n variate function

$$\hat{\theta} = \hat{\theta}(X_1, \ldots, X_n)$$

of the sample. Functions of the estimator $\hat{\theta}$ are therefore n-variate functions of the data, also. For simplicity, let $x = (x_1, \ldots, x_n)^T \in \mathbb{R}^n$, and let $x^{(1)}, x^{(2)}, \ldots$ denote a sequence of independent random samples generated from the distribution of X. Random variates from the sampling distribution of $\hat{\theta}$ can be generated by repeatedly drawing independent random samples $x^{(j)}$ and computing $\hat{\theta}^{(j)} = \hat{\theta}(x_1^{(j)}, \ldots, x_n^{(j)})$ for each sample.

6.2.1 Monte Carlo estimation and standard error

Example 6.1 (Basic Monte Carlo estimation)

Suppose that X_1, X_2 are iid from a standard normal distribution. Estimate the mean difference $E|X_1 - X_2|$.

To obtain a Monte Carlo estimate of $\theta = E[g(X_1, X_2)] = E|X_1 - X_2|$ based on m replicates, generate random samples $x^{(j)} = (x_1^{(j)}, x_2^{(j)})$ of size 2 from the standard normal distribution, $j = 1, \ldots, m$. Then compute the replicates $\hat{\theta}^{(j)} = g_j(x_1, x_2) = |x_1^{(j)} - x_2^{(j)}|$, $j = 1, \ldots, m$, and the mean of the replicates

$$\hat{\theta} = \frac{1}{m} \sum_{i=1}^{m} \hat{\theta}^{(j)} = \overline{g(X_1, X_2)} = \frac{1}{m} \sum_{i=1}^{m} |x_1^{(j)} - x_2^{(j)}|.$$

This is easy to implement, as shown below.

```
m <- 1000
g <- numeric(m)
for (i in 1:m) {
    x <- rnorm(2)
    g[i] <- abs(x[1] - x[2])
}
est <- mean(g)
```

One run produces the following estimate.

```
> est
[1] 1.128402
```

One can derive by integration that $E|X_1 - X_2| = 2/\sqrt{\pi} \doteq 1.128379$ and $Var(|X_1 - X_2|) = 2 - 4/\pi$. In this example the standard error of the estimate is $\sqrt{(2 - 4/\pi)/m} \doteq 0.02695850$. ◇

Estimating the standard error of the mean

The standard error of a mean $\overline{X}$ of a sample size n is $\sqrt{Var(X)/n}$. When the distribution of X is unknown we can substitute for F the empirical distribution F_n of the sample $x_1, \ldots, x_n$. The "plug-in" estimate of the variance of X is

$$\widehat{Var}(x) = \frac{1}{n} \sum_{i=1}^{n} (x_i - \bar{x})^2.$$

Note that $\widehat{Var}(x)$ is the population variance of the finite pseudo population $\{x_1, \ldots, x_n\}$ with cdf F_n. The corresponding estimate of the standard error of $\bar{x}$ is

$$\widehat{se}(\bar{x}) = \frac{1}{\sqrt{n}} \left\{ \frac{1}{n} \sum_{i=1}^{n} (x_i - \bar{x})^2 \right\}^{1/2} = \frac{1}{n} \left\{ \sum_{i=1}^{n} (x_i - \bar{x})^2 \right\}^{1/2}.$$

Using the unbiased estimator of $Var(X)$ we have

$$\widehat{se}(\bar{x}) = \frac{1}{\sqrt{n}} \left\{ \frac{1}{n-1} \sum_{i=1}^{n} (x_i - \bar{x})^2 \right\}^{1/2}.$$

In a Monte Carlo experiment, the sample size is large and the two estimates of standard error are approximately equal.

In Example 6.1 the sample size is m (the number of replicates of $\hat{\theta}$), and the estimate of standard error of $\hat{\theta}$ is

```
> sqrt(sum((g - mean(g))^2)) / m
[1] 0.02708121
```

In Example 6.1 we have the exact value $se(\hat{\theta}) = \sqrt{(2 - 4/\pi)/m} \doteq 0.02695850$ for comparison.

6.2.2 Estimation of MSE

Monte Carlo methods can be applied to estimate the MSE of an estimator. Recall that the MSE of an estimator $\hat{\theta}$ for a parameter θ is defined by $MSE(\hat{\theta}) = E[(\hat{\theta} - \theta)^2]$. If m (pseudo) random samples $x^{(1)}, \ldots, x^{(m)}$ are generated from the distribution of X, then a Monte Carlo estimate of the MSE of $\hat{\theta} = \hat{\theta}(x_1, \ldots, x_n)$ is

$$\widehat{MSE} = \frac{1}{m} \sum_{j=1}^{m} (\hat{\theta}^{(j)} - \theta)^2,$$

where $\hat{\theta}^{(j)} = \hat{\theta}(x^{(j)}) = \hat{\theta}(x_1^{(j)}, \ldots, x_n^{(j)})$.

Example 6.2 (Estimating the MSE of a trimmed mean)

A trimmed mean is sometimes applied to estimate the center of a continuous symmetric distribution that is not necessarily normal. In this example, we compute an estimate of the MSE of a trimmed mean. Suppose that $X_1, \ldots, X_n$ is a random sample and $X_{(1)}, \ldots, X_{(n)}$ is the corresponding ordered sample. The trimmed sample mean is computed by averaging all but the largest and smallest sample observations. More generally, the kth level trimmed sample mean is defined by

$$\overline{X}_{[-k]} = \frac{1}{n-2k} \sum_{i=k+1}^{n-k} X_{(i)}.$$

Obtain a Monte Carlo estimate of the MSE($\overline{X}_{[-1]}$) of the first level trimmed mean assuming that the sampled distribution is standard normal.

In this example, the center of the distribution is 0 and the target parameter is $\theta = E[\overline{X}] = E[\overline{X}_{[-1]}] = 0$. We will denote the first level trimmed sample mean by T. A Monte Carlo estimate of MSE(T) based on m replicates can be obtained as follows.

1. Generate the replicates $T^{(j)}$, $j = 1 \ldots, m$ by repeating:

   (a) Generate $x_1^{(j)}, \ldots, x_n^{(j)}$, iid from the distribution of X.

   (b) Sort $x_1^{(j)}, \ldots, x_n^{(j)}$ in increasing order, to obtain $x_{(1)}^{(j)} \leq \cdots \leq x_{(n)}^{(j)}$.

   (c) Compute $T^{(j)} = \frac{1}{n-2} \sum_{i=2}^{n-1} x_{(i)}^{(j)}$.

2. Compute $\widehat{MSE}(T) = \frac{1}{m} \sum_{j=1}^{m} (T^{(j)} - \theta)^2 = \frac{1}{m} \sum_{j=1}^{m} (T^{(j)})^2$.

Then $T^{(1)}, \ldots, T^{(m)}$ are independent and identically distributed according to the sampling distribution of the level-1 trimmed mean for a standard normal distribution, and we are computing the sample mean estimate $\widehat{MSE}(T)$ of MSE(T). This procedure can be implemented by writing a `for` loop as shown below (`replicate` can replace the loop; see R note 6.1 on page 161).

```
n <- 20
m <- 1000
tmean <- numeric(m)
for (i in 1:m) {
    x <- sort(rnorm(n))
    tmean[i] <- sum(x[2:(n-1)]) / (n-2)
    }
mse <- mean(tmean^2)

> mse
[1] 0.05176437
> sqrt(sum((tmean - mean(tmean))^2)) / m      #se
[1] 0.007193428
```

The estimate of MSE for the trimmed mean in this run is approximately 0.052 ($\hat{se} \doteq 0.007$). For comparison, the MSE of the sample mean $\overline{X}$ is $Var(X)/n$, which is $1/20 = 0.05$ in this example. Note that the median is actually a trimmed mean; it trims all but one or two of the observations. The simulation is repeated for the median below.

```
n <- 20
m <- 1000
tmean <- numeric(m)
for (i in 1:m) {
    x <- sort(rnorm(n))
    tmean[i] <- median(x)
    }
mse <- mean(tmean^2)

> mse
[1] 0.07483438
> sqrt(sum((tmean - mean(tmean))^2)) / m     #se
[1] 0.008649554
```

The estimate of MSE for the sample median is approximately 0.075 and $\hat{se}(\widehat{MSE}) \doteq 0.0086$.   ◇

Example 6.3 (MSE of a trimmed mean, cont.)

Compare the MSE of level-k trimmed means for the standard normal and a "contaminated" normal distribution. The contaminated normal distribution in this example is a mixture

$$pN(0, \sigma^2 = 1) + (1 - p)N(0, \sigma^2 = 100).$$

The target parameter is the mean, $\theta = 0$. (This example is from [64, 9.7].)

Write a function to estimate $\text{MSE}(\overline{X}_{[-k]})$ for different k and p. To generate the contaminated normal samples, first randomly select σ according to the probability distribution $P(\sigma = 1) = p$; $P(\sigma = 10) = 1 - p$. Note that the normal generator `rnorm` can accept a vector of parameters for standard deviation. After generating the n values for σ, pass this vector as the `sd` argument to `rnorm` (see e.g. Example 3.12 and Example 3.13).

```
n <- 20
K <- n/2 - 1
m <- 1000
mse <- matrix(0, n/2, 6)
```

```
trimmed.mse <- function(n, m, k, p) {
    #MC est of mse for k-level trimmed mean of
    #contaminated normal pN(0,1) + (1-p)N(0,100)
    tmean <- numeric(m)
    for (i in 1:m) {
        sigma <- sample(c(1, 10), size = n,
            replace = TRUE, prob = c(p, 1-p))
        x <- sort(rnorm(n, 0, sigma))
        tmean[i] <- sum(x[(k+1):(n-k)]) / (n-2*k)
        }
    mse.est <- mean(tmean^2)
    se.mse <- sqrt(mean((tmean-mean(tmean))^2)) / sqrt(m)
    return(c(mse.est, se.mse))
}


for (k in 0:K) {
    mse[k+1, 1:2] <- trimmed.mse(n=n, m=m, k=k, p=1.0)
    mse[k+1, 3:4] <- trimmed.mse(n=n, m=m, k=k, p=.95)
    mse[k+1, 5:6] <- trimmed.mse(n=n, m=m, k=k, p=.9)
}
```

The results of the simulation are shown in Table 6.1. The results in the table are n times the estimates. This comparison suggests that a robust estimator of the mean can lead to reduced MSE for contaminated normal samples.   ◇

TABLE 6.1: Estimates of Mean Squared Error for the kth Level Trimmed Mean in Example 6.3 ($n = 20$)

	Normal		$p = 0.95$		$p = 0.90$	
k	$n\widehat{MSE}$	$n\,\widehat{se}$	$n\widehat{MSE}$	$n\,\widehat{se}$	$n\widehat{MSE}$	$n\,\widehat{se}$
0	0.976	0.140	6.229	0.140	11.485	0.140
1	1.019	0.143	1.954	0.143	4.126	0.143
2	1.009	0.142	1.304	0.142	1.956	0.142
3	1.081	0.147	1.168	0.147	1.578	0.147
4	1.048	0.145	1.280	0.145	1.453	0.145
5	1.103	0.149	1.395	0.149	1.423	0.149
6	1.316	0.162	1.349	0.162	1.574	0.162
7	1.377	0.166	1.503	0.166	1.734	0.166
8	1.382	0.166	1.525	0.166	1.694	0.166
9	1.491	0.172	1.646	0.172	1.843	0.172

6.2.3   Estimating a confidence level

One type of problem that arises frequently in statistical applications is the need to evaluate the cdf of the sampling distribution of a statistic, when the density function of the statistic is unknown or intractable. For example, many commonly used estimation procedures are derived under the assumption that the sampled population is normally distributed. In practice, it is often the case that the population is non-normal and in such cases, the true distribution of the estimator may be unknown or intractable. The following examples illustrate a Monte Carlo method to assess the confidence level in an estimation procedure.

If (U, V) is a confidence interval estimate for an unknown parameter θ, then U and V are statistics with distributions that depend on the distribution F_X of the sampled population X. The confidence level is the probability that the interval (U, V) covers the true value of the parameter θ. Evaluating the confidence level is therefore an integration problem.

Note that the sample-mean Monte Carlo approaches to evaluating an integral $\int g(x)dx$ do not require that the function $g(x)$ is specified. It is only necessary that the sample from the distribution $g(X)$ can be generated. It is often the case in statistical applications, that $g(x)$ is in fact not specified, but the variable $g(X)$ is easily generated.

Consider the confidence interval estimation procedure for variance. It is well known that this procedure is sensitive to mild departures from normality. We use Monte Carlo methods to estimate the true confidence level when the normal theory confidence interval for variance is applied to non-normal data. The classical procedure based on the assumption of normality is outlined first.

Example 6.4 (Confidence interval for variance)

If $X_1, \ldots, X_n$ is a random sample from a Normal(μ, σ^2) distribution, $n \geq 2$, and S^2 is the sample variance, then

$$V = \frac{(n-1)S^2}{\sigma^2} \sim \chi^2(n-1). \tag{6.1}$$

A one side $100(1-\alpha)\%$ confidence interval is given by $(0, (n-1)S^2/\chi_\alpha^2)$, where χ_α^2 is the α-quantile of the $\chi^2(n-1)$ distribution. If the sampled population is normal with variance σ^2, then the probability that the confidence interval contains σ^2 is $1 - \alpha$.

The calculation of the 95% upper confidence limit (UCL) for a random sample size $n = 20$ from a Normal$(0, \sigma^2 = 4)$ distribution is shown below.

```
n <- 20
alpha <- .05
x <- rnorm(n, mean=0, sd=2)
UCL <- (n-1) * var(x) / qchisq(alpha, df=n-1)
```

Several runs produce the upper confidence limits UCL = 6.628, UCL = 7.348, UCL = 9.621, etc. All of these intervals contain $\sigma^2 = 4$. In this example, the sampled population is normal with $\sigma^2 = 4$, so the confidence level is exactly

$$P\left(\frac{19S^2}{\chi^2_{.05}(19)} > 4\right) = P\left(\frac{(n-1)S^2}{\sigma^2} > \chi^2_{.05}(n-1)\right) = 0.95.$$

If the sampling and estimation is repeated a large number of times, approximately 95% of the intervals based on (6.1) should contain σ^2, assuming that the sampled population is normal with variance σ^2.   ◇

Empirical confidence level is an estimate of the confidence level obtained by simulation. For the simulation experiment, repeat the steps above a large number of times, and compute the proportion of intervals that contain the target parameter.

Monte Carlo experiment to estimate a confidence level

Suppose that $X \sim F_X$ is the random variable of interest and that θ is the target parameter to be estimated.

1. For each replicate, indexed $j = 1, \ldots, m$:

   (a) Generate the jth random sample, $X_1^{(j)}, \ldots, X_n^{(j)}$.
   (b) Compute the confidence interval C_j for the jth sample.
   (c) Compute $y_j = I(\theta \in C_j)$ for the jth sample.

2. Compute the empirical confidence level $\bar{y} = \frac{1}{m}\sum_{j=1}^{m} y_j$.

The estimator $\bar{y}$ is a sample proportion estimating the true confidence level $1 - \alpha^*$, so $Var(\bar{y}) = (1-\alpha^*)\alpha^*/m$ and an estimate of standard error is $\widehat{se}(\bar{y}) = \sqrt{(1-\bar{y})\bar{y}/m}$.

Example 6.5 (MC estimate of confidence level)

Refer to Example 6.4. In this example we have $\mu = 0$, $\sigma = 2$, $n = 20$, $m = 1000$ replicates, and $\alpha = 0.05$. The sample proportion of intervals that contain $\sigma^2 = 4$ is a Monte Carlo estimate of the true confidence level. This type of simulation can be conveniently implemented by using the `replicate` function.

```
n <- 20
alpha <- .05
UCL <- replicate(1000, expr = {
    x <- rnorm(n, mean = 0, sd = 2)
    (n-1) * var(x) / qchisq(alpha, df = n-1)
    } )
```

```
#count the number of intervals that contain sigma^2=4
sum(UCL > 4)
#or compute the mean to get the confidence level
> mean(UCL > 4)
[1] 0.956
```

The result is that 956 intervals satisfied (UCL > 4), so the empirical confidence level is 95.6% in this experiment. The result will vary but should be close to the theoretical value, 95%. The standard error of the estimate is $(0.95(1 - 0.95)/1000)^{1/2} \doteq 0.00689$.   ◇

R note 6.1 *Notice that in the* `replicate` *function, the lines to be repeatedly executed are enclosed in braces { }. Alternately, the expression argument* (**expr**) *can be a function call:*

```
calcCI <- function(n, alpha) {
    y <- rnorm(n, mean = 0, sd = 2)
    return((n-1) * var(y) / qchisq(alpha, df = n-1))
}
```

```
UCL <- replicate(1000, expr = calcCI(n = 20, alpha = .05))
```

The interval estimation procedure based on (6.1) for estimating variance is sensitive to departures from normality, so the true confidence level may be different than the stated confidence level when data are non-normal. The true confidence level depends on the cdf of the statistic S^2. The confidence level is the probability that the interval $(0, (n - 1)S^2/\chi^2_\alpha)$ contains the true value of the parameter σ^2, which is

$$P\left(\frac{(n - 1)S^2}{\chi^2_\alpha} > \sigma^2\right) = P\left(S^2 > \frac{\sigma^2 \chi^2_\alpha}{n - 1}\right) = 1 - G\left(\frac{\sigma^2 \chi^2_\alpha}{n - 1}\right),$$

where $G(\cdot)$ is the cdf of S^2. If the sampled population is non-normal, we have the problem of estimating the cdf

$$G(t) = P(S^2 \le c_\alpha) = \int_0^{c_\alpha} g(x)dx,$$

where $g(x)$ is the (unknown) density of S^2 and $c_\alpha = \sigma^2\chi^2_\alpha/(n-1)$. An approximate solution can be computed empirically using Monte Carlo integration to estimate $G(c_\alpha)$. The estimate of $G(t) = P(S^2 \le t) = \int_0^t g(x)dx$, is computed by Monte Carlo integration. It is not necessary to have an explicit formula for $g(x)$, provided that we can sample from the distribution of $g(X)$.

Example 6.6 (Empirical confidence level)

In Example 6.4, what happens if the sampled population is non-normal? For example, suppose that the sampled population is $\chi^2(2)$, which has variance 4, but is distinctly non-normal. We repeat the simulation, replacing the $N(0,4)$ samples with $\chi^2(2)$ samples.

```
n <- 20
alpha <- .05
UCL <- replicate(1000, expr = {
    x <- rchisq(n, df = 2)
    (n-1) * var(x) / qchisq(alpha, df = n-1)
    } )
> sum(UCL > 4)
[1] 773
> mean(UCL > 4)
[1] 0.773
```

In this experiment, only 773 or 77.3% of the intervals contained the population variance, which is far from the 95% coverage under normality.   ◇

Remark 6.1 *The problems in Examples 6.1– 6.6 are parametric in the sense that the distribution of the sampled population is specified. The Monte Carlo approach here is sometimes called* parametric bootstrap. *The* ordinary *boot-strap discussed in Chapter 7 is a different procedure. In "parametric" boot-strap, the pseudo random samples are generated from a given probability distribution. In the "ordinary" bootstrap, the samples are generated by resampling from an observed sample. Bootstrap methods in this book refer to* resampling *methods.*

Monte Carlo methods for estimation, including several types of bootstrap confidence interval estimates, are covered in Chapter 7. Bootstrap and jack-knife methods for estimating the bias and standard error of an estimate are also covered in Chapter 7. The remainder of this chapter focuses on hypothesis tests, which are also covered in Chapter 8.

6.3   Monte Carlo Methods for Hypothesis Tests

Suppose that we wish to test a hypothesis concerning a parameter θ that lies in a parameter space Θ. The hypotheses of interest are

$$H_0 : \theta \in \Theta_0 \quad \text{vs} \quad H_1 : \theta \in \Theta_1$$

where Θ_0 and Θ_1 partition the parameter space Θ.

Two types of error can occur in statistical hypothesis testing. A Type I error occurs if the null hypothesis is rejected when in fact the null hypothesis is true. A Type II error occurs if the null hypothesis is not rejected when in fact the null hypothesis is false.

The *significance level* of a test is denoted by α, and α is an upper bound on the probability of Type I error. The probability of rejecting the null hypothesis depends on the true value of θ. For a given test procedure, let $\pi(\theta)$ denote the probability of rejecting H_0. Then

$$\alpha = \sup_{\theta \in \Theta_0} \pi(\theta).$$

The probability of Type I error is the conditional probability that the null hypothesis is rejected given that H_0 is true. Thus, if the test procedure is replicated a large number of times under the conditions of the null hypothesis, the observed Type I error rate should be at most (approximately) α.

If T is the test statistic and T^* is the observed value of the test statistic, then T^* is *significant* if the test decision based on T^* is to reject H_0. The *significance probability* or *p*-value is the smallest possible value of α such that the observed test statistic would be significant.

6.3.1 Empirical Type I error rate

An empirical Type I error rate can be computed by a Monte Carlo experiment. The test procedure is replicated a large number of times under the conditions of the null hypothesis. The empirical Type I error rate for the Monte Carlo experiment is the sample proportion of significant test statistics among the replicates.

Monte Carlo experiment to assess Type I error rate:

1. For each replicate, indexed by $j = 1, \ldots, m$:

   (a) Generate the jth random sample $x_1^{(j)}, \ldots, x_n^{(j)}$ from the null distribution.

   (b) Compute the test statistic T_j from the jth sample.

   (c) Record the test decision $I_j = 1$ if H_0 is rejected at significance level α and otherwise $I_j = 0$.

2. Compute the proportion of significant tests $\frac{1}{m}\sum_{j=1}^{m} I_j$. This proportion is the observed Type I error rate.

For the Monte Carlo experiment above, the parameter estimated is a probability and the estimate, the observed Type I error rate, is a sample proportion. If we denote the observed Type I error rate by $\hat{p}$, then an estimate of $se(\hat{p})$ is

$$\widehat{se}(\hat{p}) = \sqrt{\frac{\hat{p}(1-\hat{p})}{m}} \leq \frac{0.5}{\sqrt{m}}.$$

The procedure is illustrated below with a simple example.

Example 6.7 (Empirical Type I error rate)

Suppose that $X_1, \ldots, X_{20}$ is a random sample from a $N(\mu, \sigma^2)$ distribution. Test $H_0 : \mu = 500$ $H_1 : \mu > 500$ at $\alpha = 0.05$. Under the null hypothesis,

$$T^* = \frac{\overline{X} - 500}{S/\sqrt{20}} \sim t(19),$$

where $t(19)$ denotes the Student t distribution with 19 degrees of freedom. Large values of T^* support the alternative hypothesis. Use a Monte Carlo method to compute an empirical probability of Type I error when $\sigma = 100$, and check that it is approximately equal to $\alpha = 0.05$.

The simulation below illustrates the procedure for the case $\sigma = 100$. The t-test is implemented by t.test in R, and we are basing the test decisions on the reported p-values returned by t.test.

```
n <- 20
alpha <- .05
mu0 <- 500
sigma <- 100

m <- 10000          #number of replicates
p <- numeric(m)     #storage for p-values
for (j in 1:m) {
    x <- rnorm(n, mu0, sigma)
    ttest <- t.test(x, alternative = "greater", mu = mu0)
    p[j] <- ttest$p.value
    }

p.hat <- mean(p < alpha)
se.hat <- sqrt(p.hat * (1 - p.hat) / m)
print(c(p.hat, se.hat))
```

    [1] 0.050600000 0.002191795

The observed Type I error rate in this simulation is 0.0506, and the standard error of the estimate is approximately $\sqrt{0.05 \times 0.95/m} \doteq 0.0022$. Estimates of Type I error probability will vary, but should be close to the nominal rate $\alpha = 0.05$ because all samples were generated under the null hypothesis from the assumed model for a t-test (normal distribution). In this experiment the empirical Type I error rate differs from $\alpha = 0.05$ by less than one standard error.

Theoretically, the probability of rejecting the null hypothesis when $\mu = 500$ is exactly $\alpha = 0.05$ in this example. The simulation really only investigates empirically whether the method of computing the p-value in t.test (a numerical algorithm) is consistent with the theoretical value $\alpha = 0.05$.   ◇

One of the simplest approaches to testing for univariate normality is the skewness test. In the following example we investigate whether a test based on the asymptotic distribution of the skewness statistic achieves the nominal significance level α under the null hypothesis of normality.

Example 6.8 (Skewness test of normality)

The skewness $\sqrt{\beta_1}$ of a random variable X is defined by

$$\sqrt{\beta_1} = \frac{E[(X - \mu_X)]^3}{\sigma_X^3},$$

where $\mu_X = E[X]$ and $\sigma_X^2 = Var(X)$. (The notation $\sqrt{\beta_1}$ is the classical notation for the signed skewness coefficient.) A distribution is symmetric if $\sqrt{\beta_1} = 0$, positively skewed if $\sqrt{\beta_1} > 0$, and negatively skewed if $\sqrt{\beta_1} < 0$. The sample coefficient of skewness is denoted by $\sqrt{b_1}$, and defined as

$$\sqrt{b_1} = \frac{\frac{1}{n} \sum_{i=1}^{n} (X_i - \overline{X})^3}{(\frac{1}{n} \sum_{i=1}^{n} (X_i - \overline{X})^2)^{3/2}}. \tag{6.2}$$

(Note that $\sqrt{b_1}$ is classical notation for the signed skewness statistic.) If the distribution of X is normal, then $\sqrt{b_1}$ is asymptotically normal with mean 0 and variance $6/n$ [59]. Normal distributions are symmetric, and a test for normality based on skewness rejects the hypothesis of normality for large values of $|\sqrt{b_1}|$. The hypotheses are

$$H_0 : \sqrt{\beta_1} = 0; \qquad H_1 : \sqrt{\beta_1} \neq 0,$$

where the sampling distribution of the skewness statistic is derived under the assumption of normality.

However, the convergence of $\sqrt{b_1}$ to its limit distribution is rather slow and the asymptotic distribution is not a good approximation for small to moderate sample sizes.

Assess the Type I error rate for a skewness test of normality at $\alpha = 0.05$ based on the asymptotic distribution of $\sqrt{b_1}$ for sample sizes $n = 10, 20, 30, 50, 100,$ and 500.

The vector of critical values cv for each of the sample sizes $n = 10, 20, 30, 50, 100,$ and 500 are computed under the normal limit distribution and stored in cv.

```
n <- c(10, 20, 30, 50, 100, 500) #sample sizes
cv <- qnorm(.975, 0, sqrt(6/n))   #crit. values for each n
```

```
asymptotic critical values:
n        10     20     30     50     100    500
cv   1.5182 1.0735 0.8765 0.6790 0.4801 0.2147
```

The asymptotic distribution of $\sqrt{b_1}$ does not depend on the mean and variance of the sampled normal distribution, so the samples can be generated from the standard normal distribution. If the sample size is `n[i]` then H_0 is rejected if $|\sqrt{b_1}| > $ `cv[i]`.

First write a function to compute the sample skewness statistic.

```
sk <- function(x) {
    #computes the sample skewness coeff.
    xbar <- mean(x)
    m3 <- mean((x - xbar)^3)
    m2 <- mean((x - xbar)^2)
    return( m3 / m2^1.5 )
}
```

In the code below, the outer loop varies the sample size n and the inner loop is the simulation for the current n. In the simulation, the test decisions are saved as 1 (reject H_0) or 0 (do not reject H_0) in the vector `sktests`. When the simulation for $n = 10$ ends, the mean of `sktests` gives the sample proportion of significant tests for $n = 10$. This result is saved in `p.reject[1]`. Then the simulation is repeated for $n = 20, 30, 50, 100, 500$, and saved in `p.reject[2:6]`.

```
#n is a vector of sample sizes
#we are doing length(n) different simulations

p.reject <- numeric(length(n)) #to store sim. results
m <- 10000                     #num. repl. each sim.

for (i in 1:length(n)) {
    sktests <- numeric(m)          #test decisions
    for (j in 1:m) {
        x <- rnorm(n[i])
        #test decision is 1 (reject) or 0
        sktests[j] <- as.integer(abs(sk(x)) >= cv[i] )
        }
    p.reject[i] <- mean(sktests) #proportion rejected
}
```

```
> p.reject
[1] 0.0129 0.0272 0.0339 0.0415 0.0464 0.0539
```

The results of the simulation are the empirical estimates of Type I error rate summarized below.

n	10	20	30	50	100	500
estimate	0.0129	0.0272	0.0339	0.0415	0.0464	0.0539

With $m = 10000$ replicates the standard error of the estimate is approximately $\sqrt{0.05 \times 0.95/m} \doteq 0.0022$.

The results of the simulation suggest that the asymptotic normal approximation for the distribution of $\sqrt{b_1}$ is not adequate for sample sizes $n \leq 50$, and questionable for sample sizes as large as $n = 500$. For finite samples one should use

$$Var(\sqrt{b_1}) = \frac{6(n-2)}{(n+1)(n+3)},$$

the exact value of the variance [93] (also see [60] or [270]). Repeating the simulation with

```
cv <- qnorm(.975, 0, sqrt(6*(n-2) / ((n+1)*(n+3))))
> round(cv, 4)
[1] 1.1355 0.9268 0.7943 0.6398 0.4660 0.2134
```

produces the simulation results summarized below.

n	10	20	30	50	100	500
estimate	0.0548	0.0515	0.0543	0.0514	0.0511	0.0479

These estimates are closer to the nominal level $\alpha = 0.05$. On skewness tests and other classical tests of normality see [58] or [270]. ◇

6.3.2 Power of a Test

In a test of hypotheses H_0 vs H_1, a Type II error occurs when H_1 is true, but H_0 is not rejected. The *power* of a test is given by the *power function* $\pi : \Theta \rightarrow [0,1]$, which is the probability $\pi(\theta)$ of rejecting H_0 given that the true value of the parameter is θ. Thus, for a given $\theta_1 \in \Theta_1$, the probability of Type II error is $1 - \pi(\theta_1)$. Ideally, we would prefer a test with low probability of error. Type I error is controlled by the choice of the significance level α. Low Type II error corresponds to high power under the alternative hypothesis. Thus, when comparing test procedures for the same hypotheses at the same significance level, we are interested in comparing the power of the tests. In general the comparison is not one problem but many; the power $\pi(\theta_1)$ of a test under the alternative hypothesis depends on the particular value of the alternative θ_1. For the t-test in Example 6.7, $\Theta_1 = (500, \infty)$. In general, however, the set Θ_1 can be more complicated.

If the power function of a test cannot be derived analytically, the power of a test against a fixed alternative $\theta_1 \in \Theta_1$ can be estimated by Monte Carlo methods. Note that the power function is defined for all $\theta \in \Theta$, but the significance level α controls $\pi(\theta) \leq \alpha$ for all $\theta \in \Theta_0$.

Monte Carlo experiment to estimate power of a test against a fixed alternative

1. Select a particular value of the parameter $\theta_1 \in \Theta$.

2. For each replicate, indexed by $j = 1, \ldots, m$:

   (a) Generate the jth random sample $x_1^{(j)}, \ldots, x_n^{(j)}$ under the conditions of the alternative $\theta = \theta_1$.

   (b) Compute the test statistic T_j from the jth sample.

   (c) Record the test decision: set $I_j = 1$ if H_0 is rejected at significance level α, and otherwise set $I_j = 0$.

3. Compute the proportion of significant tests $\hat{\pi}(\theta_1) = \frac{1}{m} \sum_{j=1}^{m} I_j$.

Example 6.9 (Empirical power)

Use simulation to estimate power and plot an empirical power curve for the t-test in Example 6.7. (For a numerical approach that does not involve simulation, see the remark below.)

To plot the curve, we need the empirical power for a sequence of alternatives θ along the horizontal axis. Each point corresponds to a Monte Carlo experiment. The outer `for` loop varies the points θ (mu) and the inner `replicate` loop (see R Note 6.1) estimates the power at the current θ.

```
n <- 20
m <- 1000
mu0 <- 500
sigma <- 100
mu <- c(seq(450, 650, 10))   #alternatives
M <- length(mu)
power <- numeric(M)
for (i in 1:M) {
    mu1 <- mu[i]
    pvalues <- replicate(m, expr = {
        #simulate under alternative mu1
        x <- rnorm(n, mean = mu1, sd = sigma)
        ttest <- t.test(x,
                    alternative = "greater", mu = mu0)
        ttest$p.value } )
    power[i] <- mean(pvalues <= .05)
}
```

The estimated power $\hat{\pi}(\theta)$ values are now stored in the vector `power`. Next, plot the empirical power curve, adding vertical error bars at $\hat{\pi}(\theta) \pm \hat{se}(\hat{\pi}(\theta))$ using the `errbar` function in the `Hmisc` package [132].

```
library(Hmisc)  #for errbar
plot(mu, power)
abline(v = mu0, lty = 1)
abline(h = .05, lty = 1)

#add standard errors
se <- sqrt(power * (1-power) / m)
errbar(mu, power, yplus = power+se, yminus = power-se,
    xlab = bquote(theta))
lines(mu, power, lty=3)
detach(package:Hmisc)
```

The power curve is shown in Figure 6.1. Note that the empirical power $\hat{\pi}(\theta)$ is small when θ is close to $\theta_0 = 500$, and increasing as θ moves farther away from θ_0, approaching 1 as $\theta \to \infty$.   ◇

Remark 6.2 *The non-central t distribution arises in power calculations for t-tests. The general non-central t with parameters (ν, δ) is defined as the distribution of $T(\nu, \delta) = (Z + \delta)/\sqrt{V/\nu}$ where $Z \sim N(0, 1)$ and $V \sim \chi^2(\nu)$ are independent.*

Suppose $X_1, X_2, \ldots, X_n$ is a random sample from a $N(\mu, \sigma^2)$ distribution, and the t-statistic $T = (\overline{X} - \mu_0)/(S/\sqrt{n})$ is applied to test $H_0 : \mu = \mu_0$. Under the null hypothesis, T has the central $t(n-1)$ distribution, but if $\mu \neq \mu_0$, T has the non-central t distribution with $n - 1$ degrees of freedom and non-centrality parameter $\delta = (\mu - \mu_0)\sqrt{n}/\sigma$. A numerical approach to evaluating the cdf of the non-central t distribution, based on an algorithm of Lenth [175], is implemented in the R function pt. Also see power.t.test.   ◇

Example 6.10 (Power of the skewness test of normality)

The skewness test of normality was described in Example 6.8. In this example, we estimate by simulation the power of the skewness test of normality against a contaminated normal (normal scale mixture) alternative described in Example 6.3. The contaminated normal distribution is denoted by

$$(1 - \varepsilon)N(\mu = 0, \sigma^2 = 1) + \varepsilon N(\mu = 0, \sigma^2 = 100), \qquad 0 \leq \varepsilon \leq 1.$$

When $\varepsilon = 0$ or $\varepsilon = 1$ the distribution is normal, but the mixture is non-normal for $0 < \varepsilon < 1$. We can estimate the power of the skewness test for a sequence of alternatives indexed by ε and plot a power curve for the power of the skewness test against *this type of alternative*. For this experiment, the significance level is $\alpha = 0.1$ and the sample size is $n = 30$. The skewness statistic sk is implemented in Example 6.8.

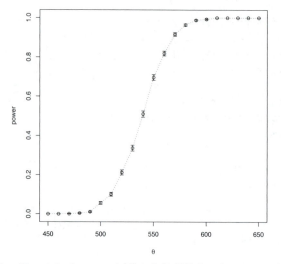

FIGURE 6.1: Empirical power $\hat{\pi}(\theta)\pm\hat{se}(\hat{\pi}(\theta))$ for the t-test of $H_0 : \theta = 500$ vs $H_1 : \theta > 500$ in Example 6.9.

```
alpha <- .1
n <- 30
m <- 2500
epsilon <- c(seq(0, .15, .01), seq(.15, 1, .05))
N <- length(epsilon)
pwr <- numeric(N)
#critical value for the skewness test
cv <- qnorm(1-alpha/2, 0, sqrt(6*(n-2) / ((n+1)*(n+3))))

for (j in 1:N) {           #for each epsilon
    e <- epsilon[j]
    sktests <- numeric(m)
    for (i in 1:m) {       #for each replicate
        sigma <- sample(c(1, 10), replace = TRUE,
            size = n, prob = c(1-e, e))
        x <- rnorm(n, 0, sigma)
        sktests[i] <- as.integer(abs(sk(x)) >= cv)
        }
    pwr[j] <- mean(sktests)
    }
#plot power vs epsilon
plot(epsilon, pwr, type = "b",
    xlab = bquote(epsilon), ylim = c(0,1))
abline(h = .1, lty = 3)
se <- sqrt(pwr * (1-pwr) / m)   #add standard errors
lines(epsilon, pwr+se, lty = 3)
lines(epsilon, pwr-se, lty = 3)
```

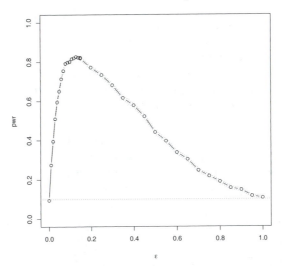

FIGURE 6.2:   Empirical power $\hat{\pi}(\varepsilon) \pm \widehat{se}(\hat{\pi}(\varepsilon))$ for the skewness test of normality against ε-contaminated normal scale mixture alternative in Example 6.10.

The empirical power curve is shown in Figure 6.2. Note that the power curve crosses the horizontal line corresponding to $\alpha = 0.10$ at both endpoints, $\varepsilon = 0$ and $\varepsilon = 1$ where the alternative is normally distributed. For $0 < \varepsilon < 1$ the empirical power of the test is greater than 0.10 and highest when ε is about 0.15. ◇

6.3.3   Power comparisons

Monte Carlo methods are often applied to compare the performance of different test procedures. A skewness test of normality was introduced in Example 6.8. There are many tests of normality in the literature (see [58] and [270]). In the following example three tests of univariate normality are compared.

Example 6.11 (Power comparison of tests of normality)

Compare the empirical power of the skewness test of univariate normality with the Shapiro-Wilk [248] test. Also compare the power of the *energy* test [263], which is based on distances between sample elements.

Let $\mathcal{N}$ denote the family of univariate normal distributions. Then the test hypotheses are

$$H_0 : F_X \in \mathcal{N} \qquad H_1 : F_X \notin \mathcal{N}.$$

The Shapiro-Wilk test is based on the regression of the sample order statistics on their expected values under normality, so it falls in the general category of tests based on regression and correlation. The approximate critical values of the statistic are determined by a transformation of the statistic W to normality [235, 236, 237] for sample sizes $7 \leq n \leq 2000$. The Shapiro-Wilk test is implemented by the R function `shapiro.test`.

The *energy* test is based on an energy distance between the sampled distribution and normal distribution, so large values of the statistic are significant. The *energy* test is a test of multivariate normality [263], so the test considered here is the special case $d = 1$. As a test of univariate normality, *energy* performs very much like the Anderson-Darling test [9]. The energy statistic for testing normality is

$$
Q_n = n \left[\frac{2}{n} \sum_{i=1}^{n} E\|x_i - X\| - E\|X - X'\| - \frac{1}{n^2} \sum_{i,j=1}^{n} \|x_i - x_j\| \right], \qquad (6.3)
$$

where X, X' are iid. Large values of Q_n are significant. In the univariate case, the following computing formula is equivalent:

$$
Q_n = n \left[\frac{2}{n} \sum_{i=1}^{n} (2Y_i\ \Phi(Y_i) + 2\phi(Y_i)) - \frac{2}{\sqrt{\pi}} - \frac{2}{n^2} \sum_{k=1}^{n} (2k - 1 - n)Y_{(k)} \right],
$$

(6.4)

where $Y_i = \frac{X_i - \mu_X}{\sigma_X}$, $Y_{(k)}$ is the kth order statistic of the standardized sample, Φ is the standard normal cdf and ϕ is the standard normal density. If the parameters are unknown, substitute the sample mean and sample standard deviation to to compute $Y_1, \ldots, Y_n$. A computing formula for the multivariate case is given in [263]. The energy test for univariate and multivariate normality is implemented in `mvnorm.etest` in the **energy** package [226].

The skewness test of normality was introduced in Examples 6.8 and 6.10. The sample skewness function `sk` is given in Example 6.8 on page 166.

For this comparison we set significance level $\alpha = 0.1$. The example below compares the power of the tests against the contaminated normal alternatives described in Example 6.3. The alternative is the normal mixture denoted by

$$
(1 - \varepsilon)N(\mu = 0, \sigma^2 = 1) + \varepsilon N(\mu = 0, \sigma^2 = 100), \qquad 0 \leq \varepsilon \leq 1.
$$

When $\varepsilon = 0$ or $\varepsilon = 1$ the distribution is normal, and in this case the empirical Type I error rate should be controlled at approximately the nominal rate $\alpha = 0.1$. If $0 < \varepsilon < 1$ the distributions are non-normal, and we are interested in comparing the empirical power of the tests against these alternatives.

```
# initialize input and output
library(energy)
alpha <- .1
n <- 30
m <- 2500          #try smaller m for a trial run
epsilon <- .1
test1 <- test2 <- test3 <- numeric(m)

#critical value for the skewness test
cv <- qnorm(1-alpha/2, 0, sqrt(6*(n-2) / ((n+1)*(n+3))))

# estimate power
for (j in 1:m) {
    e <- epsilon
    sigma <- sample(c(1, 10), replace = TRUE,
        size = n, prob = c(1-e, e))
    x <- rnorm(n, 0, sigma)
    test1[j] <- as.integer(abs(sk(x)) >= cv)
    test2[j] <- as.integer(
                shapiro.test(x)$p.value <= alpha)
    test3[j] <- as.integer(
                mvnorm.etest(x, R=200)$p.value <= alpha)
}
print(c(epsilon, mean(test1), mean(test2), mean(test3)))
detach(package:energy)
```

The simulation was repeated for several choices of ε and results saved in a matrix sim. Simulation results for $n = 30$ are summarized in Table 6.2 and in Figure 6.3. The plot is obtained as follows.

```
# plot the empirical estimates of power
plot(sim[,1], sim[,2], ylim = c(0, 1), type = "l",
    xlab = bquote(epsilon), ylab = "power")
lines(sim[,1], sim[,3], lty = 2)
lines(sim[,1], sim[,4], lty = 4)
abline(h = alpha, lty = 3)
legend("topright", 1, c("skewness", "S-W", "energy"),
    lty = c(1,2,4), inset = .02)
```

Standard error of the estimates is at most $0.5/\sqrt{m} = 0.01$. Estimates for empirical Type I error rate correspond to $\varepsilon = 0$ and $\varepsilon = 1$. All tests achieve approximately the nominal significance level $\alpha = 0.10$ within one standard error. The tests are at approximately the same significance level, so it is meaningful to compare the results for power.

The simulation results suggest that the Shapiro-Wilk and energy tests are about equally powerful against this type of alternative when $n = 30$ and $\varepsilon < 0.5$. Both have higher power than the skewness test overall and energy appears to have highest power for $0.5 \leq \varepsilon \leq 0.8$.

◇

6.4   Application: "Count Five" Test for Equal Variance

The examples in this section illustrate the Monte Carlo method for a simple two sample test of equal variance.

The two sample "Count Five" test for equality of variance introduced by McGrath and Yeh [193] counts the number of extreme points of each sample relative to the range of the other sample. Suppose the means of the two samples are equal and the sample sizes are equal. An observation in one sample is considered extreme if it is not within the range of the other sample. If either sample has five or more extreme points, the hypothesis of equal variance is rejected.

Example 6.12 (Count Five test statistic)

The computation of the test statistic is illustrated with a numerical example. Compare the side-by-side boxplots in Figure 6.4 and observe that there are some extreme points in each sample with respect to the other sample.

```
x1 <- rnorm(20, 0, sd = 1)
x2 <- rnorm(20, 0, sd = 1.5)
y <- c(x1, x2)

group <- rep(1:2, each = length(x1))
boxplot(y ~ group, boxwex = .3, xlim = c(.5, 2.5), main = "")
points(group, y)

# now identify the extreme points
> range(x1)
[1] -2.782576  1.728505
> range(x2)
[1] -1.598917  3.710319
```

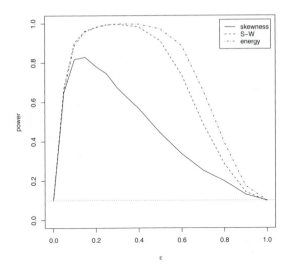

FIGURE 6.3: Empirical power of three tests of normality against a contaminated normal alternative in Example 6.11 ($n = 30$, $\alpha = 0.1$, $se \leq 0.01$)

TABLE 6.2: Empirical Power of Three Tests of Normality against a Contaminated Normal Alternative in Example 6.11 ($n = 30$, $\alpha = 0.1$, $se \leq 0.01$)

ε	skewness test	Shapiro-Wilk	energy test
0.00	0.0984	0.1076	0.1064
0.05	0.6484	0.6704	0.6560
0.10	0.8172	0.9008	0.8896
0.15	0.8236	0.9644	0.9624
0.20	0.7816	0.9816	0.9800
0.25	0.7444	0.9940	0.9924
0.30	0.6724	0.9960	0.9980
0.40	0.5672	0.9828	0.9964
0.50	0.4424	0.9112	0.9724
0.60	0.3368	0.7380	0.8868
0.70	0.2532	0.4900	0.6596
0.80	0.1980	0.2856	0.3932
0.90	0.1296	0.1416	0.1724
1.00	0.0992	0.0964	0.0980

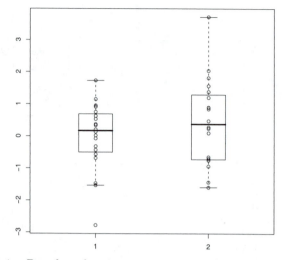

FIGURE 6.4:   Boxplots showing extreme points for the Count Five statistic in Example 6.12.

```
> i <- which(x1 < min(x2))
> j <- which(x2 > max(x1))

> x1[i]
[1] -2.782576

> x2[j]
[1] 2.035521 1.809902 3.710319
```

The Count Five statistic is the maximum number of extreme points, $\max(1, 3)$, so the Count Five test will not reject the hypothesis of equal variance. Note that we only need the number of extreme points, and the extreme count can be determined without reference to a boxplot as follows.

```
out1 <- sum(x1 > max(x2)) + sum(x1 < min(x2))
out2 <- sum(x2 > max(x1)) + sum(x2 < min(x1))
> max(c(out1, out2))
[1] 3
```

◇

Example 6.13 (Count Five test statistic, cont.)

Consider the case of two independent random samples from the same normal distribution. Estimate the sampling distribution of the maximum number of extreme points, and find the 0.80, 0.90, and 0.95 quantiles of the sampling distribution.

The function `maxout` below counts the maximum number of extreme points of each sample with respect to the range of the other sample. The sampling distribution of the extreme count statistic can be estimated by a Monte Carlo experiment.

```
maxout <- function(x, y) {
    X <- x - mean(x)
    Y <- y - mean(y)
    outx <- sum(X > max(Y)) + sum(X < min(Y))
    outy <- sum(Y > max(X)) + sum(Y < min(X))
    return(max(c(outx, outy)))
}

n1 <- n2 <- 20
mu1 <- mu2 <- 0
sigma1 <- sigma2 <- 1
m <- 1000

# generate samples under H0
stat <- replicate(m, expr={
    x <- rnorm(n1, mu1, sigma1)
    y <- rnorm(n2, mu2, sigma2)
    maxout(x, y)
    })
print(cumsum(table(stat)) / m)
print(quantile(stat, c(.8, .9, .95)))
```

The "Count Five" test criterion looks reasonable for normal distributions. The empirical cdf and quantiles are

1	2	3	4	5	6	7	8	9	10	11
0.149	0.512	0.748	0.871	0.945	0.974	0.986	0.990	0.996	0.999	1.000

80%	90%	95%
4	5	6

Notice that the `quantile` function gives 6 as the 0.95 quantile. However, if $\alpha = 0.05$ is the desired significance level, the critical value 5 appears to be the best choice. The `quantile` function is not always the best way to estimate a critical value. If `quantile` is used, compare the result to the empirical cdf. ◇

The "Count Five" test criterion can be applied for independent random samples when the random variables are similarly distributed and sample sizes are equal. (Random variables X and Y are called *similarly distributed* if Y has the same distribution as $(X - a)/b$ where a and $b > 0$ are constants.) When the data are centered by their respective population means, McGrath and Yeh

[193] show that the Count Five test on the centered data has significance level at most 0.0625.

   In practice, the populations means are generally unknown and each sample would be centered by subtracting its sample mean. Also, the sample sizes may be unequal.

Example 6.14 (Count Five test)

Use Monte Carlo methods to estimate the significance level of the test when each sample is centered by subtracting its sample mean. Here again we consider normal distributions. The function count5test returns the value 1 (reject H_0) or 0 (do not reject H_0).

```
count5test <- function(x, y) {
    X <- x - mean(x)
    Y <- y - mean(y)
    outx <- sum(X > max(Y)) + sum(X < min(Y))
    outy <- sum(Y > max(X)) + sum(Y < min(X))
    # return 1 (reject) or 0 (do not reject H0)
    return(as.integer(max(c(outx, outy)) > 5))
}
```

```
n1 <- n2 <- 20
mu1 <- mu2 <- 0
sigma1 <-  sigma2 <- 1
m <- 10000
tests <- replicate(m, expr = {
    x <- rnorm(n1, mu1, sigma1)
    y <- rnorm(n2, mu2, sigma2)
    x <- x - mean(x)   #centered by sample mean
    y <- y - mean(y)
    count5test(x, y)
    } )
```

```
alphahat <- mean(tests)
> print(alphahat)
[1] 0.0565
```

   If the samples are centered by the population mean, we should expect an empirical Type I error rate of about 0.055, from our previous simulation to estimate the quantiles of the maxout statistic. In the simulation, each sample was centered by subtracting the sample mean, and the empirical Type I error rate was 0.0565 (se $\doteq 0.0022$).   ◇

Example 6.15 (Count Five test, cont.)

Repeating the previous example, we are estimating the empirical Type I error rate when sample sizes differ and the "Count Five" test criterion is applied. Each sample is centered by subtracting the sample mean.

```
n1 <- 20
n2 <- 30
mu1 <- mu2 <- 0
sigma1 <- sigma2 <- 1
m <- 10000

alphahat <- mean(replicate(m, expr={
    x <- rnorm(n1, mu1, sigma1)
    y <- rnorm(n2, mu2, sigma2)
    x <- x - mean(x)   #centered by sample mean
    y <- y - mean(y)
    count5test(x, y)
    }))

print(alphahat)
[1] 0.1064
```

The simulation result suggests that the "Count Five" criterion does not necessarily control Type I error at $\alpha \leq 0.0625$ when the sample sizes are unequal. Repeating the simulation above with $n_1 = 20$ and $n_2 = 50$, the empirical Type I error rate was 0.2934. See [193] for a method to adjust the test criterion for unequal sample sizes. ◇

Example 6.16 (Count Five, cont.)

Use Monte Carlo methods to estimate the power of the Count Five test, where the sampled distributions are $N(\mu_1 = 0, \sigma_1^2 = 1)$, $N(\mu_2 = 0, \sigma_2^2 = 1.5^2)$, and the sample sizes are $n_1 = n_2 = 20$.

```
# generate samples under H1 to estimate power
sigma1 <- 1
sigma2 <- 1.5

power <- mean(replicate(m, expr={
    x <- rnorm(20, 0, sigma1)
    y <- rnorm(20, 0, sigma2)
    count5test(x, y)
    }))
```

```
> print(power)
[1] 0.3129
```

The empirical power of the test is 0.3129 ($se \leq 0.005$) against the alternative ($\sigma_1 = 1$, $\sigma_2 = 1.5$) with $n_1 = n_2 = 20$. See [193] for power comparisons with other tests for equal variance and applications.   ◇

Exercises

6.1 Estimate the MSE of the level k trimmed means for random samples of size 20 generated from a standard Cauchy distribution. (The target parameter θ is the center or median; the expected value does not exist.) Summarize the estimates of MSE in a table for $k = 1, 2, \ldots, 9$.

6.2 Plot the empirical power curve for the t-test in Example 6.9, changing the alternative hypothesis to $H_1 : \mu \neq 500$, and keeping the significance level $\alpha = 0.05$.

6.3 Plot the power curves for the t-test in Example 6.9 for sample sizes 10, 20, 30, 40, and 50, but omit the standard error bars. Plot the curves on the same graph, each in a different color or different line type, and include a legend. Comment on the relation between power and sample size.

6.4 Suppose that $X_1, \ldots, X_n$ are a random sample from a from a lognormal distribution with unknown parameters. Construct a 95% confidence interval for the parameter μ. Use a Monte Carlo method to obtain an empirical estimate of the confidence level.

6.5 Suppose a 95% symmetric t-interval is applied to estimate a mean, but the sample data are non-normal. Then the probability that the confidence interval covers the mean is not necessarily equal to 0.95. Use a Monte Carlo experiment to estimate the coverage probability of the t-interval for random samples of $\chi^2(2)$ data with sample size $n = 20$. Compare your t-interval results with the simulation results in Example 6.4. (The t-interval should be more robust to departures from normality than the interval for variance.)

6.6 Estimate the 0.025, 0.05, 0.95, and 0.975 quantiles of the skewness $\sqrt{b_1}$ under normality by a Monte Carlo experiment. Compute the standard error of the estimates from (2.14) using the normal approximation for the density (with exact variance formula). Compare the estimated quantiles with the quantiles of the large sample approximation $\sqrt{b_1} \approx N(0, 6/n)$.

6.7 Estimate the power of the skewness test of normality against symmetric Beta(α, α) distributions and comment on the results. Are the results different for heavy-tailed symmetric alternatives such as $t(\nu)$?

6.8   Refer to Example 6.16. Repeat the simulation, but also compute the F test of equal variance, at significance level $\hat{\alpha} \doteq 0.055$. Compare the power of the Count Five test and F test for small, medium, and large sample sizes. (Recall that the F test is not applicable for non-normal distributions.)

6.9   Let X be a non-negative random variable with $\mu = E[X] < \infty$. For a random sample $x_1, \ldots, x_n$ from the distribution of X, the Gini ratio is defined by

$$G = \frac{1}{2n^2\mu} \sum_{j=1}^{n} \sum_{i=1}^{n} |x_i - x_j|.$$

The Gini ratio is applied in economics to measure inequality in income distribution (see e.g. [163]). Note that G can be written in terms of the order statistics $x_{(i)}$ as

$$G = \frac{1}{n^2\mu} \sum_{i=1}^{n} (2i - n - 1) x_{(i)}.$$

If the mean is unknown, let $\hat{G}$ be the statistic G with μ replaced by $\bar{x}$. Estimate by simulation the mean, median and deciles of $\hat{G}$ if X is standard lognormal. Repeat the procedure for the uniform distribution and Bernoulli(0.1). Also construct density histograms of the replicates in each case.

6.10   Construct an approximate 95% confidence interval for the Gini ratio $\gamma = E[G]$ if X is lognormal with unknown parameters. Assess the coverage rate of the estimation procedure with a Monte Carlo experiment.

Projects

6.A   Use Monte Carlo simulation to investigate whether the empirical Type I error rate of the t-test is approximately equal to the nominal significance level α, when the sampled population is non-normal. The t-test is robust to mild departures from normality. Discuss the simulation results for the cases where the sampled population is (i) $\chi^2(1)$, (ii) Uniform(0,2), and (iii) Exponential(rate=1). In each case, test $H_0 : \mu = \mu_0$ vs $H_0 : \mu \neq \mu_0$, where μ_0 is the mean of $\chi^2(1)$, Uniform(0,2), and Exponential(1), respectively.

6.B   Tests for association based on Pearson product moment correlation ρ, Spearman's rank correlation coefficient ρ_s, or Kendall's coefficient τ, are implemented in cor.test. Show (empirically) that the nonparametric tests based on ρ_s or τ are less powerful than the correlation test when the sampled distribution is bivariate normal. Find an example of an alternative (a bivariate distribution (X, Y) such that X and Y are dependent) such that at least one of the nonparametric tests have better empirical power than the correlation test against this alternative.

6.C Repeat Examples 6.8 and 6.10 for Mardia's multivariate skewness test. Mardia [187] proposed tests of multivariate normality based on multivariate generalizations of skewness and kurtosis. If X and Y are iid, the multivariate population skewness $\beta_{1,d}$ is defined by Mardia as

$$\beta_{1,d} = E\left[(X - \mu)^T \Sigma^{-1} (Y - \mu)\right]^3.$$

Under normality, $\beta_{1,d} = 0$. The multivariate skewness statistic is

$$b_{1,d} = \frac{1}{n^2} \sum_{i,j=1}^{n} ((X_i - \bar{X})^T \widehat{\Sigma}^{-1} (X_j - \bar{X}))^3, \qquad (6.5)$$

where $\widehat{\Sigma}$ is the maximum likelihood estimator of covariance. Large values of $b_{1,d}$ are significant. The asymptotic distribution of $nb_{1,d}/6$ is chisquared with $d(d+1)(d+2)/6$ degrees of freedom.

6.D Repeat Example 6.11 for multivariate tests of normality. Mardia [187] defines multivariate kurtosis as

$$\beta_{2,d} = E\left[(X - \mu)^T \Sigma^{-1} (X - \mu)\right]^2.$$

For d-dimensional multivariate normal distributions the kurtosis coefficient is $\beta_{2,d} = d(d+2)$. The multivariate kurtosis statistic is

$$b_{2,d} = \frac{1}{n} \sum_{i=1}^{n} ((X_i - \bar{X})^T \widehat{\Sigma}^{-1} (X_i - \bar{X}))^2. \qquad (6.6)$$

The large sample test of multivariate normality based on $b_{2,d}$ rejects the null hypothesis at significance level α if

$$\left| \frac{b_{2,d} - d(d+2)}{\sqrt{8d(d+2)/n}} \right| \geq \Phi^{-1}(1 - \alpha/2).$$

However, $b_{2,d}$ converges very slowly to the normal limiting distribution. Compare the empirical power of Mardia's skewness and kurtosis tests of multivariate normality with the energy test of multivariate normality mvnorm.etest (energy) (6.3) [226, 263]. Consider multivariate normal location mixture alternatives where the two samples are generated from mlbench.twonorm in the mlbench package [174].

Chapter 7

Bootstrap and Jackknife

7.1 The Bootstrap

The bootstrap was introduced in 1979 by Efron [80], with further developments in 1981 [82, 81], 1982 [83], and numerous other publications including the monograph of Efron and Tibshirani [84]. Chernick [45] has an extensive bibliography. Davison and Hinkley [63] is a comprehensive reference with many applications. Also see Barbe and Bertail [19], Shao and Tu [247], and Mammen [186].

Bootstrap methods are a class of nonparametric Monte Carlo methods that estimate the distribution of a population by resampling. Resampling methods treat an observed sample as a finite population, and random samples are generated (resampled) from it to estimate population characteristics and make inferences about the sampled population. Bootstrap methods are often used when the distribution of the target population is not specified; the sample is the only information available.

The term "bootstrap" can refer to nonparametric bootstrap or parametric bootstrap. Monte Carlo methods that involve sampling from a fully specified probability distribution, such as methods of Chapter 6 are sometimes called parametric bootstrap. Nonparametric bootstrap is the subject of this chapter. In nonparametric bootstrap, the distribution is not specified.

The distribution of the finite population represented by the sample can be regarded as a pseudo-population with similar characteristics as the true population. By repeatedly generating random samples from this pseudo-population (resampling), the sampling distribution of a statistic can be estimated. Properties of an estimator such as bias or standard error can be estimated by resampling.

Bootstrap estimates of a sampling distribution are analogous to the idea of density estimation. We construct a histogram of a sample to obtain an estimate of the shape of the density function. The histogram is not the density, but in a nonparametric problem, can be viewed as a reasonable estimate of the density. We have methods to generate random samples from completely specified densities; bootstrap generates random samples from the empirical distribution of the sample.

Suppose that $x = (x_1, \ldots, x_n)$ is an observed random sample from a distribution with cdf $F(x)$. If X^* is selected at random from x, then

$$P(X^* = x_i) = \frac{1}{n}, \qquad i = 1, \ldots, n.$$

Resampling generates a random sample $X_1^*, \ldots, X_n^*$ by sampling with replacement from x. The random variables X_i^* are iid, uniformly distributed on the set $\{x_1, \ldots, x_n\}$.

The empirical distribution function (ecdf) $F_n(x)$ is an estimator of $F(x)$. It can be shown that $F_n(x)$ is a sufficient statistic for $F(x)$; that is, all the information about $F(x)$ that is contained in the sample is also contained in $F_n(x)$. Moreover, $F_n(x)$ is itself the distribution function of a random variable; namely the random variable that is uniformly distributed on the set $\{x_1, \ldots, x_n\}$. Hence the empirical cdf F_n is the cdf of X^*. Thus in bootstrap, there are two approximations. The ecdf F_n is an approximation to the cdf F_X. The ecdf F_m^* of the bootstrap replicates is an approximation to the ecdf F_n. Resampling from the sample x is equivalent to generating random samples from the distribution $F_n(x)$. The two approximations can be represented by the diagram

$$F \to X \to F_n$$
$$F_n \to X^* \to F_n^*.$$

To generate a bootstrap random sample by resampling x, generate n random integers $\{i_1, \ldots, i_n\}$ uniformly distributed on $\{1, \ldots, n\}$ and select the bootstrap sample $x^* = (x_{i_1}, \ldots, x_{i_n})$.

Suppose θ is the parameter of interest (θ could be a vector), and $\hat{\theta}$ is an estimator of θ. Then the bootstrap estimate of the distribution of $\hat{\theta}$ is obtained as follows.

1. For each bootstrap replicate, indexed $b = 1, \ldots, B$:

   (a) Generate sample $x^{*(b)} = x_1^*, \ldots, x_n^*$ by sampling with replacement from the observed sample $x_1, \ldots, x_n$.

   (b) Compute the bth replicate $\hat{\theta}^{(b)}$ from the bth bootstrap sample.

2. The bootstrap estimate of $F_{\hat{\theta}}(\cdot)$ is the empirical distribution of the replicates $\hat{\theta}^{(1)}, \ldots, \hat{\theta}^{(B)}$.

The bootstrap is applied to estimate the standard error and the bias of an estimator in the following sections. First let us see an example to illustrate the relation between the ecdf F_n and the distribution of the bootstrap replicates.

Example 7.1 (F_n and bootstrap samples)

Suppose that we have observed the sample

$$x = \{2, 2, 1, 1, 5, 4, 4, 3, 1, 2\}.$$

Resampling from x we select 1, 2, 3, 4, or 5 with probabilities 0.3, 0.3, 0.1, 0.2, and 0.1 respectively, so the cdf F_{X*} of a randomly selected replicate is exactly the ecdf $F_n(x)$:

$$F_{X*}(x) = F_n(x) = \begin{cases} 0, & x < 1; \\ 0.3, & 1 \leq x < 2; \\ 0.6, & 2 \leq x < 3; \\ 0.7, & 3 \leq x < 4; \\ 0.9, & 4 \leq x < 5; \\ 1, & x \geq 5. \end{cases}$$

Note that if F_n is not close to F_X then the distribution of the replicates will not be close to F_X. The sample x above is actually a sample from a Poisson(2) distribution. Resampling from x a large number of replicates produces a good estimate of F_n but not a good estimate of F_X, because regardless of how many replicates are drawn, the bootstrap samples will never include 0.   ◇

7.1.1 Bootstrap Estimation of Standard Error

The bootstrap estimate of standard error of an estimator $\hat{\theta}$ is the sample standard deviation of the bootstrap replicates $\hat{\theta}^{(1)}, \ldots, \hat{\theta}^{(B)}$.

$$\widehat{se}(\hat{\theta}^*) = \sqrt{\frac{1}{B-1} \sum_{b=1}^{B} (\hat{\theta}^{(b)} - \overline{\hat{\theta}^*})^2}, \tag{7.1}$$

where $\overline{\hat{\theta}^*} = \frac{1}{B} \sum_{b=1}^{B} \hat{\theta}^{(b)}$ [84, (6.6)].

According to Efron and Tibshirani [84, p. 52], the number of replicates needed for good estimates of standard error is not large; $B = 50$ is usually large enough, and rarely is $B > 200$ necessary. (Much larger B will be needed for confidence interval estimation.)

Example 7.2 (Bootstrap estimate of standard error)

The law school data set `law` in the `bootstrap` [271] package is from Efron and Tibshirani [84]. The data frame contains LSAT (average score on law school admission test score) and GPA (average undergraduate grade-point average) for 15 law schools.

LSAT 576 635 558 578 666 580 555 661 651 605 653 575 545 572 594
GPA　339 330 281 303 344 307 300 343 336 313 312 274 276 288 296

This data set is a random sample from the universe of 82 law schools in `law82` (bootstrap). Estimate the correlation between LSAT and GPA scores, and compute the bootstrap estimate of the standard error of the sample correlation.

1. For each bootstrap replicate, indexed $b = 1, \ldots, B$:

    (a) Generate sample $x^{*(b)} = x_1^*, \ldots, x_n^*$ by sampling with replacement from the observed sample $x_1, \ldots, x_n$.

    (b) Compute the bth replicate $\hat{\theta}^{(b)}$ from the bth bootstrap sample, where $\hat{\theta}$ is the sample correlation R between (LSAT, GPA).

2. The bootstrap estimate of se(R) is the sample standard deviation of the replicates $\hat{\theta}^{(1)}, \ldots, \hat{\theta}^{(B)} = R^{(1)}, \ldots, R^{(B)}$.

```
library(bootstrap)      #for the law data
print(cor(law$LSAT, law$GPA))
[1] 0.7763745
print(cor(law82$LSAT, law82$GPA))
[1] 0.7599979
```

The sample correlation is $R = 0.7763745$. The correlation for the universe of 82 law schools is $R = 0.7599979$. Use bootstrap to estimate the standard error of the correlation statistic computed from the sample of scores in `law`.

```
#set up the bootstrap
B <- 200                #number of replicates
n <- nrow(law)          #sample size
R <- numeric(B)         #storage for replicates

#bootstrap estimate of standard error of R
for (b in 1:B) {
    #randomly select the indices
    i <- sample(1:n, size = n, replace = TRUE)
    LSAT <- law$LSAT[i]         #i is a vector of indices
    GPA <- law$GPA[i]
    R[b] <- cor(LSAT, GPA)
}
#output
> print(se.R <- sd(R))
[1] 0.1358393
> hist(R, prob = TRUE)
```

The bootstrap estimate of se(R) is 0.1358393. The normal theory estimate for standard error of R is 0.115. The jackknife-after-bootstrap method of estimating $\widehat{se}(\widehat{se}(\hat{\theta}))$ is covered in Section 7.3. The histogram of the replicates of R is shown in Figure 7.1. ◇

In the next example, the `boot` function in recommended package `boot` [34] is applied to run the bootstrap. See Appendix B.1 for a note about how to write the function for the `statistic` argument in `boot`.

Example 7.3 (Bootstrap estimate of standard error: `boot` function)

Example 7.2 is repeated, using the `boot` function in `boot`. First, write a function that returns $\hat{\theta}^{(b)}$, where the first argument to the function is the sample data, and the second argument is the vector $\{i_1, \ldots, i_n\}$ of indices. If the data is x and the vector of indices is i, we need `x[i,1]` to extract the first resampled variable, and `x[i,2]` to extract the second resampled variable. The code and output is shown below.

```
r <- function(x, i) {
    #want correlation of columns 1 and 2
    cor(x[i,1], x[i,2])
}
```

The printed summary of output from the `boot` function is obtained by the command `boot` or the result can be saved in an object for further analysis. Here we save the result in `obj` and print the summary.

```
library(boot)       #for boot function
> obj <- boot(data = law, statistic = r, R = 2000)
> obj

ORDINARY NONPARAMETRIC BOOTSTRAP

Call: boot(data = law, statistic = r, R = 2000)
Bootstrap Statistics :
        original        bias      std. error
t1* 0.7763745 -0.004795305   0.1303343
```

The observed value $\hat{\theta}$ of the correlation statistic is labeled `t1*`. The bootstrap estimate of standard error of the estimate is $\widehat{se}(\hat{\theta}) \doteq 0.13$, based on 2000 replicates. To compare with formula (7.1), extract the replicates in `$t`.

```
> y <- obj$t
> sd(y)
[1] 0.1303343
```

◇

R note 7.1 *The syntax and options for the* `boot` *(boot) function and the* `bootstrap` *(bootstrap) function are different. Note that the* `bootstrap` *package [271] is a collection of functions and data for the book by Efron and Tibshirani [84], and the* `boot` *package [34] is a collection of functions and data for the book by Davison and Hinkley [63].*

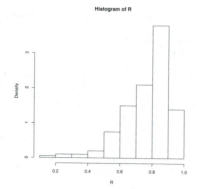

FIGURE 7.1: Bootstrap replicates for law school data in Example 7.2.

7.1.2   Bootstrap Estimation of Bias

If $\hat{\theta}$ is an unbiased estimator of θ, $E[\hat{\theta}] = \theta$. The bias of an estimator $\hat{\theta}$ for θ is

$$bias(\hat{\theta}) = E[\hat{\theta} - \theta] = E[\hat{\theta}] - \theta.$$

Thus, every statistic is an unbiased estimator of its expected value, and in particular, the sample mean of a random sample is an unbiased estimator of the mean of the distribution. An example of a biased estimator is the maximum likelihood estimator of variance, $\hat{\sigma}^2 = \frac{1}{n}\Sigma_{i=1}^n (X_i - \overline{X})^2$, which has expected value $(1 - 1/n)\sigma^2$. Thus, $\hat{\sigma}^2$ underestimates σ^2, and the bias is $-\sigma^2/n$.

The bootstrap estimation of bias uses the bootstrap replicates of $\hat{\theta}$ to estimate the sampling distribution of $\hat{\theta}$. For the finite population $x = (x_1, \dots, x_n)$, the parameter is $\hat{\theta}(x)$ and there are B independent and identically distributed estimators $\hat{\theta}^{(b)}$. The sample mean of the replicates $\{\hat{\theta}^{(b)}\}$ is unbiased for its expected value $E[\hat{\theta}^*]$, so the bootstrap estimate of bias is

$$\widehat{bias}(\hat{\theta}) = \overline{\hat{\theta}^*} - \hat{\theta}, \tag{7.2}$$

where $\overline{\hat{\theta}^*} = \frac{1}{B}\Sigma_{b=1}^B \hat{\theta}^{(b)}$, and $\hat{\theta} = \hat{\theta}(x)$ is the estimate computed from the original observed sample. (In bootstrap F_n is sampled in place of F_X, so we replace θ with $\hat{\theta}$ to estimate the bias.) Positive bias indicates that $\hat{\theta}$ on average tends to overestimate θ.

Example 7.4 (Boostrap estimate of bias)

In the `law` data of Example 7.2, compute the bootstrap estimate of bias in the sample correlation.

```
#sample estimate for n=15
theta.hat <- cor(law$LSAT, law$GPA)

#bootstrap estimate of bias
B <- 2000    #larger for estimating bias
n <- nrow(law)
theta.b <- numeric(B)

for (b in 1:B) {
    i <- sample(1:n, size = n, replace = TRUE)
    LSAT <- law$LSAT[i]
    GPA <- law$GPA[i]
    theta.b[b] <- cor(LSAT, GPA)
}
bias <- mean(theta.b - theta.hat)
> bias
[1] -0.005797944
```

The estimate of bias is -0.005797944. Note that this is close to the estimate of bias returned by the `boot` function in Example 7.3. See Section 7.3 for the jackknife-after-bootstrap method to estimate the standard error of the bootstrap estimate of bias. ◇

Example 7.5 (Bootstrap estimate of bias of a ratio estimate)

The `patch` (`bootstrap`) data from Efron and Tibshirani [84, 10.3] contains measurements of a certain hormone in the bloodstream of eight subjects after wearing a medical patch. The parameter of interest is

$$\theta = \frac{E(new) - E(old)}{E(old) - E(placebo)}.$$

If $|\theta| \leq 0.20$, this indicates bioequivalence of the old and new patches. The statistic is $\overline{Y}/\overline{Z}$. Compute a bootstrap estimate of bias in the bioequivalence ratio statistic.

```
data(patch, package = "bootstrap")
> patch
  subject placebo oldpatch newpatch     z     y
1       1    9243    17649    16449  8406 -1200
2       2    9671    12013    14614  2342  2601
3       3   11792    19979    17274  8187 -2705
4       4   13357    21816    23798  8459  1982
5       5    9055    13850    12560  4795 -1290
6       6    6290     9806    10157  3516   351
7       7   12412    17208    16570  4796  -638
8       8   18806    29044    26325 10238 -2719

n <- nrow(patch)   #in bootstrap package
B <- 2000
theta.b <- numeric(B)
theta.hat <- mean(patch$y) / mean(patch$z)

#bootstrap
for (b in 1:B) {
    i <- sample(1:n, size = n, replace = TRUE)
    y <- patch$y[i]
    z <- patch$z[i]
    theta.b[b] <- mean(y) / mean(z)
    }
bias <- mean(theta.b) - theta.hat
se <- sd(theta.b)
print(list(est=theta.hat, bias = bias,
           se = se, cv = bias/se))

$est [1] -0.0713061
$bias [1] 0.007901101
$se [1] 0.1046453
$cv [1] 0.07550363
```

If $|bias|/se \leq 0.25$, it is not usually necessary to adjust for bias [84, 10.3]. The bias is small relative to standard error ($cv < 0.08$), so in this example it is not necessary to adjust for bias.   ◇

7.2   The Jackknife

The *jackknife* is another resampling method, proposed by Quenouille [215, 216] for estimating bias, and by Tukey [274] for estimating standard error, a few decades earlier than the bootstrap. Efron [83] is a good introduction to the jackknife.

The jackknife is like a "leave-one-out" type of cross-validation. Let $x = (x_1, \ldots, x_n)$ be an observed random sample, and define the ith jackknife sample $x_{(i)}$ to be the subset of x that leaves out the ith observation x_i. That is,

$$x_{(i)} = (x_1, \ldots, x_{i-1}, x_{i+1}, \ldots, x_n).$$

If $\hat{\theta} = T_n(x)$, define the ith jackknife replicate $\hat{\theta}_{(i)} = T_{n-1}(x_{(i)})$, $i = 1, \ldots, n$.

Suppose the parameter $\theta = t(F)$ is a function of the distribution F. Let F_n be the ecdf of a random sample from the distribution F. The "plug-in" estimate of θ is $\hat{\theta} = t(F_n)$. A "plug-in" $\hat{\theta}$ is smooth in the sense that small changes in the data correspond to small changes in $\hat{\theta}$. For example, the sample mean is a plug-in estimate for the population mean, but the sample median is not a plug-in estimate for the population median.

The Jackknife Estimate of Bias

If $\hat{\theta}$ is a smooth (plug-in) statistic, then $\hat{\theta}_{(i)} = t(F_{n-1}(x_{(i)}))$, and the jackknife estimate of bias is

$$\widehat{bias}_{jack} = (n-1)(\overline{\hat{\theta}_{(\cdot)}} - \hat{\theta}), \tag{7.3}$$

where $\overline{\hat{\theta}_{(\cdot)}} = \frac{1}{n}\sum_{i=1}^{n}\hat{\theta}_{(i)}$ is the mean of the estimates from the leave-one-out samples, and $\hat{\theta} = \hat{\theta}(x)$ is the estimate computed from the original observed sample.

To see why the jackknife estimator (7.3) has the factor $n-1$, consider the case where θ is the population variance. If $x_1, \ldots, x_n$ is a random sample from the distribution of X, the plug-in estimate of the variance of X is

$$\hat{\theta} = \frac{1}{n}\sum_{i=1}^{n}(x_i - \bar{x})^2.$$

The estimator $\hat{\theta}$ is biased for σ_X^2 with

$$bias(\hat{\theta}) = E[\hat{\theta} - \sigma_X^2] = \frac{n-1}{n}\sigma_X^2 - \sigma_X^2 = -\frac{\sigma_X^2}{n}.$$

Each jackknife replicate computes the estimate $\hat{\theta}_{(i)}$ on a sample size $n-1$, so that the bias in the jackknife replicate is $-\sigma_X^2/(n-1)$. Thus, for $i = 1, \ldots, n$ we have

$$E[\hat{\theta}_{(i)} - \hat{\theta}] = E[\hat{\theta}_{(i)} - \theta] - E[\hat{\theta} - \theta]$$
$$= bias(\hat{\theta}_{(i)}) - bias(\hat{\theta})$$
$$= -\frac{\sigma_X^2}{n-1} - \left(-\frac{\sigma_X^2}{n}\right) = -\frac{\sigma_X^2}{n(n-1)} = \frac{bias(\hat{\theta})}{n-1}.$$

Thus, the jackknife estimate (7.3) with factor $(n-1)$ gives the correct estimate of bias in the plug-in estimator of variance, which is also the maximum likelihood estimator of variance.

R note 7.2 *(leave-one-out) The [ ] operator provides a very simple way to leave out the ith element of a vector.*

```
x <- 1:5
for (i in 1:5)
    print(x[-i])

[1] 2 3 4 5
[1] 1 3 4 5
[1] 1 2 4 5
[1] 1 2 3 5
[1] 1 2 3 4
```

Note that the jackknife requires only n replications to estimate the bias; the bootstrap estimate of bias typically requires several hundred replicates.

Example 7.6 (Jackknife estimate of bias)

Compute the jackknife estimate of bias for the `patch` data in Example 7.5.

```
data(patch, package = "bootstrap")
n <- nrow(patch)
y <- patch$y
z <- patch$z
theta.hat <- mean(y) / mean(z)
print (theta.hat)

#compute the jackknife replicates, leave-one-out estimates
theta.jack <- numeric(n)
for (i in 1:n)
    theta.jack[i] <- mean(y[-i]) / mean(z[-i])
bias <- (n - 1) * (mean(theta.jack) - theta.hat)

> print(bias)  #jackknife estimate of bias
[1] 0.008002488
```

◇

The jackknife estimate of standard error

A jackknife estimate of standard error [274], [84, (11.5)] is

$$
\widehat{se}_{jack} = \sqrt{\frac{n-1}{n} \sum_{i=1}^{n} \left(\hat{\theta}_{(i)} - \overline{\hat{\theta}_{(\cdot)}} \right)^2}, \tag{7.4}
$$

for smooth statistics $\hat{\theta}$.

To see why the jackknife estimator of standard error (7.4) has the factor $(n-1)/n$, consider the case where θ is the population mean and $\hat{\theta} = \overline{X}$. The standard error of the mean of X is $\sqrt{Var(X)/n}$. A factor of $(n-1)/n$ under the radial makes $\widehat{se}_{jack}$ an unbiased estimator of the standard error of the mean.

We can also consider the plug-in estimate of the standard error of the mean. In the case of a continuous random variable X, the plug-in estimate of the variance of a random sample is the variance of Y, where Y is uniformly distributed on the sample $x_1, \ldots, x_n$. That is,

$$
\widehat{Var}(Y) = \frac{1}{n} E[Y - E[Y]]^2 = \frac{1}{n} E[Y - \overline{X}]^2
$$

$$
= \frac{1}{n} \sum_{i=1}^{n} (X_i - \overline{X})^2 \cdot \frac{1}{n}
$$

$$
= \frac{n-1}{n^2} S_X^2 = \frac{n-1}{n} [\widehat{se}(\overline{X})]^2.
$$

Thus, for the jackknife estimator of standard error, a factor of $((n-1)/n)^2$ gives the plug-in estimate of variance. The factors $((n-1)/n)^2$ and $((n-1)/n)$ are approximately equal if n is not small. Efron and Tibshirani [84] remark that the choice of the factor $(n-1)/n$ instead of $((n-1)/n)^2$ is somewhat arbitrary.

Example 7.7 (Jackknife estimate of standard error)

To compute the jackknife estimate of standard error for the `patch` data in Example 7.5, use the jackknife replicates from Example 7.6.

```
se <- sqrt((n-1) *
    mean((theta.jack - mean(theta.jack))^2))
> print(se)
[1] 0.1055278
```

The jackknife estimate of standard error is 0.1055278. From the previous result for the bias, we have the estimated coefficient of variation

```
> .008002488/.1055278
[1] 0.07583298
```

$\diamond$

When the Jackknife Fails

The jackknife can fail when the statistic $\hat{\theta}$ is not "smooth." The statistic is a function of the data. Smoothness means that small changes in the data correspond to small changes in the statistic. The median is an example of a statistic that is not smooth.

Example 7.8 (Failure of jackknife)

In this example the jackknife estimate of standard error of the median is computed for a random sample of 10 integers from 1, 2 ..., 100.

```
n <- 10
x <- sample(1:100, size = n)

#jackknife estimate of se
M <- numeric(n)
for (i in 1:n) {           #leave one out
    y <- x[-i]
    M[i] <- median(y)
}
Mbar <- mean(M)
print(sqrt((n-1)/n * sum((M - Mbar)^2)))

#bootstrap estimate of se
Mb <- replicate(1000, expr = {
        y <- sample(x, size = n, replace = TRUE)
        median(y) })
print(sd(Mb))

# details and results:
# the sample, x:        29 79 41 86 91  5 50 83 51 42
# jackknife medians:    51 50 51 50 50 51 51 50 50 51
# jackknife est. of se: 1.5
# bootstrap medians:    46 50 46 79 79 51 81 65 ...
# bootstrap est. of se: 13.69387
```

Clearly something is wrong here, because the bootstrap estimate and the jackknife estimate are far apart. The jackknife fails, because the median is not smooth. ◇

In this case, when the statistic is not smooth, the delete-d jackknife (leave d observations out on each replicate) can be applied (see Efron and Tibshirani [84, 11.7]). If $\sqrt{n}/d \to 0$ and $n - d \to \infty$ then the delete-d jackknife is consistent for the median. The computing time increases because there are a large number of jackknife replicates when n and d are large.

7.3 Jackknife-after-Bootstrap

In this chapter, bootstrap estimates of standard error and bias have been introduced. These estimates are random variables. If we are interested in the variance of these estimates, one idea is to try the jackknife.

Recall that $\widehat{se}(\hat{\theta})$ is the sample standard deviation of B bootstrap replicates of $\hat{\theta}$. Now, if we leave out the ith observation, the algorithm for estimation of standard error is to resample B replicates from the $n-1$ remaining observations – for each i. In other words, we would replicate the bootstrap itself. Fortunately, there is a way to avoid replicating the bootstrap.

The *jackknife-after-bootstrap* computes an estimate for each "leave-one-out" sample. Let $J(i)$ denote the indices of bootstrap samples that do not contain x_i, and let $B(i)$ denote number of bootstrap samples that do not contain x_i. Then we can compute the jackknife replication leaving out the $B - B(i)$ samples that contain x_i [84, p. 277]. The jackknife estimate of standard error is computed by the formula (7.4). Compute

$$\widehat{se}(\hat{\theta}) = \widehat{se}_{jack}(\widehat{se}_{B(1)}, \dots, \widehat{se}_{B(n)}),$$

where

$$\widehat{se}_{B(i)} = \sqrt{\frac{1}{B(i)} \sum_{j \in J(i)} \left[\hat{\theta}_{(j)} - \overline{\hat{\theta}_{(J(i))}} \right]^2}, \tag{7.5}$$

and

$$\overline{\hat{\theta}_{(J(i))}} = \frac{1}{B(i)} \sum_{j \in J(i)} \hat{\theta}_{(j)}$$

is the sample mean of the estimates from the leave-x_i-out jackknife samples.

Example 7.9 (Jackknife-after-bootstrap)

Use the jackknife-after-bootstrap procedure to estimate the standard error of $\widehat{se}(\hat{\theta})$ for the `patch` data in Example 7.7.

```
# initialize
data(patch, package = "bootstrap")
n <- nrow(patch)
y <- patch$y
z <- patch$z
B <- 2000
theta.b <- numeric(B)
# set up storage for the sampled indices
indices <- matrix(0, nrow = B, ncol = n)
```

```
# jackknife-after-bootstrap step 1: run the bootstrap
for (b in 1:B) {
    i <- sample(1:n, size = n, replace = TRUE)
    y <- patch$y[i]
    z <- patch$z[i]
    theta.b[b] <- mean(y) / mean(z)
    #save the indices for the jackknife
    indices[b, ] <- i
    }

#jackknife-after-bootstrap to est. se(se)
se.jack <- numeric(n)
for (i in 1:n) {
    #in i-th replicate omit all samples with x[i]
    keep <- (1:B)[apply(indices, MARGIN = 1,
                FUN = function(k) {!any(k == i)})]
    se.jack[i] <- sd(theta.b[keep])
}

> print(sd(theta.b))
[1] 0.1027102
> print(sqrt((n-1) * mean((se.jack - mean(se.jack))^2)))
[1] 0.03050501
```

The bootstrap estimate of standard error is 0.1027102 and jackknife-after-bootstrap estimate of its standard error is 0.03050501.   ◇

Jackknife-after-bootstrap: Empirical influence values

The empirical influence values in jackknife-after-bootstrap are empirical quantities that measure the difference between each jackknife replicate and the observed statistic. There are several methods for estimating the influence values. One approach uses the usual jackknife differences $\hat{\theta}_{(i)} - \hat{\theta}$, $i = 1, \ldots, n$. The empinf function in the boot package computes empirical influence values by four methods. The jack.after.boot function in the boot package [34] produces a plot of empirical influence values. The plots can be used as a diagnostic tool to see the effect or influence of individual observations. See [63, Ch. 3] for examples and a discussion of how to interpret the plots.

7.4 Bootstrap Confidence Intervals

In this section several approaches to obtaining approximate confidence intervals for the target parameter in a bootstrap are discussed. The methods include the *standard normal bootstrap* confidence interval, the *basic bootstrap* confidence interval, the *bootstrap percentile* confidence interval, and the *bootstrap t confidence interval*. Readers are referred to [63] and [84] for theoretical properties and discussion of empirical performance of methods for bootstrap confidence interval estimates.

7.4.1 The Standard Normal Bootstrap Confidence Interval

The standard normal bootstrap confidence interval is the simplest approach, but not necessarily the best. Suppose that $\hat{\theta}$ is an estimator of parameter θ, and assume the standard error of the estimator is $se(\hat{\theta})$. If $\hat{\theta}$ is a sample mean and the sample size is large, then the Central Limit Theorem implies that

$$Z = \frac{\hat{\theta} - E[\hat{\theta}]}{se(\hat{\theta})} \tag{7.6}$$

is approximately standard normal. Hence, if $\hat{\theta}$ is unbiased for θ, then an approximate $100(1 - \alpha)\%$ confidence interval for θ is the Z-interval

$$\hat{\theta} \pm z_{\alpha/2}se(\hat{\theta}),$$

where $z_{\alpha/2} = \Phi^{-1}(1 - \alpha/2)$. This interval is easy to compute, but we have made several assumptions. To apply the normal distribution, we assume that the distribution of $\hat{\theta}$ is normal or $\hat{\theta}$ is a sample mean and the sample size is large. We have also implicitly assumed that $\hat{\theta}$ is unbiased for θ.

Bias can be estimated and used to center the Z statistic, but the estimator is a random variable, so the transformed variable is not normal. Here we have treated $se(\hat{\theta})$ as a known parameter, but in the bootstrap $se(\hat{\theta})$ is estimated (the sample standard deviation of the replicates).

7.4.2 The Basic Bootstrap Confidence Interval

The basic bootstrap confidence interval transforms the distribution of the replicates by subtracting the observed statistic. The quantiles of the transformed sample are used to determine the confidence limits.

The $100(1-\alpha)\%$ confidence limits for the basic bootstrap confidence interval are

$$(2\hat{\theta} - \hat{\theta}_{1-\alpha/2}, \quad 2\hat{\theta} - \hat{\theta}_{\alpha/2}). \tag{7.7}$$

To see how the confidence limits in (7.7) are determined, consider first the parametric case. Suppose that T is an estimator of θ and a_α is the α quantile of $T - \theta$. Then

$$P(T - \theta > a_\alpha) = 1 - \alpha \Rightarrow P(T - a_\alpha > \theta) = 1 - \alpha.$$

Thus, a $100(1 - 2\alpha)\%$ confidence interval with equal lower and upper tail errors α is given by $(t - a_{1-\alpha}, \, t - a_\alpha)$.

In bootstrap the distribution of T is generally unknown, but quantiles can be estimated and an approximate method applied.

Compute the sample α quantiles $\hat{\theta}_\alpha$ from the ecdf of the replicates $\hat{\theta}^*$. Denote the α quantile of $\hat{\theta}^* - \hat{\theta}$ by b_α. Then $\hat{b}_\alpha = \hat{\theta}_\alpha - \hat{\theta}$ is an estimator of b_α. An approximate upper confidence limit for a $100(1 - \alpha)\%$ confidence interval for θ is given by

$$\hat{\theta} - \hat{b}_{\alpha/2} = \hat{\theta} - (\hat{\theta}_{\alpha/2} - \hat{\theta}) = 2\hat{\theta} - \hat{\theta}_{\alpha/2}.$$

Similarly an approximate lower confidence limit is given by $2\hat{\theta} - \hat{\theta}_{1-\alpha/2}$. Thus, a $100(1 - \alpha)$ basic bootstrap confidence interval for θ is given by (7.7). See Davison and Hinkley [63, 5.2] for more details.

7.4.3   The Percentile Bootstrap Confidence Interval

A bootstrap percentile interval uses the empirical distribution of the bootstrap replicates as the reference distribution. The quantiles of the empirical distribution are estimators of the quantiles of the sampling distribution of $\hat{\theta}$, so that these (random) quantiles may match the true distribution better when the distribution of $\hat{\theta}$ is not normal. Suppose that $\hat{\theta}^{(1)}, \ldots, \hat{\theta}^{(B)}$ are the bootstrap replicates of the statistic $\hat{\theta}$. From the ecdf of the replicates, compute the $\alpha/2$ quantile $\hat{\theta}_{\alpha/2}$, and the $1 - \alpha/2$ quantile $\hat{\theta}_{1-\alpha/2}$.

Efron and Tibshirani [84, 13.3] show that the percentile interval has some theoretical advantages over the standard normal interval and somewhat better coverage performance.

Adjustments to percentile methods have been proposed. For example, the *bias-corrected and accelerated* (BCa) percentile intervals (see Section 7.5) are a modified version of percentile intervals that have better theoretical properties and better performance in practice.

The boot.ci (boot) function [34] computes five types of bootstrap confidence intervals: basic, normal, percentile, studentized, and BCa. To use this function, first call boot for the bootstrap, and pass the returned boot object to boot.ci (along with other required arguments). For more details see Davison and Hinkley [63, Ch. 5] and the boot.ci help topic.

Example 7.10 (Bootstrap confidence intervals for patch ratio statistic.)

This example illustrates how to obtain the normal, basic, and percentile bootstrap confidence intervals using the boot and boot.ci functions in the boot package. The code generates 95% confidence intervals for the ratio statistic in Example 7.5.

```
library(boot)        #for boot and boot.ci
data(patch, package = "bootstrap")

theta.boot <- function(dat, ind) {
    #function to compute the statistic
    y <- dat[ind, 1]
    z <- dat[ind, 2]
    mean(y) / mean(z)
}
```

Run the bootstrap and compute confidence interval estimates for the bioequivalence ratio.

```
y <- patch$y
z <- patch$z
dat <- cbind(y, z)
boot.obj <- boot(dat, statistic = theta.boot, R = 2000)
```

The output for the bootstrap and bootstrap confidence intervals is below.

```
print(boot.obj)
ORDINARY NONPARAMETRIC BOOTSTRAP
Call: boot(data = dat, statistic = theta.boot, R = 2000)
Bootstrap Statistics :
     original      bias     std. error
t1* -0.0713061 0.01047726   0.1010179

print(boot.ci(boot.obj,
            type = c("basic", "norm", "perc")))

BOOTSTRAP CONFIDENCE INTERVAL CALCULATIONS
Based on 2000 bootstrap replicates
CALL : boot.ci(boot.out = boot.obj, type = c("basic",
    "norm", "perc"))
Intervals :
Level     Normal             Basic              Percentile
95%  (-0.2798, 0.1162 ) (-0.3045, 0.0857 ) (-0.2283, 0.1619 )
Calculations and Intervals on Original Scale
```

Recall that the old and new patches are bioequivalent if $|\theta| \leq 0.20$. Hence, the interval estimates do not support bioequivalence of the old and new patches. Next we compute the bootstrap confidence intervals according to their definitions. Compare the following results with the boot.ci output.

```
#calculations for bootstrap confidence intervals
alpha <- c(.025, .975)

#normal
print(boot.obj$t0 + qnorm(alpha * sd(boot.obj$t)))
-0.2692975  0.1266853

#basic
print(2*boot.obj$t0 -
        quantile(boot.obj$t, rev(alpha), type=1))
   97.5%        2.5%
-0.3018698  0.0857679

#percentile
print(quantile(boot.obj$t, alpha, type=6))
   2.5%        97.5%
-0.2283370  0.1618647
```

$\diamond$

R note 7.3 *The normal interval computed by* **boot.ci** *corrects for bias. Notice that the* **boot.ci** *normal interval differs from our result by the bias estimate shown in the output from* **boot***. This is confirmed by reading the source code for the function. To view the source code for this calculation, when the* **boot** *package is loaded, enter the command* **getAnywhere(norm.ci)** *at the console. Also see* **norm.inter** *and [63] for details of calculations of quantiles.*

Example 7.11 (Bootstrap confidence intervals for the correlation statistic)

Compute 95% bootstrap confidence interval estimates for the correlation statistic in the law data of Example 7.2.

```
library(boot)
data(law, package = "bootstrap")
boot.obj <- boot(law, R = 2000,
        statistic = function(x, i){cor(x[i,1], x[i,2])})
print(boot.ci(boot.obj, type=c("basic","norm","perc")))
    ...

Intervals :
Level    Normal              Basic             Percentile
95%   (0.5182, 1.0448)  (0.5916, 1.0994)  (0.4534, 0.9611)
Calculations and Intervals on Original Scale
```

All three intervals cover the correlation $\rho = .76$ of the universe of all law schools in `law82`. One reason for the difference in the percentile and normal confidence intervals could be that the sampling distribution of correlation statistic is not close to normal (see the histogram in Figure 7.1). When the sampling distribution of the statistic is approximately normal, the percentile interval will agree with the normal interval. ◇

7.4.4 The Bootstrap t interval

Even if the distribution of $\hat\theta$ is normal and $\hat\theta$ is unbiased for θ, the normal distribution is not exactly correct for the Z statistic (7.6), because we estimate $se(\hat\theta)$. Nor can we claim that it is a Student t statistic, because the distribution of the bootstrap estimator $\widehat{se}(\hat\theta)$ is unknown. The bootstrap t interval does not use a Student t distribution as the reference distribution. Instead, the sampling distribution of a "t type" statistic (a studentized statistic) is generated by resampling. Suppose $x = (x_1, \ldots, x_n)$ is an observed sample. The $100(1-\alpha)\%$ bootstrap t confidence interval is

$$(\hat\theta - t^*_{1-\alpha/2}\widehat{se}(\hat\theta), \qquad \hat\theta - t^*_{\alpha/2}\widehat{se}(\hat\theta)),$$

where $\widehat{se}(\hat\theta)$, $t^*_{\alpha/2}$ and $t^*_{1-\alpha/2}$ are computed as outlined below.

Bootstrap t interval (studentized bootstrap interval)

1. Compute the observed statistic $\hat\theta$.

2. For each replicate, indexed $b = 1, \ldots, B$:

   (a) Sample with replacement from x to get the bth sample $x^{(b)} = (x_1^{(b)}, \ldots, x_n^{(b)})$.

   (b) Compute $\hat\theta^{(b)}$ from the bth sample $x^{(b)}$.

   (c) Compute or estimate the standard error $\widehat{se}(\hat\theta^{(b)})$ (a separate estimate for each bootstrap sample; a bootstrap estimate will resample from the current bootstrap sample $x^{(b)}$, not x).

   (d) Compute the bth replicate of the "t" statistic, $t^{(b)} = \frac{\hat\theta^{(b)} - \hat\theta}{\widehat{se}(\hat\theta^{(b)})}$.

3. The sample of replicates $t^{(1)}, \ldots, t^{(B)}$ is the reference distribution for bootstrap t. Find the sample quantiles $t^*_{\alpha/2}$ and $t^*_{1-\alpha/2}$ from the ordered sample of replicates $t^{(b)}$.

4. Compute $\widehat{se}(\hat\theta)$, the sample standard deviation of the replicates $\hat\theta^{(b)}$.

5. Compute confidence limits

$$(\hat\theta - t^*_{1-\alpha/2}\widehat{se}(\hat\theta), \qquad \hat\theta - t^*_{\alpha/2}\widehat{se}(\hat\theta)).$$

One disadvantage to the bootstrap t interval is that typically the estimates of standard error $\widehat{se}(\hat{\theta}^{(b)})$ must be obtained by bootstrap. This is a bootstrap nested inside a bootstrap. If $B = 1000$, for example, the bootstrap t confidence interval method takes approximately 1000 times longer than any of the other methods.

Example 7.12 (Bootstrap t confidence interval)

This example provides a function to compute a bootstrap t confidence interval for a univariate or a multivariate sample. The required arguments to the function are the sample data x, and the function `statistic` that computes the statistic. The default confidence level is 95%, the number of bootstrap replicates defaults to 500, and the number of replicates for estimating standard error defaults to 100.

```
boot.t.ci <-
function(x, B = 500, R = 100, level = .95, statistic){
    #compute the bootstrap t CI
    x <- as.matrix(x);  n <- nrow(x)
    stat <- numeric(B); se <- numeric(B)

    boot.se <- function(x, R, f) {
        #local function to compute the bootstrap
        #estimate of standard error for statistic f(x)
        x <- as.matrix(x); m <- nrow(x)
        th <- replicate(R, expr = {
            i <- sample(1:m, size = m, replace = TRUE)
            f(x[i, ])
            })
        return(sd(th))
    }

    for (b in 1:B) {
        j <- sample(1:n, size = n, replace = TRUE)
        y <- x[j, ]
        stat[b] <- statistic(y)
        se[b] <- boot.se(y, R = R, f = statistic)
    }
    stat0 <- statistic(x)
    t.stats <- (stat - stat0) / se
    se0 <- sd(stat)
    alpha <- 1 - level
    Qt <- quantile(t.stats, c(alpha/2, 1-alpha/2), type = 1)
    names(Qt) <- rev(names(Qt))
    CI <- rev(stat0 - Qt * se0)
}
```

Note that the `boot.se` function is a local function, visible only inside the `boot.t.ci` function. The next example applies the `boot.t.ci` function. ◇

Example 7.13 (Bootstrap t confidence interval for `patch` ratio statistic.)

Compute a 95% bootstrap t confidence interval for the ratio statistic in Examples 7.5 and 7.10.

```
dat <- cbind(patch$y, patch$z)
stat <- function(dat) {
    mean(dat[, 1]) / mean(dat[, 2]) }
ci <- boot.t.ci(dat, statistic = stat, B=2000, R=200)
print(ci)
```

```
      2.5%        97.5%
-0.2547932   0.4055129
```

The upper confidence limit of the bootstrap t confidence interval is much larger than the three intervals in Example 7.10 and the bootstrap t is the widest interval in this example. ◇

7.5   Better Bootstrap Confidence Intervals

Better bootstrap confidence intervals (see [84, Sec. 14.3]) are a modified version of percentile intervals that have better theoretical properties and better performance in practice. For a $100(1-\alpha)\%$ confidence interval, the usual $\alpha/2$ and $1-\alpha/2$ quantiles are adjusted by two factors: a correction for bias and a correction for skewness. The bias correction is denoted z_0 and the skewness or "acceleration" adjustment is a. The better bootstrap confidence interval is called BCa for "bias corrected" and "adjusted for acceleration."

For a $100(1-\alpha)\%$ BCa bootstrap confidence interval compute

$$\alpha_1 = \Phi\left(\hat{z}_0 + \frac{\hat{z}_0 + z_{\alpha/2}}{1 - \hat{a}(\hat{z}_0 + z_{\alpha/2})}\right), \tag{7.8}$$

$$\alpha_2 = \Phi\left(\hat{z}_0 + \frac{\hat{z}_0 + z_{1-\alpha/2}}{1 - \hat{a}(\hat{z}_0 + z_{1-\alpha/2})}\right), \tag{7.9}$$

where $z_\alpha = \Phi^{-1}(\alpha)$, and $\hat{z}_0$, $\hat{a}$ are given by equations (7.10) and (7.11) below. The BCa interval is

$$(\hat{\theta}^*_{\alpha_1}, \hat{\theta}^*_{\alpha_2}).$$

The upper and lower confidence limits of the BCa confidence interval are the empirical α_1 and α_2 quantiles of the bootstrap replicates.

The bias correction factor is in effect measuring the median bias of the replicates $\hat{\theta}^*$ for $\hat{\theta}$. The estimate of this bias is

$$\hat{z}_0 = \Phi^{-1}\left(\frac{1}{B}\sum_{b=1}^{B}I(\hat{\theta}^{(b)} < \hat{\theta})\right), \qquad (7.10)$$

where $I(\cdot)$ is the indicator function. Note that $\hat{z}_0 = 0$ if $\hat{\theta}$ is the median of the bootstrap replicates.

The acceleration factor is estimated from jackknife replicates:

$$\hat{a} = \frac{\sum_{i=1}^{n}(\overline{\theta_{(.)}} - \theta_{(i)})^3}{6\sum_{i=1}^{n}((\overline{\theta_{(.)}} - \theta_{(i)})^2)^{3/2}}, \qquad (7.11)$$

which measures skewness.

Other methods for estimating the acceleration have been proposed (see e.g. Shao and Tu [247]). Formula (7.11) is given by Efron and Tibshirani [84, p. 186]. The acceleration factor $\hat{a}$ is so named because it estimates the rate of change of the standard error of $\hat{\theta}$ with respect to the target parameter θ (on a normalized scale). When we use a standard normal bootstrap confidence interval, we suppose that $\hat{\theta}$ is approximately normal with mean θ and constant variance $\sigma^2(\hat{\theta})$ that does not depend on the parameter θ. However, it is not always true that the variance of an estimator has constant variance with respect to the target parameter. Consider, for example, the sample proportion $\hat{p} = X/n$ as an estimator of the probability of success p in a binomial experiment, which has variance $p(1-p)/n$. The acceleration factor aims to adjust the confidence limits to account for the possibility that the variance of the estimator may depend on the true value of the target parameter.

Properties of BCa intervals

There are two important theoretical advantages to BCa bootstrap confidence intervals. The BCa confidence intervals are transformation respecting and BCa intervals have second order accuracy.

Transformation respecting means that if $(\hat{\theta}_{\alpha_1}^*, \hat{\theta}_{\alpha_2}^*)$ is a confidence interval for θ, and $t(\theta)$ is a transformation of the parameter θ, then the corresponding interval for $t(\theta)$ is $(t(\hat{\theta}_{\alpha_1}^*), t(\hat{\theta}_{\alpha_2}^*))$. A confidence interval is first order accurate if the error tends to zero at rate $1/\sqrt{n}$ for sample size n, and second order accurate if the error tends to zero at rate $1/n$.

The bootstrap t confidence interval is second order accurate but not transformation respecting. The bootstrap percentile interval is transformation respecting but only first order accurate. The standard normal confidence interval is neither transformation respecting nor second order accurate. See [63] for discussion and comparison of theoretical properties of bootstrap confidence intervals.

Example 7.14 (BCa bootstrap confidence interval)

This example implements a function to compute a BCa confidence interval. The BCa interval is $(\hat{\theta}^*_{\alpha_1}, \hat{\theta}^*_{\alpha_2})$, where $\hat{\theta}^*_{\alpha_1}$ and $\hat{\theta}^*_{\alpha_2}$ are given by equations (7.8)–(7.11). ⋄

```
boot.BCa <-
function(x, th0, th, stat, conf = .95) {
    # bootstrap with BCa bootstrap confidence interval
    # th0 is the observed statistic
    # th is the vector of bootstrap replicates
    # stat is the function to compute the statistic

    x <- as.matrix(x)
    n <- nrow(x) #observations in rows
    N <- 1:n
    alpha <- (1 + c(-conf, conf))/2
    zalpha <- qnorm(alpha)

    # the bias correction factor
    z0 <- qnorm(sum(th < th0) / length(th))

    # the acceleration factor (jackknife est.)
    th.jack <- numeric(n)
    for (i in 1:n) {
        J <- N[1:(n-1)]
        th.jack[i] <- stat(x[-i, ], J)
    }
    L <- mean(th.jack) - th.jack
    a <- sum(L^3)/(6 * sum(L^2)^1.5)

    # BCa conf. limits
    adj.alpha <- pnorm(z0 + (z0+zalpha)/(1-a*(z0+zalpha)))
    limits <- quantile(th, adj.alpha, type=6)
    return(list("est"=th0, "BCa"=limits))
}
```

Example 7.15 (BCa bootstrap confidence interval)

Compute a BCa confidence interval for the bioequivalence ratio statistic of Example 7.10 using the function `boot.BCa` provided in Example 7.14.

```
data(patch, package = "bootstrap")
n <- nrow(patch)
B <- 2000
y <- patch$y
z <- patch$z
x <- cbind(y, z)
theta.b <- numeric(B)
theta.hat <- mean(y) / mean(z)

#bootstrap
for (b in 1:B) {
    i <- sample(1:n, size = n, replace = TRUE)
    y <- patch$y[i]
    z <- patch$z[i]
    theta.b[b] <- mean(y) / mean(z)
    }
#compute the BCa interval
stat <- function(dat, index) {
    mean(dat[index, 1]) / mean(dat[index, 2])   }

boot.BCa(x, th0 = theta.hat, th = theta.b, stat = stat)
```

In the result shown below, notice that the probabilities $\alpha/2 = 0.025$ and $1 - \alpha/2 = 0.975$ have been adjusted to 0.0339, and 0.9824.

```
$est
[1] -0.0713061


$BCa
 3.391094%  98.24405%
-0.2252715  0.1916788
```

Thus bioequivalence ($|\theta| \leq 0.20$) is not supported by the BCa confidence interval estimate of θ.   ◇

R note 7.4 (Empirical influence values) *By default, the* `type="bca"` *option of the* `boot.ci` *function computes empirical influence values by a regression method. The method in example 7.14 corresponds to the "usual jackknife" method of computing empirical jackknife values. See [63, Ch. 5] and the code for* `empinf, usual.jack`.

Example 7.16 (BCa bootstrap confidence interval using `boot.ci`)

Compute a BCa confidence interval for the bioequivalence ratio statistic of Examples 7.5 and 7.10, using the function `boot.ci` provided in the `boot` package [34].

```
boot.obj <- boot(x, statistic = stat, R=2000)
boot.ci(boot.obj, type=c("perc", "bca"))
```

The percentile confidence interval is also given for comparison.

```
BOOTSTRAP CONFIDENCE INTERVAL CALCULATIONS
Based on 2000 bootstrap replicates

CALL : boot.ci(boot.out = boot.obj, type = c("perc", "bca"))

Intervals :
Level      Percentile              BCa
95%   (-0.2368,  0.1824 )    (-0.2221,  0.2175 )
Calculations and Intervals on Original Scale
```

◇

7.6   Application: Cross Validation

Cross validation is a data partitioning method that can be used to assess the stability of parameter estimates, the accuracy of a classification algorithm, the adequacy of a fitted model, and in many other applications. The jackknife could be considered a special case of cross validation, because it is primarily used to estimate bias and standard error of an estimator.

In building a classifier, a researcher can partition the data into training and test sets. The model is estimated using the data in the training set only, and the misclassification rate is estimated by running the classifier on the test set. Similarly, the fit of any model can be assessed by holding back a test set from the model estimation, and then using the test set to see how well the model fits the new test data.

Another version of cross validation is the "n-fold" cross validation, which partitions the data into n test sets (now test points). This "leave-one-out" procedure is like the jackknife. The data could be divided into any number K partitions, so that there are K test sets. Then the model fitting leaves out one test set in turn, so that the models are fitted K times.

Example 7.17 (Model selection)

The `ironslag` (DAAG) data [185] has 53 measurements of iron content by two methods, `chemical` and `magnetic` (see "iron.dat" in [126]). A scatterplot of the data in Figure 7.2 suggests that the chemical and magnetic variables are positively correlated, but the relation may not be linear. From the plot, it appears that a quadratic polynomial, or possibly an exponential or logarithmic model might fit the data better than a line.

There are several steps to model selection, but we will focus on the prediction error. The prediction error can be estimated by cross validation, without making strong distributional assumptions about the error variable.

The proposed models for predicting magnetic measurement (Y) from chemical measurement (X) are:

1. Linear: $Y = \beta_0 + \beta_1 X + \varepsilon$.

2. Quadratic: $Y = \beta_0 + \beta_1 X + \beta_2 X^2 + \varepsilon$.

3. Exponential: $\log(Y) = \log(\beta_0) + \beta_1 X + \varepsilon$.

4. Log-Log: $\log(Y) = \beta_0 + \beta_1 \log(X) + \varepsilon$.

The code to estimate the parameters of the four models follows. Plots of the predicted response with the data are also constructed for each model and shown in Figure 7.2. To display four plots use `par(mfrow=c(2,2))`.

```
library(DAAG); attach(ironslag)
a <- seq(10, 40, .1)       #sequence for plotting fits

L1 <- lm(magnetic ~ chemical)
plot(chemical, magnetic, main="Linear", pch=16)
yhat1 <- L1$coef[1] + L1$coef[2] * a
lines(a, yhat1, lwd=2)

L2 <- lm(magnetic ~ chemical + I(chemical^2))
plot(chemical, magnetic, main="Quadratic", pch=16)
yhat2 <- L2$coef[1] + L2$coef[2] * a + L2$coef[3] * a^2
lines(a, yhat2, lwd=2)

L3 <- lm(log(magnetic) ~ chemical)
plot(chemical, magnetic, main="Exponential", pch=16)
logyhat3 <- L3$coef[1] + L3$coef[2] * a
yhat3 <- exp(logyhat3)
lines(a, yhat3, lwd=2)

L4 <- lm(log(magnetic) ~ log(chemical))
plot(log(chemical), log(magnetic), main="Log-Log", pch=16)
logyhat4 <- L4$coef[1] + L4$coef[2] * log(a)
lines(log(a), logyhat4, lwd=2)
```

◇

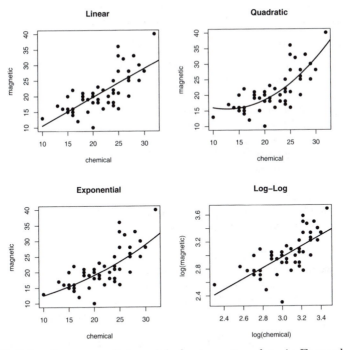

FIGURE 7.2: Four proposed models for `ironslag` data in Example 7.17.

Once the model is estimated, we want to assess the fit. Cross validation can be used to estimate the prediction errors.

Procedure to estimate prediction error by n-fold (leave-one-out) cross validation

1. For $k = 1, \ldots, n$, let observation (x_k, y_k) be the test point and use the remaining observations to fit the model.

   (a) Fit the model(s) using only the $n - 1$ observations in the training set, (x_i, y_i), $i \neq k$.

   (b) Compute the predicted response $\hat{y}_k = \hat{\beta}_0 + \hat{\beta}_1 x_k$ for the test point.

   (c) Compute the prediction error $e_k = y_k - \hat{y}_k$.

2. Estimate the mean of the squared prediction errors $\hat{\sigma}_\varepsilon^2 = \frac{1}{n} \sum_{k=1}^{n} e_k^2$.

Example 7.18 (Model selection: Cross validation)

Cross validation is applied to select a model in Example 7.17.

```
n <- length(magnetic)   #in DAAG ironslag
e1 <- e2 <- e3 <- e4 <- numeric(n)

# for n-fold cross validation
# fit models on leave-one-out samples
for (k in 1:n) {
    y <- magnetic[-k]
    x <- chemical[-k]

    J1 <- lm(y ~ x)
    yhat1 <- J1$coef[1] + J1$coef[2] * chemical[k]
    e1[k] <- magnetic[k] - yhat1

    J2 <- lm(y ~ x + I(x^2))
    yhat2 <- J2$coef[1] + J2$coef[2] * chemical[k] +
             J2$coef[3] * chemical[k]^2
    e2[k] <- magnetic[k] - yhat2

    J3 <- lm(log(y) ~ x)
    logyhat3 <- J3$coef[1] + J3$coef[2] * chemical[k]
    yhat3 <- exp(logyhat3)
    e3[k] <- magnetic[k] - yhat3

    J4 <- lm(log(y) ~ log(x))
    logyhat4 <- J4$coef[1] + J4$coef[2] * log(chemical[k])
    yhat4 <- exp(logyhat4)
    e4[k] <- magnetic[k] - yhat4
}
```

The following estimates for prediction error are obtained from the n-fold cross validation.

```
> c(mean(e1^2), mean(e2^2), mean(e3^2), mean(e4^2))
[1] 19.55644 17.85248 18.44188 20.45424
```

According to the prediction error criterion, Model 2, the quadratic model, would be the best fit for the data.

```
> L2
Call:
lm(formula = magnetic ~ chemical + I(chemical^2))
Coefficients:
(Intercept)      chemical  I(chemical^2)
   24.49262      -1.39334        0.05452
```

The fitted regression equation for Model 2 is

$$\hat{Y} = 24.49262 - 1.39334X + 0.05452X^2.$$

The residual plots for Model 2 are shown in Figure 7.3. An easy way to get several residual plots is by `plot(L2)`. Alternately, similar plots can be displayed as follows.

```
par(mfrow = c(2, 2))      #layout for graphs
plot(L2$fit, L2$res)      #residuals vs fitted values
abline(0, 0)              #reference line
qqnorm(L2$res)            #normal probability plot
qqline(L2$res)            #reference line
par(mfrow = c(1, 1))      #restore display
```

Part of the summary for the fitted quadratic model is below.

```
Residuals:
    Min      1Q  Median      3Q     Max
-8.4335 -2.7006 -0.2754  2.5446 12.2665

Residual standard error: 4.098 on 50 degrees of freedom
Multiple R-Squared: 0.5931, Adjusted R-squared: 0.5768
```

In the quadratic model the predictors X and X^2 are highly correlated. See `poly` for another approach with orthogonal polynomials. ◇

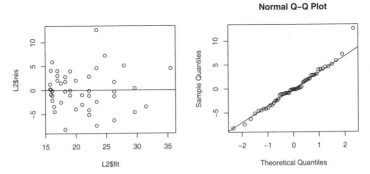

FIGURE 7.3: Residuals of the quadratic model for `ironslag` data, from Example 7.17.

Exercises

7.1 Compute a jackknife estimate of the bias and the standard error of the correlation statistic in Example 7.2.

7.2 Refer to the `law` data (`bootstrap`). Use the jackknife-after-bootstrap method to estimate the standard error of the bootstrap estimate of $se(R)$.

7.3 Obtain a bootstrap t confidence interval estimate for the correlation statistic in Example 7.2 (`law` data in `bootstrap`).

7.4 Refer to the air-conditioning data set `aircondit` provided in the `boot` package. The 12 observations are the times in hours between failures of air-conditioning equipment [63, Example 1.1]:

$$3, 5, 7, 18, 43, 85, 91, 98, 100, 130, 230, 487.$$

Assume that the times between failures follow an exponential model $\text{Exp}(\lambda)$. Obtain the MLE of the hazard rate λ and use bootstrap to estimate the bias and standard error of the estimate.

7.5 Refer to Exercise 7.4. Compute 95% bootstrap confidence intervals for the mean time between failures $1/\lambda$ by the standard normal, basic, percentile, and BCa methods. Compare the intervals and explain why they may differ.

7.6 Efron and Tibshirani discuss the `scor` (`bootstrap`) test score data on 88 students who took examinations in five subjects [84, Table 7.1], [188, Table 1.2.1]. The first two tests (mechanics, vectors) were closed book and the last three tests (algebra, analysis, statistics) were open book. Each row of the data frame is a set of scores $(x_{i1}, \ldots, x_{i5})$ for the ith student. Use a panel display to display the scatter plots for each pair of test scores. Compare the plot with the sample correlation matrix. Obtain bootstrap estimates of the standard errors for each of the following estimates: $\hat{\rho}_{12} = \hat{\rho}(\text{mec, vec})$, $\hat{\rho}_{34} = \hat{\rho}(\text{alg, ana})$, $\hat{\rho}_{35} = \hat{\rho}(\text{alg, sta})$, $\hat{\rho}_{45} = \hat{\rho}(\text{ana, sta})$.

7.7 Refer to Exercise 7.6. Efron and Tibshirani discuss the following example [84, Ch. 7]. The five-dimensional scores data have a 5×5 covariance matrix Σ, with positive eigenvalues $\lambda_1 > \cdots > \lambda_5$. In principal components analysis,

$$\theta = \frac{\lambda_1}{\sum_{j=1}^{5} \lambda_j}$$

measures the proportion of variance explained by the first principal component. Let $\hat{\lambda}_1 > \cdots > \hat{\lambda}_5$ be the eigenvalues of $\hat{\Sigma}$, where $\hat{\Sigma}$ is the MLE of Σ. Compute the sample estimate

$$\hat{\theta} = \frac{\hat{\lambda}_1}{\sum_{j=1}^{5} \hat{\lambda}_j}$$

of θ. Use bootstrap to estimate the bias and standard error of $\hat{\theta}$.

7.8   Refer to Exercise 7.7. Obtain the jackknife estimates of bias and standard error of $\hat{\theta}$.

7.9   Refer to Exercise 7.7. Compute 95% percentile and BCa confidence intervals for $\hat{\theta}$.

7.10   In Example 7.18, leave-one-out (n-fold) cross validation was used to select the best fitting model. Repeat the analysis replacing the Log-Log model with a cubic polynomial model. Which of the four models is selected by the cross validation procedure? Which model is selected according to maximum adjusted R^2?

7.11   In Example 7.18, leave-one-out (n-fold) cross validation was used to select the best fitting model. Use leave-two-out cross validation to compare the models.

Projects

7.A   Conduct a Monte Carlo study to estimate the coverage probabilities of the standard normal bootstrap confidence interval, the basic bootstrap confidence interval, and the percentile confidence interval. Sample from a normal population and check the empirical coverage rates for the sample mean. Find the proportion of times that the confidence intervals miss on the left, and the porportion of times that the confidence intervals miss on the right.

7.B   Repeat Project 7.A for the sample skewness statistic. Compare the coverage rates for normal populations (skewness 0) and $\chi^2(5)$ distributions (positive skewness).

Chapter 8

Permutation Tests

8.1 Introduction

Permutation tests are based on resampling, but unlike the ordinary bootstrap, the samples are drawn *without replacement*. Permutation tests are often applied as a nonparametric test of the general hypothesis

$$H_0 : F = G \qquad vs \qquad H_1 : F \neq G, \tag{8.1}$$

where F and G are two unspecified distributions. Under the null hypothesis, two samples from F and G, and the pooled sample, are all random samples from the same distribution F. Replicates of a two sample test statistic that compares the distributions are generated by resampling without replacement from the pooled sample. Nonparametric tests of independence, association, location, common scale, etc. can also be implemented as permutation tests. For example, in a test of multivariate independence

$$H_0 : F_{X,Y} = F_X F_Y \qquad vs \qquad H_1 : F_{X,Y} \neq F_X F_Y \tag{8.2}$$

under the null hypothesis the data in a sample need not be matched, and all pairs of samples obtained by permutations of the row labels (observations) of either sample are equally likely. Any statistic that measures dependence can be applied in a permutation test.

Permutation tests also can be applied to multi-sample problems, with similar methodology. For example, to test

$$H_0 : F_1 = \cdots = F_k \qquad vs \qquad H_1 : F_i \neq F_j \text{ for some } i, j \tag{8.3}$$

the samples are drawn without replacement from the k pooled samples. Any test statistic for the multi-sample problem can then be applied in a permutation test.

This chapter covers several applications of permutation tests for the general hypotheses (8.1) and (8.2). See Efron and Tibshirani [84, Ch. 15] or Davison and Hinkley for background, examples, and further discussion of permutation tests.

Permutation Distribution

Suppose that two independent random samples $X_1, \ldots, X_n$ and $Y_1, \ldots, Y_m$ are observed from the distributions F_X and F_Y, respectively. Let Z be the ordered set $\{X_1, \ldots, X_n, Y_1, \ldots, Y_m\}$, indexed by

$$\nu = \{1, \ldots, n, n+1, \ldots, n+m\} = \{1, \ldots, N\}.$$

Then $Z_i = X_i$ if $1 \leq i \leq n$ and $Z_i = Y_{i-n}$ if $n+1 \leq i \leq n+m$. Let $Z^* = (X^*, Y^*)$ represent a partition of the pooled sample $Z = X \cup Y$, where X^* has n elements and Y^* has $N - n = m$ elements. Then Z^* corresponds to a permutation π of the integers ν, where $Z_i^* = Z_{\pi(i)}$. The number of possible partitions is equal to the number $\binom{N}{n}$ of different ways to select the first n indices of $\pi(\nu)$, hence there are $\binom{N}{n}$ different ways to partition the pooled sample Z into two subsets of size n and m.

The Permutation Lemma [84, p. 207] states that under $H_0 : F_X = F_Y$, a randomly selected Z^* has probability

$$\frac{1}{\binom{N}{n}} = \frac{n!\,m!}{N!}$$

of equaling any of its possible values. That is, if $F_X = F_Y$ then all permutations are equally likely.

If $\hat{\theta}(X, Y) = \hat{\theta}(Z, \nu)$ is a statistic, then the *permutation distribution* of $\hat{\theta}^*$ is the distribution of the replicates

$$\{\hat{\theta}^*\} = \left\{\hat{\theta}(Z, \pi_j(\nu)), j = 1, \ldots, \binom{N}{n}\right\}$$

$$= \{\hat{\theta}^{(j)} \mid \pi_j(\nu) \text{ is a permutation of } \nu\}.$$

The cdf of $\hat{\theta}^*$ is given by

$$F_{\theta^*}(t) = P(\hat{\theta}^* \leq t) = \binom{N}{n}^{-1} \sum_{j=1}^{N} I(\hat{\theta}^{(j)} \leq t). \tag{8.4}$$

Thus, if $\hat{\theta}$ is applied to test a hypothesis and large values of $\hat{\theta}$ are significant, then the permutation test rejects the null hypothesis when $\hat{\theta}$ is large relative to the distribution of the permutation replicates. The achieved significance level (ASL) of the observed statistic $\hat{\theta}$ is the probability

$$P(\hat{\theta}^* \geq \hat{\theta}) = \binom{N}{n}^{-1} \sum_{j=1}^{N} I(\hat{\theta}^{(j)} \geq \hat{\theta}),$$

where $\hat{\theta} = \hat{\theta}(Z, \nu)$ is the statistic computed on the observed sample. The ASL for a lower-tail or two-tail test based on $\hat{\theta}$ is computed in a similar way.

In practice, unless the sample size is very small, evaluating the test statistic for all of the $\binom{N}{n}$ permutations is computationally excessive. An approximate permutation test is implemented by randomly drawing a large number of samples without replacement.

Approximate permutation test procedure

1. Compute the observed test statistic $\hat{\theta}(X, Y) = \hat{\theta}(Z, \nu)$.

2. For each replicate, indexed $b = 1, \ldots, B$:

    (a) Generate a random permutation $\pi_b = \pi(\nu)$.

    (b) Compute the statistic $\hat{\theta}^{(b)} = \hat{\theta}^*(Z, \pi_b)$.

3. If large values of $\hat{\theta}$ support the alternative, compute the ASL (the empirical p-value) by

$$\hat{p} = \frac{1 + \#\{\hat{\theta}^{(b)} \geq \hat{\theta}\}}{B+1} = \frac{\left\{1 + \sum_{b=1}^{B} I(\hat{\theta}^{(b)} \geq \hat{\theta})\right\}}{B+1}.$$

For a lower-tail or two-tail test $\hat{p}$ is computed in a similar way.

4. Reject H_0 at significance level α if $\hat{p} \leq \alpha$.

The formula for $\hat{p}$ is given by Davison and Hinkley [63, p. 159], who state that "at least 99 and at most 999 random permutations should suffice."

Methods for implementing an approximate permutation test are illustrated in the examples that follow. Although the `boot` function [34] can be used to generate the replicates, it is not necessary to use `boot`. For a multivariate permutation test using `boot` see the examples in Section 8.3.

Example 8.1 (Permutation distribution of a statistic)

The permutation distribution of a statistic is illustrated for a small sample, from the `chickwts` data in R. Weights in grams are recorded, for six groups of newly hatched chicks fed different supplements. There are six types of feed supplements. A quick graphical summary of the data can be displayed by `boxplot(formula(chickwts))`. The plot (not shown) suggests that soybean and linseed groups may be similar. The distribution of weights for these two groups are compared below.

```
attach(chickwts)
x <- sort(as.vector(weight[feed == "soybean"]))
y <- sort(as.vector(weight[feed == "linseed"]))
detach(chickwts)
```

The ordered chick weights for the two samples are

```
X:  158 171 193 199 230 243 248 248 250 267 271 316 327 329
Y:  141 148 169 181 203 213 229 244 257 260 271 309
```

The groups can be compared in several ways. For example, sample means, sample medians, or other trimmed means can be compared. More generally, one can ask whether the distributions of the two variables differ and compare the groups by any statistic that measures a distance between two samples.

Consider the sample mean. If the two samples are drawn from normal populations with equal variances, we can apply the two-sample t-test. The sample means are $\overline{X} = 246.4286$ and $\overline{Y} = 218.7500$. The two sample t statistic is $T = 1.3246$. In this problem, however, the distributions of the weights are unknown. The achieved significance level of T can be computed from the permutation distribution without requiring distributional assumptions.

The sample sizes are $n = 14$ and $m = 12$, so there are a total of

$$\binom{n+m}{n} = \binom{26}{14} = \frac{26!}{14!\,12!} = 9,657,700$$

different partitions of the pooled sample into two subsets of size 14 and 12. Thus, even for small samples, enumerating all possible partitions of the pooled sample is not practical. An alternate approach is to generate a large number of the permutation samples, to obtain the approximate permutation distribution of the replicates. Draw a random sample of n indices from 1:N without replacement, which determines a randomly selected partition (X^*, Y^*). In this way we can generate a large number of the permutation samples. Then compare the observed statistic T to the replicates T^*.

The approximate permutation test procedure is illustrated below with the two-sample t statistic.

```
R <- 999                 #number of replicates
z <- c(x, y)             #pooled sample
K <- 1:26
reps <- numeric(R)       #storage for replicates
t0 <- t.test(x, y)$statistic

for (i in 1:R) {
    #generate indices k for the first sample
    k <- sample(K, size = 14, replace = FALSE)
    x1 <- z[k]
    y1 <- z[-k]          #complement of x1
    reps[i] <- t.test(x1, y1)$statistic
    }
p <- mean(c(t0, reps) >= t0)

> p
[1] 0.101
```

The value of $\hat{p}$ is the proportion of replicates T^* that are at least as large as the observed test statistic (an approximate p-value). For a two-tail test the ASL is $2\hat{p}$ if $\hat{p} \leq 0.5$ (it is $2(1 - \hat{p})$ if $\hat{p} > 0.5$). The ASL is 0.202 so the null hypothesis is not rejected. For comparison, the two-sample t-test reports p-value $= 0.198$. A histogram of the replicates of T is displayed by

```
hist(reps, main = "", freq = FALSE, xlab = "T (p = 0.202)",
     breaks = "scott")
points(t0, 0, cex = 1, pch = 16)      #observed T
```

which is shown in Figure 8.1.   ◇

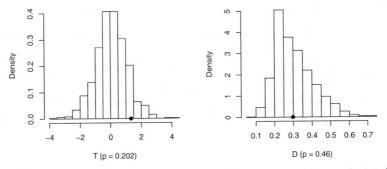

FIGURE 8.1:   Permutation distribution of replicates in Example 8.1 (left) and Example 8.2 (right).

8.2   Tests for Equal Distributions

Suppose that $X = (X_1, \ldots, X_n)$ and $Y = (Y_1, \ldots, Y_m)$ are independent random samples from distributions F and G respectively, and we wish to test the hypothesis $H_0 : F = G$ vs the alternative $H_1 : F \neq G$. Under the null hypothesis, samples X, Y, and the pooled sample $Z = X \cup Y$, are all random samples from the same distribution F. Moreover, under H_0, any subset X^* of size n from the pooled sample, and its complement Y^*, also represent independent random samples from F.

Suppose that $\hat{\theta}$ is a two-sample statistic that measures the distance in some sense between F and G. Without loss of generality, we can suppose that large values of $\hat{\theta}$ support the alternative $F \neq G$. By the permutation lemma, under the null hypothesis all values of $\hat{\theta}^* = \hat{\theta}(X^*, Y^*)$ are equally likely. The permutation distribution of $\hat{\theta}^*$ is given by (8.4), and an exact permutation test or the approximate permutation test procedure given on page 217 can be applied.

Two-sample tests for univariate data

To apply a permutation test of equal distributions, choose a test statistic that measures the difference between two distributions. For example, the two-sample Kolmogorov-Smirnov (K-S) statistic or the two-sample Cramér-von Mises statistic can be applied in the univariate case. Many other statistics are in the literature, although the K-S statistic is one of the most widely applied for univariate distributions. It is applied in the following example.

Example 8.2 (Permutation distribution of the K-S statistic)

In Example 8.1 the means of the soybean and linseed groups were compared. Suppose now that we are interested in testing for any type of difference in the two groups. The hypotheses of interest are $H_0 : F = G$ vs $H_1 : F \neq G$, where F is the distribution of weight of chicks fed soybean supplements and G is the distribution of weight of chicks fed linseed supplements. The Kolmogorov-Smirnov statistic D is the maximum absolute difference between the ecdf's of the two samples, defined by

$$D = \sup_{1 \leq i \leq N} |F_n(z_i) - G_m(z_i)|,$$

where F_n is the ecdf of the first sample $x_1, \ldots, x_n$ and G_m is the ecdf of the second sample $y_1, \ldots, y_m$. Note that $0 \leq D \leq 1$ and large values of D support the alternative $F \neq G$. The observed value of $D = D(X, Y) = 0.2976190$ can be computed using `ks.test`. To determine whether this value of D is strong evidence for the alternative, we compare D with the replicates $D^* = D(X^*, Y^*)$.

```
R <- 999              #number of replicates
z <- c(x, y)          #pooled sample
K <- 1:26
D <- numeric(R)       #storage for replicates
options(warn = -1)
D0 <- ks.test(x, y, exact = FALSE)$statistic
for (i in 1:R) {
    #generate indices k for the first sample
    k <- sample(K, size = 14, replace = FALSE)
    x1 <- z[k]
    y1 <- z[-k]        #complement of x1
    D[i] <- ks.test(x1, y1, exact = FALSE)$statistic
    }
```

```
p <- mean(c(D0, D) >= D0)
options(warn = 0)
> p
[1] 0.46
```

The approximate ASL 0.46 does not support the alternative hypothesis that distributions differ. A histogram of the replicates of D is displayed by

```
hist(D, main = "", freq = FALSE, xlab = "D (p = 0.46)",
    breaks = "scott")
points(D0, 0, cex = 1, pch = 16)        #observed D
```

which is shown in Figure 8.1.   ◇

R note 8.1 *In Example 8.2 the Kolmogorov-Smirnov test ks.test generates a warning each time it tries to compute a p-value, because there are ties in the data. We are not using the p-value, so it is safe to ignore these warnings. Display of warnings or messages at the console regarding warnings can be suppressed by options(warn = -1). The default value is warn = 0.*

Example 8.3 (Two-sample K-S test)

Test whether the distributions of chick weights for the sunflower and linseed groups differ. The K-S test can be applied as in Example 8.2.

```
attach(chickwts)
x <- sort(as.vector(weight[feed == "sunflower"]))
y <- sort(as.vector(weight[feed == "linseed"]))
detach(chickwts)
```

The sample sizes are $n = m = 12$, and the observed K-S test statistic is $D = 0.8333$. The summary statistics below suggest that the distributions of weights for these two groups may differ.

```
> summary(cbind(x, y))
       x                 y
 Min.    :226.0   Min.    :141.0
 1st Qu.:312.8    1st Qu.:178.0
 Median :328.0    Median :221.0
 Mean    :328.9   Mean    :218.8
 3rd Qu.:340.2    3rd Qu.:257.8
 Max.    :423.0   Max.    :309.0
```

Repeating the simulation in Example 8.2 with the sunflower sample replacing the soybean sample produces the following result.

```
p <- mean(c(D0, D) >= D0)
> p
[1] 0.001
```

Thus, none of the replicates are as large as the observed test statistic. Here the sample evidence supports the alternative hypothesis that the distributions differ. ◇

Another univariate test for the two-sample problem is the Cramér-von Mises test [56, 281]. The Cramér-von Mises statistic, which estimates the integrated squared distance between the distributions, is defined by

$$W_2 = \frac{mn}{(m+n)^2} \left[\sum_{i=1}^{n} (F_n(x_i) - G_m(x_i))^2 + \sum_{j=1}^{m} (F_n(y_j) - G_m(y_j))^2 \right],$$

where F_n is the ecdf of the sample $x_1, \ldots, x_n$ and G_m is the ecdf of the sample $y_1, \ldots, y_m$. Large values of W_2 are significant. The implementation of the Cramér-von Mises test is left as an exercise.

The multivariate tests discussed in the next section can also be applied for testing $H_0 : F = G$ in the univariate case.

8.3 Multivariate Tests for Equal Distributions

Classical approaches to the two-sample problem in the univariate case based on comparing empirical distribution functions, such as the Kolmogorov–Smirnov and Cramér-von Mises tests, do not have a natural distribution free extension to the multivariate case. Multivariate tests based on maximum likelihood depend on distributional assumptions about the underlying populations. Hence although likelihood tests may apply in special cases, they do not apply to the general two-sample or k-sample problem, and may not be robust to departures from these assumptions.

Many of the procedures that are available for the multivariate two-sample problem (8.1) require a computational approach for implementation. Bickel [27] constructed a consistent distribution free multivariate extension of the univariate Smirnov test by conditioning on the pooled sample. Friedman and Rafsky [101] proposed distribution free multivariate generalizations of the Wald-Wolfowitz runs test and Smirnov test for the two-sample problem, based on the minimal spanning tree of the pooled sample. A class of consistent, asymptotically distribution free tests for the multivariate problem is based on nearest neighbors [28, 139, 240]. The nearest neighbor tests apply to testing the k-sample hypothesis when all distributions are continuous. A multivariate nonparametric test for equal distributions was developed independently by Baringhaus and Franz [20] and Székely and Rizzo [261, 262], which is implemented as an approximate permutation test. We will discuss the latter two, the nearest neighbor tests and the energy test [226, 261].

In the following sections multivariate samples will be denoted by boldface type. Suppose that

$$\mathbf{X} = \{X_1, \ldots, X_{n_1}\} \in \mathbb{R}^d, \quad \mathbf{Y} = \{Y_1, \ldots, Y_{n_2}\} \in \mathbb{R}^d,$$

are independent random samples, $d \geq 1$. The pooled data matrix is $\mathbf{Z}$, an $n \times d$ matrix with observations in rows:

$$\mathbf{Z}_{n \times d} = \begin{bmatrix} x_{1,1} & x_{1,2} & \cdots & x_{1,d} \\ x_{2,1} & x_{2,2} & \cdots & x_{2,d} \\ \vdots & \vdots & & \vdots \\ x_{n_1,1} & x_{n_1,2} & \cdots & x_{n_1,d} \\ y_{1,1} & y_{1,2} & \cdots & y_{1,d} \\ y_{2,1} & y_{2,2} & \cdots & y_{2,d} \\ \vdots & \vdots & & \vdots \\ y_{n_2,1} & y_{n_2,2} & \cdots & y_{n_2,d} \end{bmatrix}, \tag{8.5}$$

where $n = n_1 + n_2$.

Nearest neighbor tests

A multivariate test for equal distributions is based on nearest neighbors. The nearest neighbor (NN) tests are a type of test based on ordered distances between sample elements, which can be applied when the distributions are continuous.

Usually the distance is the Euclidean norm $\|z_i - z_j\|$. The NN tests are based on the first through r^{th} nearest neighbor coincidences in the pooled sample. Consider the simplest case, $r = 1$. For example, if the observed samples are the weights in Example 8.3

	[,1]	[,2]	[,3]	[,4]	[,5]	[,6]	[,7]	[,8]	[,9]	[,10]	[,11]	[,12]
x	423	340	392	339	341	226	320	295	334	322	297	318
y	309	229	181	141	260	203	148	169	213	257	244	271

then the first nearest neighbor of $x_1 = 423$ is $x_3 = 392$, which are in the same sample. The first nearest neighbor of $x_6 = 226$ is $y_2 = 229$, in different samples. In general, if the sampled distributions are equal, then the pooled sample has on average less nearest neighbor coincidences than under the alternative hypothesis. In this example, most of the nearest neighbors are found in the same sample.

Let $\mathbf{Z} = \{X_1, \ldots, X_{n_1}, Y_1, \ldots, Y_{n_2}\}$ as in (8.5). Denote the first nearest neighbor of Z_i by $NN_1(Z_i)$. Count the number of first nearest neighbor coincidences by the indicator function $I_i(1)$, which is defined by

$$I_i(1) = 1 \text{ if } Z_i \text{ and } NN_1(Z_i) \text{ belong to the same sample;}$$
$$I_i(1) = 0 \text{ if } Z_i \text{ and } NN_1(Z_i) \text{ belong to different samples.}$$

The first nearest neighbor statistic is the proportion of first nearest neighbor coincidences

$$T_{n,1} = \frac{1}{n} \sum_{i=1}^{n} I_i(1),$$

where $n = n_1 + n_2$. Large values of $T_{n,1}$ support the alternative hypothesis that the distributions differ.

Similarly, denote the second nearest neighbor of a sample element Z_i by $NN_2(Z_i)$ and define the indicator function $I_i(2)$, which is 1 if $NN_2(Z_i)$ is in the same sample as Z_i and otherwise $I_i(2) = 0$. The second nearest neighbor statistic is based on the first and second nearest neighbor coincidences, defined by

$$T_{n,2} = \frac{1}{2n} \sum_{i=1}^{n} (I_i(1) + I_i(2)).$$

In general, the rth nearest neighbor of Z_i is defined to be the sample element Z_j satisfying $\|Z_i - Z_\ell\| < \|Z_i - Z_j\|$ for exactly $r - 1$ indices $1 \leq \ell \leq n$, $\ell \neq i$. Denote the rth nearest neighbor of a sample element Z_i by $NN_r(Z_i)$. For $i = 1, \ldots, n$ the indicator function $I_i(r)$ is defined by $I_i(r) = 1$ if Z_i and $NN_r(Z_i)$ belong to the same sample, and otherwise $I_i(r) = 0$. The Jth nearest neighbor statistic measures the proportion of first through J^{th} nearest neighbor coincidences:

$$T_{n,J} = \frac{1}{nJ} \sum_{i=1}^{n} \sum_{r=1}^{J} I_i(r). \tag{8.6}$$

Under the hypothesis of equal distributions, the pooled sample has on average less nearest neighbor coincidences than under the alternative hypothesis, so the test rejects the null hypothesis for large values of $T_{n,J}$. Henze [139] proved that the limiting distribution of a class of nearest neighbor statistics is normal for any distance generated by a norm on $\mathbb{R}^d$. Schilling [240] derived the mean and variance of the distribution of $T_{n,2}$ for selected values of n_1/n and d in the case of Euclidean norm. In general, the parameters of the normal distribution may be difficult to obtain analytically. If we condition on the pooled sample to implement an exact permutation test, the procedure is distribution free. The test can be implemented as an approximate permutation test, following the procedure outlined on page 217.

Remark 8.1 *Nearest neighbor statistics are functions of the ordered distances between sample elements. The sampled distributions are assumed to be continuous, so there are no ties. Thus, resampling without replacement is the correct resampling method and the permutation test rather than the ordinary bootstrap should be applied. In the ordinary bootstrap, many ties would occur in the bootstrap samples.*

Searching for nearest neighbors is not a trivial computational problem, but fast algorithms have been developed [25, 12, 13]. A fast nearest neighbor method nn is available in the knnFinder package. The algorithm uses a kd-tree. According to the package author Kemp [160], "The advantage of the kd-tree is that it runs in $O(M \log M)$ time ... where M is the number of data points using Bentley's kd-tree."

Example 8.4 (Finding nearest neighbors)

The following numerical example illustrates the usage of the nn (knnFinder) [160] function as a method to find the indices of the first through r^{th} nearest neighbors. The pooled data matrix **Z** is assumed to be in the layout (8.5).

```
library(knnFinder)  #for nn function

#generate a small multivariate data set
x <- matrix(rnorm(12), 3, 4)
y <- matrix(rnorm(12), 3, 4)

z <- rbind(x, y)
o <- rep(0, nrow(z))

DATA <- data.frame(cbind(z, o))
NN <- nn(DATA, p = nrow(z)-1)
```

In the distance matrix below, for example, the first through fifth nearest neighbors of Z_1 are Z_4, Z_2, Z_3, Z_5, Z_6.

```
> D <- dist(z)
> round(as.matrix(D), 2)

     1    2    3    4    5    6
1 0.00 2.29 2.69 1.88 2.87 3.94
2 2.29 0.00 2.60 3.27 2.45 3.69
3 2.69 2.60 0.00 2.14 0.43 3.52
4 1.88 3.27 2.14 0.00 2.52 3.66
5 2.87 2.45 0.43 2.52 0.00 3.48
6 3.94 3.69 3.52 3.66 3.48 0.00
```

The index matrix returned by the function nn identifies the nearest neighbors as follows. The ith row of $nn.idx on the next page contains the subscripts (indices) of $NN_1(Z_i)$, $NN_2(Z_i)$, ..., the nearest neighbors of Z_i. According to the first row of the index matrix $nn.idx, the indices of the first through fifth nearest neighbors of Z_1 are 4, 2, 3, 5, and 6, respectively.

```
> NN$nn.idx
   X1 X2 X3 X4 X5
1  4  2  3  5  6
2  1  5  3  4  6
3  5  4  2  1  6
4  1  3  5  2  6
5  3  2  4  1  6
6  5  3  4  2  1

> round(NN$nn.dist, 2)
     X1   X2   X3   X4   X5
1 1.88 2.29 2.69 2.87 3.94
2 2.29 2.45 2.60 3.27 3.69
3 0.43 2.14 2.60 2.69 3.52
4 1.88 2.14 2.52 3.27 3.66
5 0.43 2.45 2.52 2.87 3.48
6 3.48 3.52 3.66 3.69 3.94
```

In this small data set it is easy to compute the nearest neighbor statistics. For example, $T_{n,1} = 2/6 \doteq 0.333$ and

$$T_{n,2} = \frac{1}{2n}\sum_{i=1}^{n}(I_i(1) + I_i(2)) = \frac{1}{12}(2+1) = 0.25.$$

◇

Example 8.5 (Nearest neighbor statistic)

In this example a method of computing nearest neighbor statistics from the result of nn (knnFinder) is shown. Compute $T_{n,3}$ for the chickwts data from Example 8.3.

```
library(knnFinder)
attach(chickwts)
x <- as.vector(weight[feed == "sunflower"])
y <- as.vector(weight[feed == "linseed"])
detach(chickwts)

z <- c(x, y)
o <- rep(0, length(z))
z <- as.data.frame(cbind(z, o))
NN <- nn(z, p=3)
```

The data and the index matrix NN$nn.idx are shown on the facing page.

pooled sample			$nn.idx		
	[,1]		X1	X2	X3
[1,]	423	1	3	5	2
[2,]	340	2	4	5	9
[3,]	392	3	1	5	2
[4,]	339	4	2	5	9
[5,]	341	5	2	4	9
[6,]	226	6	14	21	23
[7,]	320	7	12	10	13
[8,]	295	8	11	13	12
[9,]	334	9	4	2	5
[10,]	322	10	7	12	9
[11,]	297	11	8	13	12
[12,]	318	12	7	10	13	I=1 if index <= 12

I=1 if index > 12

[13,]	309	13	12	7	11
[14,]	229	14	6	23	21
[15,]	181	15	20	18	21
[16,]	141	16	19	20	15
[17,]	260	17	22	24	23
[18,]	203	18	21	15	6
[19,]	148	19	16	20	15
[20,]	169	20	15	19	16
[21,]	213	21	18	6	14
[22,]	257	22	17	23	24
[23,]	244	23	22	14	17
[24,]	271	24	17	22	8

The first three nearest neighbors of each sample element Z_i are in the ith row. In the first block, count the number of entries that are between 1 and $n_1 = 12$. In the second block, count the number of entries that are between $n_1 + 1 = 13$ and $n_1 + n_2 = 24$.

```
block1 <- NN$nn.idx[1:12, ]
block2 <- NN$nn.idx[13:24, ]
i1 <- sum(block1 < 12.5)
i2 <- sum(block2 > 12.5)

> c(i1, i2)
[1] 29 29
```

Then

$$T_{n,3} = \frac{1}{3n} \sum_{i=1}^{n} \sum_{j=1}^{3} I_i(j) = \frac{1}{3(24)}(29 + 29) = \frac{58}{72} = 0.8055556.$$

◇

Example 8.6 (Nearest neighbor test)

The permutation test for $T_{n,3}$ in Example 8.5 can be applied using the `boot` function in the `boot` package [34] as follows.

```
library(boot)
Tn3 <- function(z, ix, sizes) {
    n1 <- sizes[1]
    n2 <- sizes[2]
    n <- n1 + n2
    z <- z[ix, ]
    o <- rep(0, NROW(z))
    z <- as.data.frame(cbind(z, o))
    NN <- nn(z, p=3)
    block1 <- NN$nn.idx[1:n1, ]
    block2 <- NN$nn.idx[(n1+1):n, ]
    i1 <- sum(block1 < n1 + .5)
    i2 <- sum(block2 > n1 + .5)
    return((i1 + i2) / (3 * n))
}
N <- c(12, 12)
boot.obj <- boot(data = z, statistic = Tn3,
    sim = "permutation", R = 999, sizes = N)
```

Note: The permutation samples can also be generated by the `sample` function. The result of the simulation is

```
> boot.obj
DATA PERMUTATION
Call: boot(data = z, statistic = Tn3, R = 999,
    sim = "permutation", sizes = N)


Bootstrap Statistics :
     original       bias     std. error
t1* 0.8055556  -0.3260066   0.07275428
```

The output from `boot` does not include a p-value, of course, because `boot` has no way of knowing what hypotheses are being tested. What is printed at the console is a summary of the `boot` object. The `boot` object itself is a list that contains several things including the permutation replicates of the test statistic. The test decision can be obtained from the observed statistic in `$t0` and the replicates in `$t`.

```
> tb <- c(boot.obj$t, boot.obj$t0)
> mean(tb >= boot.obj$t0)
[1] 0.001
```

The ASL is $\hat{p} = 0.001$, so the hypothesis of equal distributions is rejected. The histogram of replicates of $T_{n,3}$ is shown in Figure 8.2.

```
hist(tb, freq=FALSE, main="",
     xlab="replicates of T(n,3) statistic")
points(boot.obj$t0, 0, cex=1, pch=16)
```

◇

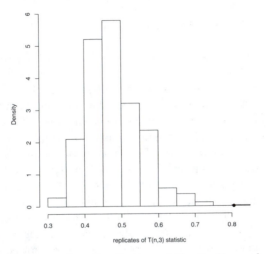

FIGURE 8.2: Permutation distribution of $T_{n,3}$ in Example 8.6.

The multivariate r^{th} nearest neighbor test can be implemented by an approximate permutation test. The steps are to write a general function that computes the statistic $T_{n,r}$ for any given (n_1, n_2, r) and permutation of the row indices of the pooled sample. Then apply **boot** or generate permutations using **sample**, similar to the implementation of the permutation test shown in Example 8.6.

Energy test for equal distributions

The *energy* distance or *e*-distance statistic $\mathcal{E}_n$ is defined by

$$\mathcal{E}_n = e(\mathbf{X}, \mathbf{Y}) = \frac{n_1 n_2}{n_1 + n_2} \left(\frac{2}{n_1 n_2} \sum_{i=1}^{n_1} \sum_{j=1}^{n_2} \|X_i - Y_j\| \right.$$

$$\left. - \frac{1}{n_1^2} \sum_{i=1}^{n_1} \sum_{j=1}^{n_1} \|X_i - X_j\| - \frac{1}{n_2^2} \sum_{i=1}^{n_2} \sum_{j=1}^{n_2} \|Y_i - Y_j\| \right). \quad (8.7)$$

On the name "energy" and concept of energy statistics in general see [258, 259]. The non-negativity of $e(\mathbf{X}, \mathbf{Y})$ is a special case of the following inequality. If X, X', Y, Y' are independent random vectors in $\mathbb{R}^d$ with finite expectations, $X \overset{D}{=} X'$ and $Y \overset{D}{=} Y'$, then

$$2E\|X - Y\| - E\|X - X'\| - E\|Y - Y'\| \geq 0, \tag{8.8}$$

and equality holds if and only if X and Y are identically distributed [262, 263]. The $\mathcal{E}$ distance between the distribution of X and Y is

$$\mathcal{E}(X, Y) = 2E\|X - Y\| - E\|X - X'\| - E\|Y - Y'\|$$

and the empirical distance $\mathcal{E}_n = e(\mathbf{X}, \mathbf{Y})$ is a constant times the plug-in estimator of $\mathcal{E}(X, Y)$.

Clearly large e-distance corresponds to different distributions, and measures the distance between distributions in a similar sense as the univariate empirical distribution function (edf) statistics. In contrast to edf statistics, however, e-distance does not depend on the notion of a sorted list, and e-distance is by definition a multivariate measure of distance between distributions.

If X and Y are not identically distributed, and $n = n_1 + n_2$, then $E[\mathcal{E}_n]$ is asymptotically a positive constant times n. As the sample size n tends to infinity, under the null hypothesis $E[\mathcal{E}_n]$ tends to a positive constant, while under the alternative hypothesis $E[\mathcal{E}_n]$ tends to infinity. Not only the expected value of $\mathcal{E}_n$, but $\mathcal{E}_n$ itself, converges (in distribution) under the null hypothesis, and tends to infinity (stochastically) otherwise. A test for equal distributions based on $\mathcal{E}_n$ is universally consistent against all alternatives with finite first moments [261, 262]. The asymptotic distribution of $\mathcal{E}_n$ is a quadratic form of centered Gaussian random variables, with coefficients that depend on the distributions of X and Y.

To implement the test, suppose that $\mathbf{Z}$ is the $n \times d$ data matrix of the pooled sample as in (8.5). The permutation operation is applied to the row indices of $\mathbf{Z}$. The calculation of the test statistic has $O(n^2)$ time complexity, where $n = n_1 + n_2$ is the size of the pooled sample. (In the univariate case the statistic can be written as a linear combination of the order statistics, with $O(n \log n)$ complexity.)

Example 8.7 (Two-sample energy statistic)

The approximate permutation energy test is implemented in `eqdist.etest` in the `energy` package [226]. However, in order to illustrate the details of the implementation for a multivariate permutation test, we provide an R version below. Note that the `energy` implementation is considerably faster than the example below, because in `eqdist.etest` the calculation of the test statistic is implemented in an external C library.

The $\mathcal{E}_n$ statistic is a function of the pairwise distances between sample elements. The distances remain invariant under any permutation of the indices, so it is *not* necessary to recalculate distances for each permutation sample.

However, it *is* necessary to provide a method for looking up the correct distance in the original distance matrix given the permutation of indices.

```
edist.2 <- function(x, ix, sizes) {
    # computes the e-statistic between 2 samples
    # x:          Euclidean distances of pooled sample
    # sizes:      vector of sample sizes
    # ix:         a permutation of row indices of x

    dst <- x
    n1 <- sizes[1]
    n2 <- sizes[2]
    ii <- ix[1:n1]
    jj <- ix[(n1+1):(n1+n2)]
    w <- n1 * n2 / (n1 + n2)

    # permutation applied to rows & cols of dist. matrix
    m11 <- sum(dst[ii, ii]) / (n1 * n1)
    m22 <- sum(dst[jj, jj]) / (n2 * n2)
    m12 <- sum(dst[ii, jj]) / (n1 * n2)
    e <- w * ((m12 + m12) - (m11 + m22))
    return (e)
}
```

Below, the simulated samples in $\mathbb{R}^d$ are generated from distributions that differ in location. The first distribution is centered at $\mu_1 = (0, \ldots, 0)^T$ and the second distribution is centered at $\mu_2 = (a, \ldots, a)^T$.

```
d <- 3
a <- 2 / sqrt(d)
x <- matrix(rnorm(20 * d), nrow = 20, ncol = d)
y <- matrix(rnorm(10 * d, a, 1), nrow = 10, ncol = d)
z <- rbind(x, y)
dst <- as.matrix(dist(z))

> edist.2(dst, 1:30, sizes = c(20, 10))
[1] 9.61246
```

The observed value of the test statistic is $\mathcal{E}_n = 9.61246$. ◇

The function edist.2 is designed to be used with the boot (boot) function [34] to perform the permutation test. Alternately, generate the permutation vectors ix using the sample function. The boot method is shown in the following example.

Example 8.8 (Two-sample energy test)

This example shows how to apply the `boot` function to perform an approximate permutation test using a multivariate test statistic function. Apply the permutation test to the data matrix `z` in Example 8.7.

```
library(boot)  #for boot function
dst <- as.matrix(dist(z))
N <- c(20, 10)

boot.obj <- boot(data = dst, statistic = edist.2,
    sim = "permutation", R = 999, sizes = N)

> boot.obj

DATA PERMUTATION

Call: boot(data = dst, statistic = edist.2, R = 999,
    sim = "permutation", sizes = N)

Bootstrap Statistics :
    original     bias      std. error
t1*  9.61246  -7.286621    1.025068
```

The permutation vectors generated by `boot` will have the same length as the `data` argument. If `data` is a vector then the permutation vector generated by `boot` will have length equal to the `data` vector. If `data` is a matrix, then the permutation vector will have length equal to the number of rows of the matrix. For this reason, it is necessary to convert the `dist` object to an $n \times n$ distance matrix.

The ASL is computed from the replicates in the bootstrap object.

```
e <- boot.obj$t0
tb <- c(e, boot.obj$t)
mean(tb >= e)
[1] 0.001

hist(tb, main = "", breaks="scott", freq=FALSE,
    xlab="Replicates of e")
points(e, 0, cex=1, pch=16)
```

None of the replicates exceed the observed value 9.61246 of the test statistic. The approximate achieved significance level is 0.001, and we reject the hypothesis of equal distributions. Replicates of $\mathcal{E}_n$ are shown in Figure 8.3(a).

The large estimate of bias reported by the boot function gives an indication that the test statistic is large, because $\mathcal{E}(X, Y) \geq 0$ and $\mathcal{E}(X, Y) = 0$ if and only if the sampled distributions are equal.

Finally, let us check the result of the test when the sampled distributions are identical.

```
d <- 3
a <- 0
x <- matrix(rnorm(20 * d), nrow = 20, ncol = d)
y <- matrix(rnorm(10 * d, a, 1), nrow = 10, ncol = d)
z <- rbind(x, y)
dst <- as.matrix(dist(z))

N <- c(20, 10)
dst <- as.matrix(dist(z))
boot.obj <- boot(data = dst, statistic = edist.2,
    sim="permutation", R=999, sizes=N)
> boot.obj
...
Bootstrap Statistics :
    original    bias    std. error
t1* 1.664265 0.7325929   1.051064

e <- boot.obj$t0
E <- c(boot.obj$t, e)

mean(E >= e)
[1] 0.742
    hist(E, main = "", breaks="scott",
        xlab="Replicates of e", freq=FALSE)
    points(e, 0, cex=1, pch=16)
```

In the second example the approximate achieved significance level is 0.742 and the hypothesis of equal distributions is not rejected. Notice that the estimate of bias here is small. The histogram of replicates is shown in Figure 8.3(b). ◇

The $\mathcal{E}$ distance and two-sample e-statistic $\mathcal{E}_n$ are easily generalized to the k-sample problem. See e.g. the function edist (energy), which returns a dissimilarity object like the dist object.

Example 8.9 (*k*-sample energy distances)

The function edist.2 in Example 8.7 is a two-sample version of the function edist in the energy package [226], which summarizes the empirical $\mathcal{E}$-distances between $k \geq 2$ samples. The syntax is

```
edist(x, sizes, distance=FALSE, ix=1:sum(sizes), alpha=1)
```

The argument alpha is an exponent $0 < \alpha \leq 2$ on the Euclidean distance.

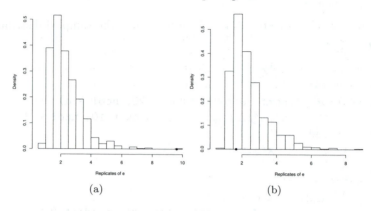

(a)   (b)

FIGURE 8.3:   Permutation distribution of the two-sample e-statistic replicates in Example 8.7.

It can be shown that for all $0 < \alpha < 2$ the corresponding $e^{(\alpha)}$-distance
determines a statistically consistent test of equal distributions for all random
vectors with finite first moments [262].
   Consider the four-dimensional iris data. Compute the e-distance matrix for
the three species of iris.

```
library(energy)   #for edist
z <- iris[ , 1:4]
dst <- dist(z)

> edist(dst, sizes = c(50, 50, 50), distance = TRUE)
          1          2
2 123.55381
3 195.30396   38.85415
```

A test for the k-sample hypothesis of equal distributions is based on k-sample
e-distances with a suitable weight function.   ◇

Comparison of nearest neighbor and energy tests

Example 8.10 (Power comparison)

In a simulation experiment, we compared the empirical power of the third
nearest neighbor test based on $T_{n,3}$ (8.6) and the energy test based on $\mathcal{E}_n$
(8.7). The distributions compared,

$$F_1 = N_2(\mu = (0,0)^2, \Sigma = I_2), \qquad F_2 = N_2(\mu = (0,\delta)^T, \Sigma = I_2),$$

differ in location. The empirical power was estimated for $\delta = 0, 0.5, 0.75, 1$,
from a simulation of permutation tests on $10,000$ pairs of samples. Each per-

mutation test decision was based on 499 permutation replicates (each entry in the table required $5 \cdot 10^6$ calculations of the test statistic). Empirical results are given below for selected alternatives, sample sizes, and dimension, at significance level $\alpha = 0.1$. Both the $\mathcal{E}_n$ and $T_{n,3}$ statistics achieved approximately correct empirical significance in our simulations (see case $\delta = 0$ in Table 8.1), although the Type I error rate for $T_{n,3}$ may be slightly inflated when n is small.

TABLE 8.1: Significant Tests (nearest whole percent at $\alpha = 0.1$, $se \leq 0.5\%$) of Bivariate Normal Location Alternatives $F_1 = N_2((0,0)^T, I_2)$, $F_2 = N_2((0,\delta)^T, I_2)$

		$\delta = 0$		$\delta = 0.5$		$\delta = 0.75$		$\delta = 1$	
n_1	n_2	$\mathcal{E}_n$	$T_{n,3}$	$\mathcal{E}_n$	$T_{n,3}$	$\mathcal{E}_n$	$T_{n,3}$	$\mathcal{E}_n$	$T_{n,3}$
10	10	10	12	23	19	40	29	58	42
15	15	9	11	30	21	53	34	75	52
20	20	10	12	37	23	64	38	86	58
25	25	10	11	43	25	73	42	93	65
30	30	10	11	48	25	81	47	96	70
40	40	11	10	59	28	90	52	99	78
50	50	10	11	69	29	95	58	100	82
75	75	10	11	85	37	99	69	100	93
100	100	10	10	92	40	100	79	100	100

These alternatives differ in location only, and the empirical evidence summarized in Table 8.1 suggests that $\mathcal{E}_n$ is more powerful than $T_{n,3}$ against this class of alternatives. $\diamond$

8.4 Application: Distance Correlation

A test of independence of random vectors $X \in \mathbb{R}^p$ and $Y \in \mathbb{R}^q$

$$H_0 : F_{XY} = F_X F_Y \qquad vs \qquad H_1 : F_{XY} \neq F_X F_Y$$

can be implemented as a permutation test. The permutation test does not require distributional assumptions, or any type of model specification for the dependence structure. Not many universally consistent nonparametric tests exist for the general hypothesis above. In this section we will discuss a new multivariate nonparametric test of independence based on distance correlation [265] that is consistent against all dependent alternatives with finite first moments. The test will be implemented as a permutation test.

Distance Correlation

Distance correlation is a new measure of dependence between random vectors introduced by Székely, Rizzo, and Bakirov [265]. For all distributions with finite first moments, distance correlation $\mathcal{R}$ generalizes the idea of correlation in two fundamental ways:

1. $\mathcal{R}(X, Y)$ is defined for X and Y in arbitrary dimension.

2. $\mathcal{R}(X, Y) = 0$ characterizes independence of X and Y.

Distance correlation satisfies $0 \leq \mathcal{R} \leq 1$, and $\mathcal{R} = 0$ only if X and Y are independent. Distance covariance $\mathcal{V}$ provides a new approach to the problem of testing the joint independence of random vectors. The formal definitions of the population coefficients $\mathcal{V}$ and $\mathcal{R}$ are given in [265]. The definitions of the empirical coefficients are as follows.

Definition 8.1 *The empirical distance covariance $\mathcal{V}_n(\mathbf{X}, \mathbf{Y})$ is the nonnegative number defined by*

$$\mathcal{V}_n^2(\mathbf{X}, \mathbf{Y}) = \frac{1}{n^2} \sum_{k,l=1}^{n} A_{kl} B_{kl}, \qquad (8.9)$$

where A_{kl} and B_{kl} are defined in equations (8.11-8.12) below. Similarly, $\mathcal{V}_n(\mathbf{X})$ is the nonnegative number defined by

$$\mathcal{V}_n^2(\mathbf{X}) = \mathcal{V}_n^2(\mathbf{X}, \mathbf{X}) = \frac{1}{n^2} \sum_{k,l=1}^{n} A_{kl}^2 . \qquad (8.10)$$

The formulas for A_{kl} and B_{kl} in (8.9–8.10) are given by

$$A_{kl} = a_{kl} - \bar{a}_{k.} - \bar{a}_{.l} + \bar{a}_{..} ; \qquad (8.11)$$
$$B_{kl} = b_{kl} - \bar{b}_{k.} - \bar{b}_{.l} + \bar{b}_{..} , \qquad (8.12)$$

where

$$a_{kl} = \|X_k - X_l\|_p, \qquad b_{kl} = \|Y_k - Y_l\|_q, \qquad k, l = 1, \dots, n,$$

and the subscript "." denotes that the mean is computed for the index that it replaces. Note that these formulas are similar to computing formulas in analysis of variance, so the distance covariance statistic is very easy to compute. Although it may not be obvious that $\mathcal{V}_n^2(\mathbf{X}, \mathbf{Y}) \geq 0$, this fact as well as the motivation for the definition of $\mathcal{V}_n$ is explained in [265].

Definition 8.2 *The empirical distance correlation $\mathcal{R}_n(\mathbf{X}, \mathbf{Y})$ is the square root of*

$$\mathcal{R}_n^2(\mathbf{X}, \mathbf{Y}) = \begin{cases} \frac{\mathcal{V}_n^2(\mathbf{X}, \mathbf{Y})}{\sqrt{\mathcal{V}_n^2(\mathbf{X}) \mathcal{V}_n^2(\mathbf{Y})}}, & \mathcal{V}_n^2(\mathbf{X}) \mathcal{V}_n^2(\mathbf{Y}) > 0; \\ 0, & \mathcal{V}_n^2(\mathbf{X}) \mathcal{V}_n^2(\mathbf{Y}) = 0. \end{cases} \qquad (8.13)$$

The asymptotic distribution of $n\mathcal{V}_n^2$ is a quadratic form of centered Gaussian random variables, with coefficients that depend on the distributions of X and Y. For the general problem of testing independence when the distributions of X and Y are unknown, the test based on $n\mathcal{V}_n^2$ can be implemented as a permutation test.

Before proceeding to the details of the permutation test, we implement the calculation of the distance covariance statistic (dCov).

Example 8.11 (Distance covariance statistic)

In the distance covariance function dCov, operations on the rows and columns of the distance matrix generate the matrix with entries A_{kl}. Note that each term

$$A_{kl} = a_{kl} - \bar{a}_{k.} - \bar{a}_{.l} + \bar{a}_{..}; \qquad a_{kl} = \|X_k - X_l\|$$

is a function of the distance matrix of the X sample. In the function Akl, the sweep operator is used twice. The first sweep subtracts $\bar{a}_{.l}$, the row means, from the distances a_{kl}. The second sweep subtracts $\bar{a}_{k.}$, the column means, from the result of the first sweep. (The column means and row means are equal because the distance matrix is symmetric.) If the samples are x and y, then the matrix $A = (A_{kl})$ is returned by Akl(x) and the matrix $B = (B_{kl})$ is returned by Akl(y). The remaining calculations are simple functions of these two matrices.

```
dCov <- function(x, y) {
    x <- as.matrix(x)
    y <- as.matrix(y)
    n <- nrow(x)
    m <- nrow(y)
    if (n != m || n < 2) stop("Sample sizes must agree")
    if (! (all(is.finite(c(x, y)))))
        stop("Data contains missing or infinite values")

    Akl <- function(x) {
        d <- as.matrix(dist(x))
        m <- rowMeans(d)
        M <- mean(d)
        a <- sweep(d, 1, m)
        b <- sweep(a, 2, m)
        return(b + M)
    }
    A <- Akl(x)
    B <- Akl(y)
    dCov <- sqrt(mean(A * B))
    dCov
}
```

A simple example to try out the dCov function is the following. Compute $\mathcal{V}_n$ for the bivariate distributions of iris setosa (petal length, petal width) and (sepal length, sepal width).

```
z <- as.matrix(iris[1:50, 1:4])
x <- z[ , 1:2]
y <- z[ , 3:4]
# compute the observed statistic
> dCov(x, y)
[1] 0.06436159
```

The returned value is $\mathcal{V}_n = 0.06436159$. Here $n = 50$ so the test statistic for a test of independence is $n\mathcal{V}_n^2 \doteq 0.207$.   ◇

Example 8.12 (Distance correlation statistic)

The distance covariance must be computed to get the distance correlation statistic. Rather than call the distance covariance function three times, which means repeated calculation of the distances and the A and B matrices, it is more efficient to combine all operations in one function.

```
DCOR <- function(x, y) {
    x <- as.matrix(x)
    y <- as.matrix(y)
    n <- nrow(x)
    m <- nrow(y)
    if (n != m || n < 2) stop("Sample sizes must agree")
    if (! (all(is.finite(c(x, y)))))
        stop("Data contains missing or infinite values")
    Akl <- function(x) {
        d <- as.matrix(dist(x))
        m <- rowMeans(d)
        M <- mean(d)
        a <- sweep(d, 1, m)
        b <- sweep(a, 2, m)
        return(b + M)
    }
    A <- Akl(x)
    B <- Akl(y)
    dCov <- sqrt(mean(A * B))
    dVarX <- sqrt(mean(A * A))
    dVarY <- sqrt(mean(B * B))
    dCor <- sqrt(dCov / sqrt(dVarX * dVarY))
    list(dCov=dCov, dCor=dCor, dVarX=dVarX, dVarY=dVarY)
}
```

Applying the function `DCOR` to the `iris` data we obtain all of the distance dependence statistics in one step.

```
z <- as.matrix(iris[1:50, 1:4])
x <- z[ , 1:2]
y <- z[ , 3:4]


> unlist(DCOR(x, y))
        dCov        dCor       dVarX       dVarY
  0.06436159  0.61507138  0.28303069  0.10226284
```

$\diamond$

Permutation tests of independence

A permutation test of independence is implemented as follows. Suppose that $X \in \mathbb{R}^p$ and $Y \in \mathbb{R}^q$ and $Z = (X, Y)$. Then Z is a random vector in $\mathbb{R}^{p+q}$. In the following, we suppose that a random sample is in an $n \times (p+q)$ data matrix $\mathbf{Z}$ with observations in rows:

$$\mathbf{Z}_{n \times d} = \begin{bmatrix} x_{1,1} & x_{1,2} & \cdots & x_{1,p} & y_{1,1} & y_{1,2} & \cdots & y_{1,q} \\ x_{2,1} & x_{2,2} & \cdots & x_{2,p} & y_{2,1} & y_{2,2} & \cdots & y_{2,q} \\ \vdots & \vdots & & \vdots & & & & \\ x_{n,1} & x_{n,2} & \cdots & x_{n,p} & y_{n,1} & y_{n,2} & \cdots & x_{n,q} \end{bmatrix}.$$

Let ν_1 be the row labels of the X sample and let ν_2 be the row labels of the Y sample. Then (Z, ν_1, ν_2) is the sample from the joint distribution of X and Y. If X and Y are dependent, the samples must be paired and the ordering of labels ν_2 cannot be changed independently of ν_1. Under independence, the samples X and Y need not be matched. Any permutation of the row labels of the X or Y sample generates a permutation replicate. The permutation test procedure for independence permutes the row indices of one of the samples (it is not necessary to permute both ν_1 and ν_2).

Approximate permutation test procedure for independence

Let $\hat{\theta}$ be a two sample statistic for testing multivariate independence.

1. Compute the observed test statistic $\hat{\theta}(X, Y) = \hat{\theta}(Z, \nu_1, \nu_2)$.
2. For each replicate, indexed $b = 1, \ldots, B$:
   (a) Generate a random permutation $\pi_b = \pi(\nu_2)$.
   (b) Compute the statistic $\hat{\theta}^{(b)} = \hat{\theta}^*(Z, \pi_b) = \hat{\theta}(X, Y^*, \pi(\nu_2))$.
3. If large values of $\hat{\theta}$ support the alternative, compute the ASL by

$$\hat{p} = \frac{1 + \#\{\hat{\theta}^{(b)} \geq \hat{\theta}\}}{B+1} = \frac{\left\{1 + \sum_{b=1}^{B} I(\hat{\theta}^{(b)} \geq \hat{\theta})\right\}}{B+1}.$$

The ASL for a lower-tail or two-tail test based on $\hat{\theta}$ is computed in a similar way.

4. Reject H_0 at significance level α if $\hat{p} \leq \alpha$.

Example 8.13 (Distance covariance test)

This example tests whether the bivariate distributions (petal length, petal width) and (sepal length, sepal width) of iris setosa are independent. To implement a permutation test, write a function to compute the replicates of the test statistic $n\mathcal{V}_n^2$ that takes as its first argument the data matrix and as its second argument the permutation vector.

```
ndCov2 <- function(z, ix, dims) {
    #dims contains dimensions of x and y
    p <- dims[1]
    q1 <- dims[2] + 1
    d <- p + dims[2]
    x <- z[ , 1:p]      #leave x as is
    y <- z[ix, q1:d]    #permute rows of y
    return(nrow(z) * dCov(x, y)^2)
}

library(boot)
z <- as.matrix(iris[1:50, 1:4])
boot.obj <- boot(data = z, statistic = ndCov2, R = 999,
        sim = "permutation", dims = c(2, 2))

tb <- c(boot.obj$t0, boot.obj$t)
hist(tb, nclass="scott", xlab="", main="",
        freq=FALSE)
points(boot.obj$t0, 0, cex=1, pch=16)

> mean(tb >= boot.obj$t0)
[1] 0.066
> boot.obj
DATA PERMUTATION
Call: boot(data = z, statistic = ndCov2, R = 999,
        sim = "permutation", dims = c(2, 2))
Bootstrap Statistics :
     original      bias      std. error
t1* 0.2071207 -0.05991699    0.0353751
```

The achieved significance level is 0.066 so the null hypothesis of independence is rejected at $\alpha = 0.10$. The histogram of replicates of the dCov statistic is shown in Figure 8.4. ◇

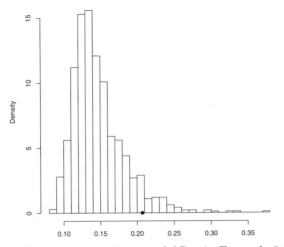

FIGURE 8.4: Permutation replicates of dCov in Example 8.13.

One of the advantages of the dCov test is that it is sensitive to all types of dependence structures in data. Procedures based on the classical definition of covariance, or measures of association based on ranks are generally less effective against non-monotone types of dependence. An alternative with non-monotone dependence is tested in the following example.

Example 8.14 (Power of dCov)

Consider the data generated by the following nonlinear model. Suppose that

$$Y_{ij} = X_{ij}\varepsilon_{ij}, \qquad i = 1,\ldots,n, \; j = 1,\ldots,5,$$

where $X \sim N_5(0, I_5)$ and $\varepsilon \sim N_5(0, \sigma^2 I_5)$ are independent. Then X and Y are dependent, but if the parameter σ is large, the dependence can be hard to detect. We compared the permutation test implementation of *dCov* with the parametric Wilks Lambda (W) likelihood ratio test [296] using Bartlett's approximation for the critical value (see e.g. [188, Sec. 5.3.2b]). Recall that Wilks Lambda tests whether the covariance $\Sigma_{12} = Cov(X, Y)$ is the zero matrix.

From a power comparison with 10,000 test decisions for each of the sample sizes we have obtained the results shown in Table 8.2 and Figure 8.5. Figure 8.5 shows a plot of power vs sample size. Table 8.2 reports the empirical power for a subset of the cases in the plot.

The dCov test is clearly more powerful in this empirical comparison. This example illustrates that the parametric Wilks Lambda test based on product-moment correlation is not always powerful against non-monotone types of dependence. The dCov test is statistically consistent with power approaching 1 as $n \to \infty$ (theoretically and empirically). ◇

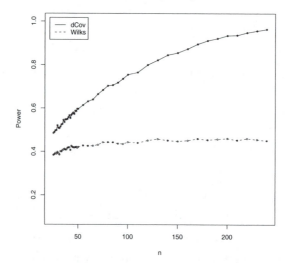

FIGURE 8.5:   Empirical power comparison of the distance covariance test dCov and Wilks Lambda W in Example 8.14.

TABLE 8.2:   Example 8.14: Percent of Significant Tests of Independence of $Y = X\varepsilon$ at $\alpha = 0.1$ ($se \leq 0.5\%$)

n	$dCov$	W	n	$dCov$	W	n	$dCov$	W
25	48.56	38.43	55	61.39	42.74	100	75.40	44.36
30	50.89	39.16	60	63.09	42.60	120	79.97	45.20
35	54.56	40.86	65	63.96	42.64	140	84.51	45.21
40	55.79	41.88	70	66.43	43.08	160	87.31	45.17
45	57.93	41.91	75	68.32	44.28	180	91.13	45.46
50	59.63	42.05	80	70.27	44.34	200	93.43	46.12

For properties of distance covariance and distance correlation, proofs of convergence and consistency, and more empirical results, see [265]. The distance correlation and covariance statistics and the corresponding permutation tests are provided in the **energy** package [226].

Exercises

8.1   Implement the two-sample Cramér-von Mises test for equal distributions as a permutation test. Apply the test to the data in Examples 8.1 and 8.2.

8.2   Implement the bivariate Spearman rank correlation test for independence [255] as a permutation test. The Spearman rank correlation test statistic can be obtained from function `cor` with `method = "spearman"`. Compare the achieved significance level of the permutation test with the p-value reported by `cor.test` on the same samples.

8.3   The Count 5 test for equal variances in Section 6.4 is based on the maximum number of extreme points. Example 6.15 shows that the Count 5 criterion is not applicable for unequal sample sizes. Implement a permutation test for equal variance based on the maximum number of extreme points that applies when sample sizes are not necessarily equal.

8.4   Complete the steps to implement a rth-nearest neighbors test for equal distributions. Write a function to compute the test statistic. The function should take the data matrix as its first argument, and an index vector as the second argument. The number of nearest neighbors r should follow the index argument.

Projects

8.A   Replicate the power comparison in Example 8.10, reducing the number of permutation tests from 10000 to 2000 and number of replicates from 499 to 199. Use the `eqdist.etest` (energy) version of the energy test.

8.B   The `aml` (boot) [34] data contains estimates of the times to remission for two groups of patients with acute myelogenous leukaemia (AML). One group received maintenance chemotherapy treament and the other group did not. See the description in the `aml` data help topic. Following Davison and Hinkley [63, Example 4.12], compute the log-rank statistic and apply a permutation test procedure to test whether the survival distributions of the two groups are equal.

Chapter 9

Markov Chain Monte Carlo Methods

9.1 Introduction

Markov Chain Monte Carlo (MCMC) methods encompass a general frame-work of methods introduced by Metropolis et al. [197] and Hastings [138] for Monte Carlo integration. Recall (see Section 5.2) that Monte Carlo integra-tion estimates the integral

$$\int_A g(t)dt$$

with a sample mean, by restating the integration problem as an expectation with respect to some density function $f(\cdot)$. The integration problem then is reduced to finding a way to generate samples from the target density $f(\cdot)$.

The MCMC approach to sampling from $f(\cdot)$ is to construct a Markov chain with stationary distribution $f(\cdot)$, and run the chain for a sufficiently long time until the chain converges (approximately) to its stationary distribution.

This chapter is a brief introduction to MCMC methods, with the goal of understanding the main ideas and how to implement some of the methods in R. In the following sections, methods of constructing the Markov chains are illustrated, such as the Metropolis and Metropolis-Hastings algorithms, and the Gibbs sampler, with applications. Methods of checking for convergence are briefly discussed. In addition to references listed in Section 5.1, see e.g. Casella and George [40], Chen, Shao, and Ibrahim [44], Chib and Greenberg [47], Gamerman [103], Gelman et al. [108], or Tierney [272]. For a thorough, accessible treatment with applications, see Gilks, Richardson, and Spiegelhal-ter [120]. For reference on Monte Carlo methods including extensive treatment of MCMC methods see Robert and Casella [228].

9.1.1 Integration problems in Bayesian inference

Many applications of Markov Chain Monte Carlo methods are problems that arise in Bayesian inference. From a Bayesian perspective, in a statistical model both the observables and the parameters are random. Given observed data $x = \{x_1, \ldots, x_n\}$, and parameters θ, x depends on the *prior* distribution

$f_\theta(\theta)$. This dependence is expressed by the likelihood $f(x_1, \ldots, x_n | \theta)$. The joint distribution of (x, θ) is therefore

$$f_{x,\theta}(x, \theta) = f_{x|\theta}(x_1, \ldots, x_n | \theta) f_\theta(\theta).$$

One can then update the distribution of θ conditional on the information in the sample $x = \{x_1, \ldots, x_n\}$, so that by Bayes Theorem the *posterior* distribution of θ is given by

$$f_{\theta|x}(\theta|x) = \frac{f_{x|\theta}(x_1, \ldots, x_n|\theta) f_\theta(\theta)}{\int f_{x|\theta}(x_1, \ldots, x_n|\theta) f_\theta(\theta) d\theta} = \frac{f_{x|\theta}(x) f_\theta(\theta)}{\int f_{x|\theta}(x) f_\theta(\theta) d\theta}.$$

Then the conditional expectation of a function $g(\theta)$ with respect to the posterior density is

$$E[g(\theta|x)] = \int g(\theta) f_{\theta|x}(\theta) \, d\theta = \frac{\int g(\theta) f_{x|\theta}(x) f_\theta(\theta) d\theta}{\int f_{x|\theta}(x) f_\theta(\theta) d\theta}. \tag{9.1}$$

To state the problem in more general terms,

$$E[g(Y)] = \frac{\int g(t) \pi(t) \, dt}{\int \pi(t) \, dt}, \tag{9.2}$$

where $\pi(\cdot)$ is (proportional to) a density or a likelihood. If $\pi(\cdot)$ is a density function, then (9.2) is just the usual definition $E[g(Y)] = \int g(t) f_Y(t) dt$. If $\pi(\cdot)$ is a likelihood, then the normalizing constant in the denominator is needed. In Bayesian analysis, $\pi(\cdot)$ is a posterior density. The expectation (9.2) can be evaluated even if $\pi(\cdot)$ is known only up to a constant. This simplifies the problem because in practice the normalizing constant for a posterior density $f_{\theta|x}(\theta)$ is often difficult to evaluate.

The practical problem, however, is that the integrations in (9.2) are often mathematically intractable, and difficult to compute by numerical methods, especially in higher dimensions. Markov Chain Monte Carlo provides a method for this type of integration problem.

9.1.2 Markov Chain Monte Carlo Integration

The Monte Carlo estimate of $E[g(\theta)] = \int g(\theta) f_{\theta|x}(\theta) d\theta$ is the sample mean

$$\bar{g} = \frac{1}{m} \sum_{i=1}^{m} g(x_i),$$

where $x_1, \ldots, x_m$ is a sample from the distribution with density $f_{\theta|x}$. If $x_1, \ldots, x_m$ are independent (it is a random sample) then by the laws of large numbers, the sample mean $\bar{g}$ converges in probability to $E[g(\theta)]$ as sample size n tends to infinity. In this case, one can in principle draw as large a Monte

Carlo sample as required to obtain the desired precision in the estimate $\overline{g}$. Here the first "MC" in "MCMC" is not needed; Monte Carlo integration can be used.

However, in a problem such as (9.1) it may be quite difficult to implement a method for generating independent observations from the density $f_{\theta|x}$. Nevertheless, even if the sample observations are dependent, a Monte Carlo integration can be applied if the observations can be generated so that their joint density is roughly the same as the joint density of a random sample. This is where the first "MC" comes to the rescue. Markov Chain Monte Carlo methods estimate the integral in (9.1) or (9.2) by *Monte Carlo* integration, and the *Markov Chain* provides the sampler that generates the random observations from the target distribution.

By a generalization of the strong law of large numbers, if $\{X_0, X_1, X_2, \dots\}$ is a realization of an irreducible, ergodic Markov Chain with stationary distribution π, then

$$\overline{g(X)}_m = \frac{1}{m} \sum_{t=0}^{m} g(X_t)$$

converges with probability one to $E[g(X)]$ as $m \to \infty$, where X has the stationary distribution π and the expectation is taken with respect to π (provided the expectation exists).

For a brief review of discrete-time discrete-state-space Markov Chains see Section 2.8. For an introduction to Markov chains and stochastic processes see Ross [234].

9.2 The Metropolis-Hastings Algorithm

The *Metropolis-Hastings algorithms* are a class of Markov Chain Monte Carlo methods including the special cases of the Metropolis sampler, the Gibbs sampler, the independence sampler, and the random walk. The main idea is to generate a Markov Chain $\{X_t | t = 0, 1, 2, \dots\}$ such that its stationary distribution is the target distribution. The algorithm must specify, for a given state X_t, how to generate the next state X_{t+1}. In all of the Metropolis-Hastings (M-H) sampling algorithms, there is a candidate point Y generated from a proposal distribution $g(\cdot|X_t)$. If this candidate point is accepted, the chain moves to state Y at time $t+1$ and $X_{t+1} = Y$; otherwise the chain stays in state X_t and $X_{t+1} = X_t$. Note that the proposal distribution can depend on the previous state X_t. For example, if the proposal distribution is normal, one choice for $g(\cdot|X_t)$ might be Normal$(\mu_t = X_t, \sigma^2)$ for some fixed σ^2.

The choice of proposal distribution is very flexible, but the chain generated by this choice must satisfy certain regularity conditions. The proposal distribution must be chosen so that the generated chain will converge to a stationary distribution – the target distribution f. Required conditions for the generated chain are irreducibility, positive recurrence, and aperiodicity (see [229]). A proposal distribution with the same support set as the target distribution will usually satisfy these regularity conditions. Refer to [121, Ch. 7-8], [228, Ch. 7] or [229] for further details on the choice of proposal distribution.

9.2.1 Metropolis-Hastings Sampler

The *Metropolis-Hastings sampler* generates the Markov chain $\{X_0, X_1, \dots\}$ as follows.

1. Choose a proposal distribution $g(\cdot|X_t)$ (subject to regularity conditions stated above).

2. Generate X_0 from a distribution g.

3. Repeat (until the chain has converged to a stationary distribution according to some criterion):

   (a) Generate Y from $g(\cdot|X_t)$.
   (b) Generate U from Uniform(0,1).
   (c) If
   $$U \leq \frac{f(Y)g(X_t|Y)}{f(X_t)g(Y|X_t)}$$
   accept Y and set $X_{t+1} = Y$; otherwise set $X_{t+1} = X_t$.
   (d) Increment t.

Observe that in step (3c) the candidate point Y is accepted with probability

$$\alpha(X_t, Y) = \min\left(1, \frac{f(Y)g(X_t|Y)}{f(X_t)g(Y|X_t)}\right), \tag{9.3}$$

so that it is only necessary to know the density of the target distribution f up to a constant.

Assuming that the proposal distribution satisfies the regularity conditions, the Metropolis-Hastings chain will converge to a unique stationary distribution π. The algorithm is designed so that the stationary distribution of the Metropolis-Hastings chain is indeed the target distribution, f.

Suppose (r, s) are two elements of the state space of the chain, and without loss of generality suppose that $f(s)g(r|s) \geq f(r)g(s|r)$. Thus, $\alpha(r, s) = 1$ and the joint density of (X_t, X_{t+1}) at (r, s) is $f(r)g(s|r)$. The joint density of (X_t, X_{t+1}) at (s, r) is

$$f(s)g(r|s)\,\alpha(s, r) = f(s)g(r|s)\left(\frac{f(r)g(s|r)}{f(s)g(r|s)}\right) = f(r)g(s|r).$$

The transition kernel is

$$K(r, s) = \alpha(r, s)g(s|r) + I(s = r)\left[1 - \int_\alpha (r, s)g(s|r)ds\right].$$

(The second term in $K(r, s)$ arises when the candidate point is rejected and $X_{t+1} = X_t$.) Hence we have the system of equations

$$\alpha(r, s)f(r)g(s|r) = \alpha(s, r)f(s)g(r|s),$$

$$I(s = r)\left[1 - \int_{\alpha}(r, s)g(s|r)ds\right]f(r) = I(r = s)\left[1 - \int_{\alpha}(s, r)g(r|s)ds\right]f(s)$$

for the Metropolis-Hastings chain, and f satisfies the detailed balance condition $K(s, r)f(s) = K(r, s)f(r)$. Therefore f is the stationary distribution of the chain. See Theorems 6.46 and 7.2 in [228].

Example 9.1 (Metropolis-Hastings sampler)

Use the Metropolis-Hastings sampler to generate a sample from a Rayleigh distribution. The Rayleigh density [156, (18.76)] is

$$f(x) = \frac{x}{\sigma^2}e^{-x^2/(2\sigma^2)}, \qquad x \geq 0, \sigma > 0.$$

The Rayleigh distribution is used to model lifetimes subject to rapid aging, because the hazard rate is linearly increasing. The mode of the distribution is at σ, $E[X] = \sigma\sqrt{\pi/2}$ and $Var(X) = \sigma^2(4 - \pi)/2$.

For the proposal distribution, try the chisquared distribution with degrees of freedom X_t. Implementation of a Metropolis-Hastings sampler for this example is as follows. Note that the base of the array in R is 1, so we initialize the chain at X_0 in x[1].

1. Set $g(\cdot|X)$ to the density of $\chi^2(X)$.

2. Generate X_0 from distribution $\chi^2(1)$ and store in x[1].

3. Repeat for $i = 2, \ldots, N$:

   (a) Generate Y from $\chi^2(df = X_t) = \chi^2(df=x[i-1])$.
   (b) Generate U from Uniform(0, 1).
   (c) With $X_t = x[i-1]$, compute

   $$r(X_t, Y) = \frac{f(Y)g(X_t|Y)}{f(X_t)g(Y|X_t)},$$

   where f is the Rayleigh density with parameter σ, $g(Y|X_t)$ is the $\chi^2(df = X_t)$ density evaluated at Y, and $g(X_t|Y)$ is the $\chi^2(df = Y)$ density evaluated at X_t.

   If $U \leq r(X_t, Y)$ accept Y and set $X_{t+1} = Y$; otherwise set $X_{t+1} = X_t$. Store X_{t+1} in x[i].

   (d) Increment t.

The constants in the densities cancel, so

$$r(x_t, y) = \frac{f(y)g(x_t|y)}{f(x_t)g(y|x_t)} = \frac{ye^{-y^2/2\sigma^2}}{x_t e^{-x_t^2/2\sigma^2}} \times \frac{\Gamma(\frac{x_t}{2})2^{x_t/2}x_t^{y/2-1}e^{-x_t/2}}{\Gamma(\frac{y}{2})2^{y/2}y^{x_t/2-1}e^{-y/2}}.$$

This ratio can be simplified further, but in the following simulation for clarity we will evaluate the Rayleigh and chisquare densities separately. The following function evaluates the Rayleigh(σ) density.

```
f <- function(x, sigma) {
    if (any(x < 0)) return (0)
    stopifnot(sigma > 0)
    return((x / sigma^2) * exp(-x^2 / (2*sigma^2)))
}
```

In the simulation below, a Rayleigh($\sigma = 4$) sample is generated using the chisquare proposal distribution. At each transition, the candidate point Y is generated from $\chi^2(\nu = X_{i-1})$

```
xt <- x[i-1]
y <- rchisq(1, df = xt)
```

and for each y, the numerator and denominator of $r(X_{i-1}, Y)$ are computed in num and den. The counter k records the number of rejected candidate points.

```
m <- 10000
sigma <- 4
x <- numeric(m)
x[1] <- rchisq(1, df=1)
k <- 0
u <- runif(m)

for (i in 2:m) {
    xt <- x[i-1]
    y <- rchisq(1, df = xt)
    num <- f(y, sigma) * dchisq(xt, df = y)
    den <- f(xt, sigma) * dchisq(y, df = xt)
    if (u[i] <= num/den) x[i] <- y else {
        x[i] <- xt
        k <- k+1      #y is rejected
        }
    }

> print(k)
[1] 4009
```

In this example, approximately 40% of the candidate points are rejected, so the chain is somewhat inefficient.

To see the generated sample as a realization of a stochastic process, we can plot the sample vs the time index. The following code will display a partial plot starting at time index 5000.

```
index <- 5000:5500
y1 <- x[index]
plot(index, y1, type="l", main="", ylab="x")
```

The plot is shown in Figure 9.1. Note that at times the candidate point is rejected and the chain does not move at these time points; this corresponds to the short horizontal paths in the graph.   ◇

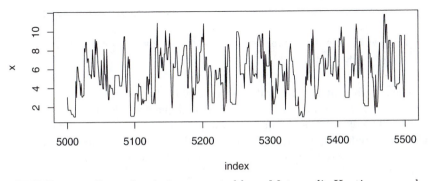

FIGURE 9.1: Part of a chain generated by a Metropolis-Hastings sampler of a Rayleigh distribution in Example 9.1.

Example 9.1 is a simple example intended to illustrate how to implement a Metropolis-Hastings sampler. There are better ways to generate samples from Rayleigh distributions. In fact, an explicit formula for the quantiles of the Rayleigh distribution are given by

$$x_q = F^{-1}(q) = \sigma\{-2\log(1-q)\}^{1/2}, \qquad 0 < q < 1. \tag{9.4}$$

Using F^{-1} one could write a simple generator for Rayleigh using the inverse transform method of Section 3.2.1 with antithetic sampling (Section 5.4).

Example 9.2 (Example 9.1, cont.)

The following code compares the quantiles of the target Rayleigh($\sigma = 4$) distribution with the quantiles of the generated chain in a quantile-quantile plot (QQ plot).

```
b <- 2001        #discard the burnin sample
y <- x[b:m]
a <- ppoints(100)
QR <- sigma * sqrt(-2 * log(1 - a))   #quantiles of Rayleigh
Q <- quantile(x, a)

qqplot(QR, Q, main="",
    xlab="Rayleigh Quantiles", ylab="Sample Quantiles")

hist(y, breaks="scott", main="", xlab="", freq=FALSE)
lines(QR, f(QR, 4))
```

The histogram of the generated sample with the Rayleigh($\sigma = 4$) density superimposed is shown in Figure 9.2(a) and the QQ plot is shown in Figure 9.2(b). The QQ plot is an informal approach to assessing the goodness-of-fit of the generated sample with the target distribution. From the plot, it appears that the sample quantiles are in approximate agreement with the theoretical quantiles.   ◇

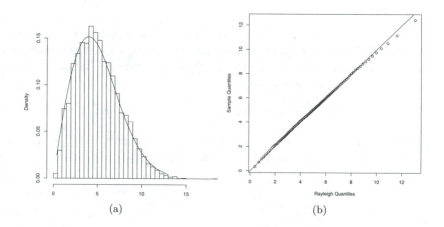

(a)   (b)

FIGURE 9.2: Histogram with target Rayleigh density and QQ plot for a Metropolis-Hastings chain in Example 9.1.

9.2.2 The Metropolis Sampler

The Metropolis-Hastings sampler [138, 197] is a generalization of the *Metropolis sampler* [197]. In the Metropolis algorithm, the proposal distribution is symmetric. That is, the proposal distribution $g(\cdot|X_t)$ satisfies

$$g(X|Y) = g(Y|X),$$

so that in (9.3) the proposal distribution g cancels from

$$r(X_t, Y) = \frac{f(Y)g(X_t|Y)}{f(X_t)g(Y|X_t)},$$

and the candidate point Y is accepted with probability

$$\alpha(X_t, Y) = \min\left(1, \frac{f(Y)}{f(X_t)}\right).$$

9.2.3 Random Walk Metropolis

The *random walk Metropolis* sampler is an example of a Metropolis sampler. Suppose the candidate point Y is generated from a symmetric proposal distribution $g(Y|X_t) = g(|X_t - Y|)$. Then at each iteration, a random increment Z is generated from $g(\cdot)$, and Y is defined by $Y = X_t + Z$. For example, the random increment might be normal with zero mean, so that the candidate point is $Y|X_t \sim Normal(X_t, \sigma^2)$ for some fixed $\sigma^2 > 0$.

Convergence of the random walk Metropolis is often sensitive to the choice of scale parameter. When variance of the increment is too large, most of the candidate points are rejected and the algorithm is very inefficient. If the variance of the increment is too small, the candidate points are almost all accepted, so the random walk Metropolis generates a chain that is almost like a true random walk, which is also inefficient. One approach to selecting the scale parameter is to monitor the acceptance rates, which should be in the range [0.15, 0.5] [230].

Example 9.3 (Random walk Metropolis)

Implement the random walk version of the Metropolis sampler to generate the target distribution Student t with ν degrees of freedom, using the proposal distribution $Normal(X_t, \sigma^2)$. In order to see the effect of different choices of variance of the proposal distribution, try repeating the simulation with different choices of σ.

The $t(\nu)$ density is proportional to $(1 + x^2/\nu)^{-(\nu+1)/2}$, so

$$r(x_t, y) = \frac{f(Y)}{f(X_t)} = \frac{\left(1 + \frac{y^2}{\nu}\right)^{-(\nu+1)/2}}{\left(1 + \frac{x_t^2}{\nu}\right)^{-(\nu+1)/2}}.$$

In this simulation below, the t densities in $r(x_{i-1}, y)$ will be computed by the dt function. Then y is accepted or rejected and X_i generated by

```
if (u[i] <= dt(y, n) / dt(x[i-1], n))
    x[i] <- y
else
    x[i] <- x[i-1]
```

These steps are combined into a function to generate the chain, given the parameters n and σ, initial value X_0, and the length of the chain, N.

```
rw.Metropolis <- function(n, sigma, x0, N) {
    x <- numeric(N)
    x[1] <- x0
    u <- runif(N)
    k <- 0
    for (i in 2:N) {
        y <- rnorm(1, x[i-1], sigma)
            if (u[i] <= (dt(y, n) / dt(x[i-1], n)))
            x[i] <- y  else {
                x[i] <- x[i-1]
                k <- k + 1
            }
    }
    return(list(x=x, k=k))
    }
```

Four chains are generated for different variances σ^2 of the proposal distribution.

```
n <- 4  #degrees of freedom for target Student t dist.
N <- 2000
sigma <- c(.05, .5, 2,   16)

x0 <- 25
rw1 <- rw.Metropolis(n, sigma[1], x0, N)
rw2 <- rw.Metropolis(n, sigma[2], x0, N)
rw3 <- rw.Metropolis(n, sigma[3], x0, N)
rw4 <- rw.Metropolis(n, sigma[4], x0, N)

#number of candidate points rejected
> print(c(rw1$k, rw2$k, rw3$k, rw4$k))
[1]    14   136   891 1798
```

Only the third chain has a rejection rate in the range $[0.15, 0.5]$. The plots in Figure 9.3 show that the random walk Metropolis sampler is very sensitive to the variance of the proposal distribution. Recall that the variance of the

$t(\nu)$ distribution is $\nu/(\nu - 2)$, $\nu > 2$. Here $\nu = 4$ and the standard deviation of the target distribution is $\sqrt{2}$.

In the first plot of Figure 9.3 with $\sigma = 0.05$, the ratios $r(X_t, Y)$ tend to be large and almost every candidate point is accepted. The increments are small and the chain is almost like a true random walk. Chain 1 has not converged to the target in 2000 iterations. The chain in the second plot generated with $\sigma = 0.5$ is converging very slowly and requires a much longer burn-in period. In the third plot ($\sigma = 2$) the chain is mixing well and converging to the target distribution after a short burn-in period of about 500. Finally, in the fourth plot, where $\sigma = 16$, the ratios $r(X_t, Y)$ are smaller and most of the candidate points are rejected. The fourth chain converges, but it is inefficient.

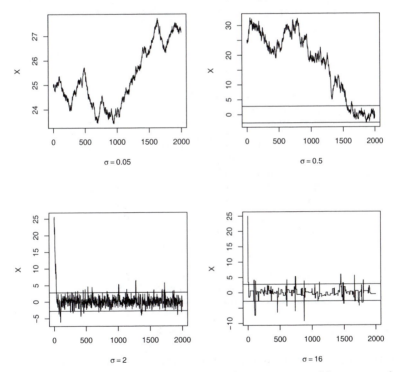

FIGURE 9.3: Random walk Metropolis chains generated by proposal distributions with different variances in Example 9.3.

◇

Example 9.4 (Example 9.3, cont.)

Usually in MCMC problems one does not have the theoretical quantiles of the target distribution available for comparison, but in this case the output of the random walk Metropolis chains in Example 9.3 can be compared with the theoretical quantiles of the target distribution. Discard the burn-in values in the first 500 rows of each chain. The quantiles are computed by the `apply` function (applying `quantile` to the columns of the matrix). The quantiles of the target distribution and the sample quantiles of the four chains `rw1`, `rw2`, `rw3`, and `rw4` are in Table 9.1.

```
a <- c(.05, seq(.1, .9, .1), .95)
Q <- qt(a, n)
rw <- cbind(rw1$x, rw2$x, rw3$x, rw4$x)
mc <- rw[501:N, ]
Qrw <- apply(mc, 2, function(x) quantile(x, a))
print(round(cbind(Q, Qrw), 3))            #not shown
xtable::xtable(round(cbind(Q, Qrw), 3)) #latex format
```

◇

TABLE 9.1:   Quantiles of Target
Distribution and Chains in Example 9.4

	Q	rw1	rw2	rw3	rw4
5%	−2.13	23.66	−1.16	−1.92	−2.40
10%	−1.53	23.77	−0.39	−1.47	−1.35
20%	−0.94	23.99	0.67	−1.01	−0.90
30%	−0.57	24.29	4.15	−0.63	−0.64
40%	−0.27	24.68	9.81	−0.25	−0.47
50%	0.00	25.29	17.12	0.01	−0.15
60%	0.27	26.14	18.75	0.27	0.06
70%	0.57	26.52	21.79	0.59	0.25
80%	0.94	26.93	25.42	0.92	0.52
90%	1.53	27.27	28.51	1.55	1.18
95%	2.13	27.39	29.78	2.37	1.90

R note 9.1 *Table 9.1 was exported to LATEXformat by the* xtable *function in the* xtable *package [61].*

Example 9.5 (Bayesian inference: A simple investment model)

In general, the returns on different investments are not independent. To reduce risk, portfolios are sometimes selected so that returns of securities are negatively correlated. Rather than the correlation of returns, here the daily performance is ranked. Suppose five stocks are tracked for 250 trading days (one year), and each day the "winner" is picked based on maximum return relative to the market. Let X_i be the number of days that security i is a winner. Then the observed vector of frequencies $(x_1, \ldots, x_5)$ is an observation from the joint distribution of $(X_1, \ldots, X_5)$. Based on historical data, suppose that the prior odds of an individual security being a winner on any given day are $[1 : (1-\beta) : (1-2\beta) : 2\beta : \beta]$, where $\beta \in (0, 0.5)$ is an unknown parameter. Update the estimate of β for the current year of winners.

According to this model, the multinomial joint distribution of $X_1, \ldots, X_5$ has the probability vector

$$p = \left(\frac{1}{3}, \frac{(1-\beta)}{3}, \frac{(1-2\beta)}{3}, \frac{2\beta}{3}, \frac{\beta}{3} \right).$$

The posterior distribution of β given $(x_1, \ldots, x_5)$ is therefore

$$Pr[\beta | (x_1, \ldots, x_5)] = \frac{250!}{x_1! x_2! x_3! x_4! x_5!} p_1^{x_1} p_2^{x_2} p_3^{x_3} p_4^{x_4} p_5^{x_5}.$$

In this example, we cannot directly simulate random variates from the posterior distribution. One approach to estimating β is to generate a chain that converges to the posterior distribution and estimate β from the generated chain. Use the random walk Metropolis sampler with a uniform proposal distribution to generate the posterior distribution of β. The candidate point Y is accepted with probability

$$\alpha(X_t, Y) = \min \left(1, \frac{f(Y)}{f(X_t)} \right).$$

The multinomial coefficient cancels from the ratio in $\alpha(X, Y)$, so that

$$\frac{f(Y)}{f(X)} = \frac{(1/3)^{x_1}((1-Y)/3)^{x_2}((1-2Y)/3)^{x_3}((2Y)/3)^{x_4}(Y/3)^{x_5}}{(1/3)^{x_1}((1-X)/3)^{x_2}((1-2X)/3)^{x_3}((2X)/3)^{x_4}(X/3)^{x_5}}.$$

The ratio can be further simplified, but the numerator and denominator are evaluated separately in the implementation below. In order to check the results, start by generating the observed frequencies from a distribution with specified β.

```
b <- .2          #actual value of beta
w <- .25         #width of the uniform support set
m <- 5000        #length of the chain
burn <- 1000     #burn-in time
days <- 250
x <- numeric(m)  #the chain
```

```
# generate the observed frequencies of winners
i <- sample(1:5, size=days, replace=TRUE,
        prob=c(1, 1-b, 1-2*b, 2*b, b))
win <- tabulate(i)
> print(win)
[1] 82 72 45 34 17
```

The tabulated frequencies in `win` are the simulated numbers of trading days that each of the stocks were the daily winner. Based on this year's observed distribution of winners, we want to estimate the parameter β.

The following function `prob` computes the target density (without the constant).

```
prob <- function(y, win) {
    # computes (without the constant) the target density
    if (y < 0 || y >= 0.5)
        return (0)
    return((1/3)^win[1] *
        ((1-y)/3)^win[2] * ((1-2*y)/3)^win[3] *
            ((2*y)/3)^win[4] * (y/3)^win[5])
}
```

Finally the random walk Metropolis chain is generated. Two sets of uniform random variates are required; one for generating the proposal distribution and another for the decision to accept or reject the candidate point.

```
u <- runif(m)           #for accept/reject step
v <- runif(m, -w, w)    #proposal distribution
x[1] <- .25
for (i in 2:m) {
    y <- x[i-1] + v[i]
    if (u[i] <= prob(y, win) / prob(x[i-1], win))
        x[i] <- y  else
            x[i] <- x[i-1]
}
```

The plot of the chains in Figure 9.4(a) shows that the chain has converged, approximately, to the target distribution. Now the generated chain provides an estimate of β, after discarding a burn-in sample. From the histogram of the sample in Figure 9.4(b) the plausible values for β are close to 0.2.

The original sample table of relative frequencies, and the MCMC estimates of the multinomial probabilities are given below.

```
> print(win)
[1] 82 72 45 34 17
> print(round(win/days, 3))
[1] 0.328 0.288 0.180 0.136 0.068
```

```
> print(round(c(1, 1-b, 1-2*b, 2*b, b)/3, 3))
[1] 0.333 0.267 0.200 0.133 0.067
> xb <- x[(burn+1):m]
> print(mean(xb))
[1] 0.2101277
```

The sample mean of the generated chain is 0.2101277 (the simulated year of winners table was generated with $\beta = 0.2$).   ◇

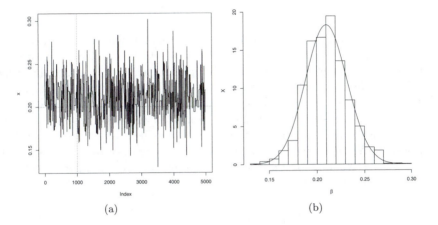

(a)   (b)

FIGURE 9.4: Random walk Metropolis chain for β in Example 9.5.

9.2.4   The Independence Sampler

Another special case of the Metropolis-Hastings sampler is the independence sampler [272]. The proposal distribution in the independence sampling algorithm does not depend on the previous value of the chain. Thus, $g(Y|X_t) = g(Y)$ and the acceptance probability (9.3) is

$$\alpha(X_t, Y) = \min\left(1, \ \frac{f(Y)g(X_t)}{f(X_t)g(Y)}\right).$$

The independence sampler is easy to implement and tends to work well when the proposal density is a close match to the target density, but otherwise does not perform well. Roberts [229] discusses convergence of the independence sampler, and comments that "it is rare for the independence sampler to be useful as a stand-alone algorithm." Nevertheless, we illustrate the procedure in the following example, because the independence sampler can be useful in hybrid MCMC methods (see e.g. [119]).

Example 9.6 (Independence sampler)

Assume that a random sample $(z_1, \ldots, z_n)$ from a two-component normal mixture is observed. The mixture is denoted by

$$pN(\mu_1, \sigma_1^2) + (1-p)N(\mu_2, \sigma_2^2),$$

and the density of the mixture (see Chapter 3) is

$$f^*(z) = pf_1(z) + (1-p)f_2(z),$$

where f_1 and f_2 are the densities of the two normal distributions, respectively. If the densities f_1 and f_2 are completely specified, the problem is to estimate the mixing parameter p given the observed sample. Generate a chain using an independence sampler that has the posterior distribution of p as the target distribution.

The proposal distribution should be supported on the set of valid probabilities p; that is, the interval $(0, 1)$. The most obvious choices are the beta distributions. With no prior information on p, one might consider the Beta(1,1) proposal distribution (Beta(1,1) is Uniform(0,1)). The candidate point Y is accepted with probability

$$\alpha(X_t, Y) = \min\left(1, \ \frac{f(Y)g(X_t)}{f(X_t)g(Y)}\right),$$

where $g(\cdot)$ is the Beta proposal density. Thus, if the proposal distribution is Beta(a, b), then $g(y) \propto y^{a-1}(1-y)^{b-1}$ and Y is accepted with probability $\min(1, f(y)g(x_t)/f(x_t)g(y))$, where

$$\frac{f(y)g(x_t)}{f(x_t)g(y)} = \frac{x_t^{a-1}(1-x_t)^{b-1}\prod_{j=1}^{n}[yf_1(z_j) + (1-y)f_2(z_j)]}{y^{a-1}(1-y)^{b-1}\prod_{j=1}^{n}[x_t f_1(z_j) + (1-x_t)f_2(z_j)]}.$$

In the following simulation the proposal distribution is Uniform(0,1). The simulated data is generated from the normal mixture

$$0.2N(0, 1) + 0.8N(5, 1).$$

The first steps are to initialize constants and generate the observed sample. Then an observed sample is generated. To generate the chain, all random numbers can be generated in advance because the candidate Y does not depend on X_t.

```
m <- 5000 #length of chain
xt <- numeric(m)
a <- 1                  #parameter of Beta(a,b) proposal dist.
b <- 1                  #parameter of Beta(a,b) proposal dist.
p <- .2                 #mixing parameter
n <- 30                 #sample size
mu <- c(0, 5)           #parameters of the normal densities
sigma <- c(1, 1)

# generate the observed sample
i <- sample(1:2, size=n, replace=TRUE, prob=c(p, 1-p))
x <- rnorm(n, mu[i], sigma[i])

# generate the independence sampler chain
u <- runif(m)
y <- rbeta(m, a, b)        #proposal distribution
xt[1] <- .5

for (i in 2:m) {
    fy <- y[i] * dnorm(x, mu[1], sigma[1]) +
            (1-y[i]) * dnorm(x, mu[2], sigma[2])
    fx <- xt[i-1] * dnorm(x, mu[1], sigma[1]) +
            (1-xt[i-1]) * dnorm(x, mu[2], sigma[2])

    r <- prod(fy / fx) *
            (xt[i-1]^(a-1) * (1-xt[i-1])^(b-1)) /
            (y[i]^(a-1) * (1-y[i])^(b-1))

    if (u[i] <= r) xt[i] <- y[i] else
        xt[i] <- xt[i-1]
    }

plot(xt, type="l", ylab="p")
hist(xt[101:m], main="", xlab="p", prob=TRUE)
print(mean(xt[101:m]))
```

The histogram of the generated sample after discarding the first 100 points is shown in Figure 9.5 on the next page. The mean of the remaining sample is 0.2516. The time plot of the generated chain is shown in Figure 9.6(a), which mixes well and converges quickly to a stationary distribution.

For comparison, we repeated the simulation with a Beta(5,2) proposal distribution. In this simulation the sample mean of the chain after discarding the burn-in sample is 0.2593, but the chain that is generated, shown in Figure 9.6(b) on the following page, is not very efficient.   ◇

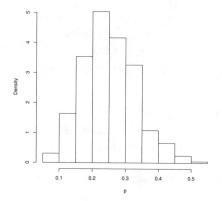

FIGURE 9.5:   Distribution of the independence sampler chain for p with proposal distribution Beta(1, 1) in Example 9.6, after discarding a burn-in sample of length 100.

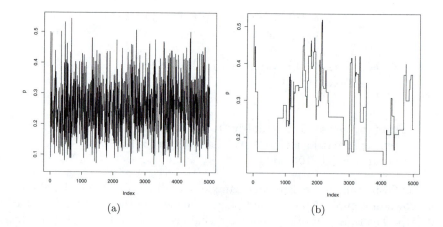

FIGURE 9.6:   Chain generated by independence sampler for p with proposal distribution Beta(1, 1) (left) and Beta(5, 2) (right) in Example 9.6.

9.3 The Gibbs Sampler

The Gibbs sampler was named by Geman and Geman [111], because of its application to analysis of Gibbs lattice distributions. However, it is a general method that can be applied to a much wider class of distributions [111, 106, 105]. It is another special case of the Metropolis-Hastings sampler. See the introduction to Gibbs sampling by Casella and George [40].

The Gibbs sampler is often applied when the target is a multivariate distribution. Suppose that all the univariate conditional densities are fully specified and it is reasonably easy to sample from them. The chain is generated by sampling from the marginal distributions of the target distribution, and every candidate point is therefore accepted.

Let $X = (X_1, \ldots, X_d)$ be a random vector in $\mathbb{R}^d$. Define the $d - 1$ dimensional random vectors

$$X_{(-j)} = (X_1, \ldots, X_{j-1}, X_{j+1}, \ldots, X_d),$$

and denote the corresponding univariate conditional density of X_j given $X_{(-j)}$ by $f(X_j | X_{(-j)})$. The Gibbs sampler generates the chain by sampling from each of the d conditional densities $f(X_j | X_{(-j)})$.

In the following algorithm for the Gibbs sampler, we denote X_t by $X(t)$.

1. Initialize $X(0)$ at time $t = 0$.

2. For each iteration, indexed $t = 1, 2, \ldots$ repeat:

   (a) Set $x_1 = X_1(t - 1)$.
   (b) For each coordinate $j = 1, \ldots, d$
       (a) Generate $X_j^*(t)$ from $f(X_j | x_{(-j)})$.
       (b) Update $x_j = X_j^*(t)$.
   (c) Set $X(t) = (X_1^*(t), \ldots, X_d^*(t))$ (every candidate is accepted).
   (d) Increment t.

Example 9.7 (Gibbs sampler: Bivariate distribution)

Generate a bivariate normal distribution with mean vector (μ_1, μ_2), variances σ_1^2, σ_2^2, and correlation ρ, using Gibbs sampling.

In the bivariate case, $X = (X_1, X_2)$, $X_{(-1)} = X_2$, $X_{(-2)} = X_1$. The conditional densities of a bivariate normal distribution are univariate normal with parameters

$$E[X_2 | x_1] = \mu_1 + \rho \frac{\sigma_2}{\sigma_1}(x_1 - \mu_1),$$
$$\text{Var}(X_2 | x_1) = (1 - \rho^2)\sigma_2^2,$$

and the chain is generated by sampling from

$$f(x_1|x_2) \sim \text{Normal}(\mu_1 + \frac{\rho\sigma_1}{\sigma_2}(x_2 - \mu_2), (1 - \rho^2)\sigma_1^2),$$

$$f(x_2|x_1) \sim \text{Normal}(\mu_2 + \frac{\rho\sigma_2}{\sigma_1}(x_1 - \mu_1), (1 - \rho^2)\sigma_2^2).$$

For a bivariate distribution (X_1, X_2), at each iteration the Gibbs sampler

1. Sets $(x_1, x_2) = X(t - 1)$;

2. Generates $X_1^*(t)$ from $f(X_1|x_2)$;

3. Updates $x_1 = X_1^*(t)$;

4. Generates $X_2^*(t)$ from $f(X_2|x_1)$;

5. Sets $X(t) = (X_1^*(t), X_2^*(t))$.

```
#initialize constants and parameters
N <- 5000                    #length of chain
burn <- 1000                 #burn-in length
X <- matrix(0, N, 2)         #the chain, a bivariate sample

rho <- -.75                  #correlation
mu1 <- 0
mu2 <- 2
sigma1 <- 1
sigma2 <- .5
s1 <- sqrt(1-rho^2)*sigma1
s2 <- sqrt(1-rho^2)*sigma2

###### generate the chain #####

X[1, ] <- c(mu1, mu2)                    #initialize

for (i in 2:N) {
    x2 <- X[i-1, 2]
    m1 <- mu1 + rho * (x2 - mu2) * sigma1/sigma2
    X[i, 1] <- rnorm(1, m1, s1)
    x1 <- X[i, 1]
    m2 <- mu2 + rho * (x1 - mu1) * sigma2/sigma1
    X[i, 2] <- rnorm(1, m2, s2)
}

b <- burn + 1
x <- X[b:N, ]
```

The first 1000 observations are discarded from the chain in matrix X and the remaining observations are in x. Summary statistics for the column means, the sample covariance, and correlation matrices are shown below.

```
# compare sample statistics to parameters
> colMeans(x)
[1] -0.03030001  2.01176134
> cov(x)
            [,1]         [,2]
[1,]   1.0022207 -0.3757518
[2,]  -0.3757518  0.2482327
> cor(x)
             [,1]          [,2]
[1,]   1.0000000 -0.7533379
[2,]  -0.7533379  1.0000000

plot(x, main="", cex=.5, xlab=bquote(X[1]),
     ylab=bquote(X[2]), ylim=range(x[,2]))
```

The sample means, variances, and correlation are close to the true parameters, and the plot in Figure 9.7 exhibits the elliptical symmetry of the bivariate normal, with negative correlation. (The version printed is a randomly selected subset of 1000 generated variates after discarding the burn-in sample.)   ◇

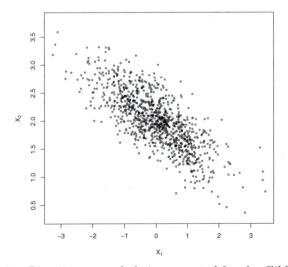

FIGURE 9.7: Bivariate normal chain generated by the Gibbs sampler in Example 9.7.

9.4   Monitoring Convergence

In several examples using various Metropolis-Hastings algorithms, we have seen that some generated chains have not converged to the target distribution. In general, for an arbitrary Metropolis-Hastings sampler the number of iterations that are sufficient for approximate convergence to the target distribution or what length burn-in sample is required are unknown. Moreover, Gelman and Rubin [110] provide examples of slow convergence that cannot be detected by examining a single chain. A single chain may appear to have converged because the generated values have a small variance within a local part of the support set of the target distribution, but in reality the chain has not explored all of the support set. By examining several parallel chains, slow convergence should be more evident, particularly if the initial values of the chain are overdispersed with respect to the target distribution. Methods have been proposed in the literature for monitoring the convergence of MCMC chains (see e.g. [33, 54, 116, 138, 227, 219]). In this section we discuss and illustrate the approach suggested by Gelman and Rubin [107, 109] for monitoring convergence of Metropolis-Hastings chains.

9.4.1   The Gelman-Rubin Method

The Gelman-Rubin [107, 109] method of monitoring convergence of a M-H chain is based on comparing the behavior of several generated chains with respect to the variance of one or more scalar summary statistics. The estimates of the variance of the statistic are analogous to estimates based on between-sample and within-sample mean squared errors in a one-way analysis of variance (ANOVA).

Let ψ be a scalar summary statistic that estimates some parameter of the target distribution. Generate k chains $\{X_{ij} : 1 \leq i \leq k, \ 1 \leq j \leq n\}$ of length n. (Here the chains are indexed with initial time $t = 1$.) Compute $\{\psi_{in} = \psi(X_{i1}, \ldots, X_{in})\}$ for each chain at time n. We expect that if the chains are converging to the target distribution as $n \to \infty$, then the sampling distribution of the statistics $\{\psi_{in}\}$ should be converging to a common distribution.

The Gelman-Rubin method uses the between-sequence variance of ψ and the within-sequence variance of ψ to estimate an upper bound and a lower bound for variance of ψ, converging to variance ψ from above and below, respectively, as the chain converges to the target distribution.

Consider the chains up to time n to represent data from a balanced one-way ANOVA on k groups with n observations. Compute the estimates of between-sample and within-sample variance analogous to the sum of squares for treatments and the sum of squares for error, and the corresponding mean squared errors as in ANOVA.

The between-sequence variance is

$$B = \frac{1}{k-1} \sum_{i=1}^{k} \sum_{j=1}^{n} (\overline{\psi}_{i\cdot} - \overline{\psi}_{\cdot\cdot})^2 = \frac{n}{k-1} \sum_{i=1}^{k} (\overline{\psi}_{i\cdot} - \overline{\psi}_{\cdot\cdot})^2,$$

where

$$\overline{\psi}_{i\cdot} = (1/n) \sum_{j=1}^{n} \psi_{ij}, \qquad \overline{\psi}_{\cdot\cdot} = (1/(nk)) \sum_{i=1}^{k} \sum_{j=1}^{n} \psi_{ij}.$$

Within the ith sequence, the sample variance is

$$s_i^2 = \frac{1}{n} \sum_{j=1}^{n} (\psi_{ij} - \overline{\psi}_{i\cdot})^2,$$

and the pooled estimate of within sample variance is

$$W = \frac{1}{nk-k} \sum_{i=1}^{k} (n-1)s_i^2 = \frac{1}{k} \sum_{i=1}^{k} s_i^2.$$

The between-sequence and within-sequence estimates of variance are combined to estimate an upper bound for $Var(\psi)$

$$\widehat{Var}(\psi) = \frac{n-1}{n} W + \frac{1}{n} B. \tag{9.5}$$

If the chains were random samples from the target distribution, (9.5) is an unbiased estimator of $Var(\psi)$. In this application (9.5) is positively biased for the variance of ψ if the initial values of the chain are over-dispersed, but converges to $Var(\psi)$ as $n \to \infty$. On the other hand, if the chains have not converged by time n, the chains have not yet mixed well across the entire support set of the target distribution so the within-sample variance W underestimates the variance of ψ. As $n \to \infty$ we have the expected value of (9.5) converging to $Var(\psi)$ from above and W converging to $Var(\psi)$ from below. If $\widehat{Var}(\psi)$ is large relative to W this suggests that the chain has not converged to the target distribution by time n.

The Gelman-Rubin statistic is the *estimated potential scale reduction*

$$\sqrt{\hat{R}} = \sqrt{\frac{\widehat{Var}(\psi)}{W}}, \tag{9.6}$$

which can be interpreted as measuring the factor by which the standard deviation of ψ could be reduced by extending the chain. The factor $\sqrt{\hat{R}}$ decreases to 1 as the length of the chain tends to infinity, so $\sqrt{\hat{R}}$ should be close to 1 if the chains have approximately converged to the target distribution. Gelman [107] suggests that $\hat{R}$ should be less than 1.1 or 1.2.

Example 9.8 (Gelman-Rubin method of monitoring convergence)

This example illustrates the Gelman-Rubin method of monitoring convergence of a Metropolis chain. The target distribution is Normal(0,1), and the proposal distribution is Normal(X_t, σ^2). The scalar summary statistic ψ_{ij} is the mean of the ith chain up to time j. After generating all chains the diagnostic statistics are computed in the Gelman.Rubin function below.

```
Gelman.Rubin <- function(psi) {
    # psi[i,j] is the statistic psi(X[i,1:j])
    # for chain in i-th row of X
    psi <- as.matrix(psi)
    n <- ncol(psi)
    k <- nrow(psi)

    psi.means <- rowMeans(psi)      #row means
    B <- n * var(psi.means)         #between variance est.
    psi.w <- apply(psi, 1, "var")   #within variances
    W <- mean(psi.w)                #within est.
    v.hat <- W*(n-1)/n + (B/n)      #upper variance est.
    r.hat <- v.hat / W              #G-R statistic
    return(r.hat)
    }
```

Since several chains are to be generated, the M-H sampler is written as a function normal.chain.

```
normal.chain <- function(sigma, N, X1) {
    #generates a Metropolis chain for Normal(0,1)
    #with Normal(X[t], sigma) proposal distribution
    #and starting value X1
    x <- rep(0, N)
    x[1] <- X1
    u <- runif(N)

    for (i in 2:N) {
        xt <- x[i-1]
        y <- rnorm(1, xt, sigma)      #candidate point
        r1 <- dnorm(y, 0, 1) * dnorm(xt, y, sigma)
        r2 <- dnorm(xt, 0, 1) * dnorm(y, xt, sigma)
        r <- r1 / r2
        if (u[i] <= r) x[i] <- y else
            x[i] <- xt
        }
    return(x)
    }
```

In the following simulation, the proposal distribution has a small variance $\sigma^2 = 0.04$. When the variance is small relative to the target distribution, the chains are usually converging slowly.

```
sigma <- .2       #parameter of proposal distribution
k <- 4            #number of chains to generate
n <- 15000        #length of chains
b <- 1000         #burn-in length

#choose overdispersed initial values
x0 <- c(-10, -5, 5, 10)

#generate the chains
X <- matrix(0, nrow=k, ncol=n)
for (i in 1:k)
    X[i, ] <- normal.chain(sigma, n, x0[i])

#compute diagnostic statistics
psi <- t(apply(X, 1, cumsum))
for (i in 1:nrow(psi))
    psi[i,] <- psi[i,] / (1:ncol(psi))
print(Gelman.Rubin(psi))

#plot psi for the four chains
par(mfrow=c(2,2))
for (i in 1:k)
    plot(psi[i, (b+1):n], type="l",
        xlab=i, ylab=bquote(psi))
par(mfrow=c(1,1)) #restore default

#plot the sequence of R-hat statistics
rhat <- rep(0, n)
for (j in (b+1):n)
    rhat[j] <- Gelman.Rubin(psi[,1:j])
plot(rhat[(b+1):n], type="l", xlab="", ylab="R")
abline(h=1.1, lty=2)
```

The plots of the four sequences of the summary statistic (the mean) ψ are shown in Figure 9.8 from time 1001 to 15000. Rather than interpret the plots, one can refer directly to the value of the factor $\hat{R}$ to monitor convergence. The value $\hat{R} = 1.447811$ at time $n = 5000$ suggests that the chain should be extended. The plot of $\hat{R}$ (Figure 9.9(a)) over time 1001 to 15000 suggests that the chain has approximately converged to the target distribution within approximately 10000 iterations ($\hat{R} = 1.1166$). The dashed line on the plot is at $\hat{R} = 1.1$. Some intermediate values are 1.2252, 1.1836, 1.1561, and 1.1337 at times 6000, 7000, 8000, and 9000, respectively. The value of $\hat{R}$ is less than 1.1 within time 11200.

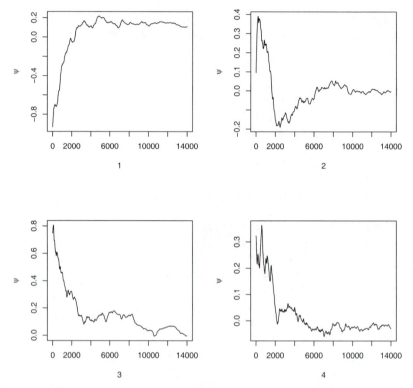

FIGURE 9.8: Sequences of the running means ψ for four Metropolis-Hastings chains in Example 9.8.

For comparison the simulation is repeated, where the variance of the proposal distribution is $\sigma^2 = 4$. The plot of $\hat{R}$ is shown in Figure 9.9(b) for time 1001 to 15000. From this plot it is evident that the chain is converging faster than when the proposal distribution had a very small variance. The value of $\hat{R}$ is below 1.2 within 2000 iterations and below 1.1 within 4000 iterations. ◇

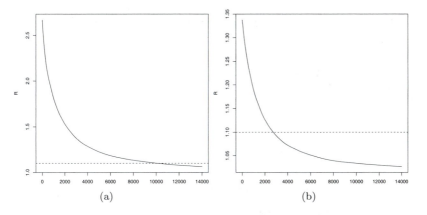

FIGURE 9.9: Sequence of the Gelman-Rubin $\hat{R}$ for four Metropolis-Hastings chains in Example 9.8 (a) $\sigma = 0.2$, (b) $\sigma = 2$.

9.5 Application: Change Point Analysis

A Poisson process is often chosen to model the frequency of rare events. Poisson processes are discussed in Section 3.7. A homogeneous Poisson process $\{X(t), t \geq 0\}$ with constant rate λ is a counting process with independent and stationary increments, such that $X(0) = 0$ and the number of events $X(t)$ in $[0, t]$ has the Poisson(λt) distribution.

Suppose that the parameter λ, which is the expected number of events that occur in a unit of time, has changed at some point in time k. That is, $X_t \sim$ Poisson(μt) for $0 < t \leq k$ and $X_t \sim$ Poisson(λt) for $k < t$. Given a sample of n observations from this process, the problem is to estimate μ, λ and k.

For a specific application, consider the following well known example. The `coal` data in the `boot` package [34] gives the dates of 191 explosions in coal mines which resulted in 10 or more fatalities from March 15, 1851 until March 22, 1962. The data are given in [126], originally from [153]. This problem has been discussed by many authors, including e.g. [36, 37, 63, 121, 171, 192]. A Bayesian model and Gibbs sampling can be applied to estimate the change point in the annual number of coal mining disasters.

Example 9.9 (Coal mining disasters)

In the `coal` data, the date of the disaster is given. The integer part of the date gives the year. For simplicity truncate the fractional part of the year. As a first step, tabulate the number of disasters per year and create a time plot.

```
library(boot)        #for coal data
data(coal)
year <- floor(coal)
y <- table(year)
plot(y)   #a time plot
```

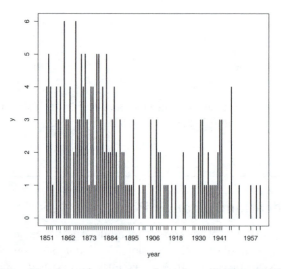

FIGURE 9.10:   Number of annual coal mining disasters in Example 9.9.

From the plot in Figure 9.10 it appears that a change in the average number of disasters per year may have occurred somewhere around the turn of the century. Note that vector of frequencies returned by `table` omits the years where there are zero counts, so for the change point analysis `tabulate` is applied.

```
y <- floor(coal[[1]])
y <- tabulate(y)
y <- y[1851:length(y)]
```

Sequence of annual number of coal mining disasters:

```
4 5 4 1 0 4 3 4 0 6 3 3 4 0 2 6 3 3 5 4 5 3 1 4 4
1 5 5 3 4 2 5 2 2 3 4 2 1 3 2 2 1 1 1 1 3 0 0 1 0
1 1 0 0 3 1 0 3 2 2 0 1 1 1 0 1 0 1 0 0 0 2 1 0 0
0 1 1 0 2 3 3 1 1 2 1 1 1 1 2 3 3 0 0 0 1 4 0 0 0
1 0 0 0 0 0 1 0 0 1 0 1
```

Let Y_i be the number of disasters in year i, where 1851 is year 1. Assume that the change point occurs at year k, and the number of disasters in year i is a Poisson random variable, where

$$Y_i \sim \text{Poisson}(\mu), \qquad i = 1, \ldots, k,$$
$$Y_i \sim \text{Poisson}(\lambda), \qquad i = k+1, \ldots, n.$$

There are $n = 112$ observations ending with year 1962.

Assume the Bayesian model with independent priors

$$k \sim \text{Uniform}\{1, 2, \ldots, n\},$$
$$\mu \sim \text{Gamma}(0.5, b_1),$$
$$\lambda \sim \text{Gamma}(0.5, b_2),$$

introducing additional parameters b_1 and b_2, independently distributed as a positive multiple of a chisquare random variable. That is,

$$b_1|Y, \mu, \lambda, b_2, k \sim \text{Gamma}(0.5, \mu + 1),$$
$$b_2|Y, \mu, \lambda, b_1, k \sim \text{Gamma}(0.5, \lambda + 1).$$

Let $S_k = \sum_{i=1}^{k} Y_i$, and $S'_k = S_n - S_k$ To apply the Gibbs sampler, the fully specified conditional distributions are needed. The conditional distributions for μ, λ, b_1, and b_2 are given by

$$\mu \,|\, y, \lambda, b_1, b_2, k \sim \text{Gamma}(0.5 + S_k, \, k + b_1);$$
$$\lambda \,|\, y, \mu, b_1, b_2, k \sim \text{Gamma}(0.5 + S'_k, \, n - k + b_2);$$
$$b_1 \,|\, y, \mu, \lambda, b_2, k \sim \text{Gamma}(0.5, \mu + 1);$$
$$b_2 \,|\, y, \mu, \lambda, b_1, k \sim \text{Gamma}(0.5, \lambda + 1),$$

and the posterior density of the change point k is

$$f(k|Y, \mu, \lambda, b_1, b_2) = \frac{L(Y; k, \mu, \lambda)}{\sum_{j=1}^{n} L(Y; j, \mu, \lambda)}, \tag{9.7}$$

where

$$L(Y; k, \mu, \lambda) = e^{k(\lambda - \mu)} \left(\frac{\mu}{\lambda}\right)^{S_k}$$

is the likelihood function.

For the change point analysis with the model specified on the previous page, the Gibbs sampler algorithm is as follows ($G(a, b)$ denotes the Gamma(shape= a, rate= b) distribution).

1. Initialize k by a random draw from 1:n, and initialize λ, μ, b_1, b_2 to 1.
2. For each iteration, indexed $t = 1, 2, \ldots$ repeat:

   (a) Generate $\mu(t)$ from $G(0.5 + S_{k(t-1)}, k(t-1) + b_1(t-1))$.
   (b) Generate $\lambda(t)$ from $G(0.5 + S'_{k(t-1)}, n - k(t-1) + b_2(t-1))$.
   (c) Generate $b_1(t)$ from $G(0.5, \mu(t) + 1)$.
   (d) Generate $b_2(t)$ from $G(0.5, \lambda(t) + 1)$.
   (e) Generate $k(t)$ from the multinomial distribution defined by (9.7) using the updated values of λ, μ, b_1, b_2.
   (f) $X(t) = (\mu(t), \lambda(t), b_1(t), b_2(t), k(t))$ (every candidate is accepted).
   (g) Increment t.

The implementation of the Gibbs sampler for this problem is shown on the facing page.

From the output of the Gibbs sampler below, the following sample means are obtained after discarding a burn-in sample of size 200. The estimated change point is $k \doteq 40$. From year $k = 1$ (1851) to $k = 40$ (1890) the estimated Poisson mean is $\hat{\mu} \doteq 3.1$, and from year $k = 41$ (1891) forward the estimated Poisson mean is $\hat{\lambda} \doteq 0.93$.

```
b <- 201
j <- k[b:m]
> print(mean(k[b:m]))
[1] 39.935
> print(mean(lambda[b:m]))
[1] 0.9341033
> print(mean(mu[b:m]))
[1] 3.108575
```

Histograms and plots of the chains are shown in Figures 9.11 and 9.12. Code to generate the plots is given on page 279. ◇

```
# Gibbs sampler for the coal mining change point

# initialization
n <- length(y)      #length of the data
m <- 1000           #length of the chain
mu <- lambda <- k <- numeric(m)
L <- numeric(n)
k[1] <- sample(1:n, 1)
mu[1] <- 1
lambda[1] <- 1
b1 <- 1
b2 <- 1



# run the Gibbs sampler
for (i in 2:m) {
    kt <- k[i-1]

    #generate mu
    r <- .5 + sum(y[1:kt])
    mu[i] <- rgamma(1, shape = r, rate = kt + b1)

    #generate lambda
    if (kt + 1 > n) r <- .5 + sum(y) else
        r <- .5 + sum(y[(kt+1):n])
    lambda[i] <- rgamma(1, shape = r, rate = n - kt + b2)

    #generate b1 and b2
    b1 <- rgamma(1, shape = .5, rate = mu[i]+1)
    b2 <- rgamma(1, shape = .5, rate = lambda[i]+1)

    for (j in 1:n) {
        L[j] <- exp((lambda[i] - mu[i]) * j) *
                    (mu[i] / lambda[i])^sum(y[1:j])
        }
    L <- L / sum(L)

    #generate k from discrete distribution L on 1:n
    k[i] <- sample(1:n, prob=L, size=1)
}
```

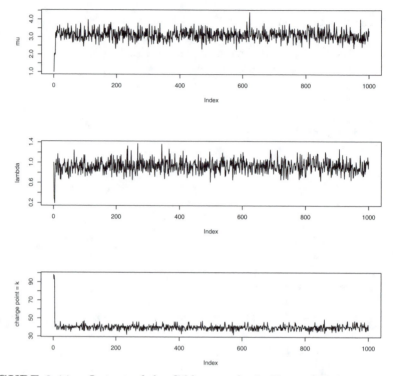

FIGURE 9.11: Output of the Gibbs sampler in Example 9.9.

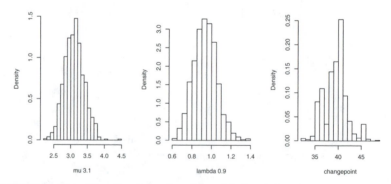

FIGURE 9.12: Distribution of μ, λ, and k from the change point analysis for coal mining disasters in Example 9.9.

Several contributed packages for R offer implementations of the methods in this chapter. See, for example, the packages mcmc and MCMCpack [117, 191]. The coda (Convergence Diagnosis and Output Analysis) package [212] provides utilities that summarize, plot, and diagnose convergence of mcmc objects created by functions in MCMCpack. Also see mcgibbsit [291]. For implementation of Bayesian methods in general, see the task view on CRAN "Bayesian Inference" for a description of several packages.

Exercises

9.1 Repeat Example 9.1 for the target distribution Rayleigh($\sigma = 2$). Compare the performance of the Metropolis-Hastings sampler for Example 9.1 and this problem. In particular, what differences are obvious from the plot corresponding to Figure 9.1?

9.2 Repeat Example 9.1 using the proposal distribution $Y \sim \text{Gamma}(X_t, 1)$ (shape parameter X_t and rate parameter 1).

9.3 Use the Metropolis-Hastings sampler to generate random variables from a standard Cauchy distribution. Discard the first 1000 of the chain, and compare the deciles of the generated observations with the deciles of the standard Cauchy distribution (see qcauchy or qt with df=1). Recall that a Cauchy(θ, η) distribution has density function

$$f(x) = \frac{1}{\theta\pi(1 + [(x - \eta)/\theta]^2)}, \qquad -\infty < x < \infty, \ \theta > 0.$$

The standard Cauchy has the Cauchy($\theta = 1, \eta = 0$) density. (Note that the standard Cauchy density is equal to the Student t density with one degree of freedom.)

9.4 Implement a random walk Metropolis sampler for generating the standard Laplace distribution (see Exercise 3.2). For the increment, simulate from a normal distribution. Compare the chains generated when different variances are used for the proposal distribution. Also, compute the acceptance rates of each chain.

9.5 What effect, if any, does the width w have on the mixing of the chain in Example 9.5? Repeat the simulation keeping the random number seed fixed, trying different proposal distributions based on the random increments from Uniform($-w, w$), varying w.

9.6 Rao [220, Sec. 5g] presented an example on genetic linkage of 197 animals in four categories (also discussed in [67, 106, 171, 266]). The group sizes are

(125, 18, 20, 34). Assume that the probabilities of the corresponding multinomial distribution are

$$\left(\frac{1}{2} + \frac{\theta}{4}, \frac{1-\theta}{4}, \frac{1-\theta}{4}, \frac{\theta}{4}\right).$$

Estimate the posterior distribution of θ given the observed sample, using one of the methods in this chapter.

9.7 Implement a Gibbs sampler to generate a bivariate normal chain (X_t, Y_t) with zero means, unit standard deviations, and correlation 0.9. Plot the generated sample after discarding a suitable burn-in sample. Fit a simple linear regression model $Y = \beta_0 + \beta_1 X$ to the sample and check the residuals of the model for normality and constant variance.

9.8 This example appears in [40]. Consider the bivariate density

$$f(x, y) \propto \binom{n}{x} y^{x+a-1}(1-y)^{n-x+b-1}, \qquad x = 0, 1, \ldots, n, \ 0 \le y \le 1.$$

It can be shown (see e.g. [23]) that for fixed a, b, n, the conditional distributions are Binomial(n, y) and Beta$(x + a, n - x + b)$. Use the Gibbs sampler to generate a chain with target joint density $f(x, y)$.

9.9 Modify the Gelman-Rubin convergence monitoring given in Example 9.8 so that only the final value of $\hat{R}$ is computed, and repeat the example, omitting the graphs.

9.10 Refer to Example 9.1. Use the Gelman-Rubin method to monitor convergence of the chain, and run the chain until the chain has converged approximately to the target distribution according to $\hat{R} < 1.2$. (See Exercise 9.9.) Also use the coda [212] package to check for convergence of the chain by the Gelman-Rubin method. Hints: See the help topics for the coda functions gelman.diag, gelman.plot, as.mcmc, and mcmc.list.

9.11 Refer to Example 9.5. Use the Gelman-Rubin method to monitor convergence of the chain, and run the chain until the chain has converged approximately to the target distribution according to $\hat{R} < 1.2$. Also use the coda [212] package to check for convergence of the chain by the Gelman-Rubin method. (See Exercises 9.9 and 9.10.)

9.12 Refer to Example 9.6. Use the Gelman-Rubin method to monitor convergence of the chain, and run the chain until the chain has converged approximately to the target distribution according to $\hat{R} < 1.2$. Also use the coda [212] package to check for convergence of the chain by the Gelman-Rubin method. (See Exercises 9.9 and 9.10.)

R Code

Code for Figure 9.3 on page 255

Reference lines are added at the $t_{0.025}(\nu)$ and $t_{0.975}(\nu)$ quantiles.

```
par(mfrow=c(2,2))  #display 4 graphs together
refline <- qt(c(.025, .975), df=n)
rw <- cbind(rw1$x, rw2$x, rw3$x,  rw4$x)
for (j in 1:4) {
    plot(rw)[,j], type="l",
        xlab=bquote(sigma == .(round(sigma[j],3))),
        ylab="X", ylim=range(rw[,j]))
    abline(h=refline)
}
par(mfrow=c(1,1)) #reset to default
```

Code for Figures 9.4(a) on page 259 and 9.4(b) on page 259

```
plot(x, type="l")
abline(h=b, v=burn, lty=3)
xb <- x[- (1:burn)]
hist(xb, prob=TRUE, xlab=bquote(beta), ylab="X", main="")
z <- seq(min(xb), max(xb), length=100)
lines(z, dnorm(z, mean(xb), sd(xb)))
```

Code for Figure 9.11 on page 276

```
# plots of the chains for Gibbs sampler output

par(mfcol=c(3,1), ask=TRUE)
plot(mu, type="l", ylab="mu")
plot(lambda, type="l", ylab="lambda")
plot(k, type="l", ylab="change point = k")
```

Code for Figure 9.12 on page 276

```
# histograms from the Gibbs sampler output

par(mfrow=c(2,3))
labelk <- "changepoint"
label1 <- paste("mu", round(mean(mu[b:m]), 1))
label2 <- paste("lambda", round(mean(lambda[b:m]), 1))

hist(mu[b:m], main="", xlab=label1,
     breaks = "scott", prob=TRUE) #mu posterior
hist(lambda[b:m], main="", xlab=label2,
     breaks = "scott", prob=TRUE) #lambda posterior
hist(j, breaks=min(j):max(j), prob=TRUE, main="",
    xlab = labelk)
par(mfcol=c(1,1), ask=FALSE)  #restore display
```

Chapter 10

Probability Density Estimation

Density estimation is a collection of methods for constructing an estimate of a probability density, as a function of an observed sample of data. In previous chapters, we have used density estimation informally to describe the distribution of data. A *histogram* is a type of density estimator. Another type of density estimator is provided in the R function `density`. As explained in the following sections, `density` computes *kernel density estimates*.

Several methods of density estimation are discussed in the literature. In this chapter we restrict attention to nonparametric density estimation. A density estimation problem requires a nonparametric approach if we have no information about the target distribution other than the observed data. In other cases we may have incomplete information about the distribution, so that traditional estimation methods are not directly applicable. For example, suppose it is known that the data arise from a location-scale family, but the family is not specified. Nonparametric density estimation may not always be the best approach, however. Perhaps the data are assumed to be a sample from a normal mixture model, which is a type of classification problem; one can apply EM or other parametric estimation procedures. For problems that require a nonparametric approach, density estimation provides a flexible and powerful tool for visualization, exploration, and analysis of data.

Readers are referred to Scott [244], Silverman [252] or Devroye [70] for an overview of univariate and multivariate density estimation methods including kernel methods. On multivariate density estimation see Scott [244].

10.1   Univariate Density Estimation

In this section univariate density estimation methods are presented, including the histogram, frequency polygon, average shifted histogram, and kernel density estimators.

10.1.1 Histograms

Several methods for computing the histogram density estimate are presented and illustrated with examples. These methods include the normal reference rule, Sturges [257], Scott [241], and Freedman-Diaconis [99] rules for determining the class boundaries.

Introduced in elementary statistics courses, and available in all popular statistics packages, the probability histogram is the most widely used density estimate in descriptive statistics. However, even in the elementary data analysis projects we are faced with tricky questions such as how to determine the best number of bins, the boundaries and width of class intervals, or how to handle unequal class interval widths. In many software packages, these decisions are made automatically, but sometimes produce undesirable results. With R software, the user has control over several options described below.

The histogram is a piecewise constant approximation of the density function. Because data, in general, is contaminated by noise, the estimator that presents too much detail (fitting more closely with the data) is not necessarily "better." The choice of bin width for a histogram is a choice of smoothing parameter. A narrow bin width may *undersmooth* the data, presenting too much detail, while wider bin width may *oversmooth* the data, obscuring important features. Several rules are commonly applied that suggest an optimal choice of bin width. These rules are discussed below. The choice of smoothing parameter and bin center is a challenging problem that continues to attract much attention in research.

Suppose that a random sample $X_1, \ldots, X_n$ is observed. To construct a frequency or probability histogram of the sample, the data must be sorted into bins, and the binning operation is determined by the boundaries of the class intervals. Although in principle any class boundaries can be used, some choices are more reasonable than others in terms of the quality of information about the population density.

In this book we only discuss uniform bin width. Among the commonly applied rules for determining the boundaries of class intervals of a histogram are Sturges' rule [257], Scott's normal reference rule [241], the Freedman-Diaconis (FD) rule [99], and various modifications of these rules.

Given class intervals of equal width h, the histogram density estimate based on a sample size n is

$$\hat{f}(x) = \frac{\nu_k}{nh}, \qquad t_k \leq x < t_{k+1}, \tag{10.1}$$

where ν_k is the number of sample points in the class interval $[t_k, t_{k+1})$. If the bin width is exactly 1, then the density estimate is the relative frequency of the class containing the point x.

The bias of a histogram density estimator (10.1) is proportional to the bin width h. The bias in a histogram density estimate is determined by f', the first order derivative of the density. For other density estimators such as the frequency polygon, ASH, and kernel density estimators, the bias is determined by f'', the second order derivative of the density. Estimators of higher order are not usually applied because the density estimates can be negative.

Sturges' Rule

Although Sturges' rule [257] tends to oversmooth the data and either Scott's rule or FD are generally preferable, Sturges' rule is the default in many statistical packages. In this section we present the motivation for this rule and also use it to illustrate the behavior of the `hist` histogram plotting function and how to change the default behavior. Sturges' rule is based on the implicit assumption that the sampled population is normally distributed. In this case, it is natural to choose a family of discrete distributions that converge in distribution to normal as the number of classes (and sample size n) tend to infinity. The most obvious candidate is the binomial distribution with probability of success $1/2$. For example, if the sample size is $n = 64$, one could select seven class intervals such that the frequency histogram corresponding to a Binomial$(6, 1/2)$ sample has expected class frequencies

$$\binom{6}{0}, \binom{6}{1}, \binom{6}{2}, \ldots \binom{6}{6} = 1, 6, 15, 20, 15, 6, 1,$$

which sum to $n = 64$. Now consider sample sizes $n = 2^k$, $k = 1, 2, \ldots$. For large k (large n) the distribution of Binomial$(k, 1/2)$ is approximately Normal$(\mu = n/2, \sigma^2 = n/4)$. Here $k = \log_2 n$ and we have $k + 1$ bins with expected class frequencies

$$\binom{\log_2 n}{j}, \qquad j = 0, 1, \ldots, k.$$

According to Sturges, the optimal [257] width of class intervals is given by

$$\frac{R}{1 + \log_2 n},$$

where R is the sample range. The number of bins depends only on the sample size n, and not on the distribution. This choice of class interval is designed for data sampled from symmetric, unimodal populations, but is not a good choice for skewed distributions or distributions with more than one mode. For large samples, Sturges' rule tends to oversmooth (see Table 10.1).

Example 10.1 (Histogram density estimates using Sturges' Rule)

Although `breaks = "Sturges"` is the default in the `hist` function in R, this default value is a suggestion only unless a vector of class boundaries is given. For example, compare the following default behavior of `hist` for number of classes with Sturges' Rule.

```
n <- 25
x <- rnorm(n)
# calc breaks according to Sturges' Rule
nclass <- ceiling(1 + log2(n))
cwidth <- diff(range(x) / nclass)
breaks <- min(x) + cwidth * 0:nclass
h.default <- hist(x, freq = FALSE, xlab = "default",
    main = "hist: default")
z <- qnorm(ppoints(1000))
lines(z, dnorm(z))
h.sturges <- hist(x, breaks = breaks, freq = FALSE,
    main = "hist: Sturges")
lines(z, dnorm(z))
```

The corresponding numerical values of breaks and counts are shown below, and the histograms produced by each method are displayed in Figure 10.1(a). The default method is a modification of Sturges' Rule that selects "nice" break points.

```
> print(h.default$breaks)
[1] -2.0 -1.5 -1.0 -0.5  0.0  0.5  1.0  1.5  2.0
> print(h.default$counts)
[1] 3 0 4 6 2 7 2 1
> print(round(h.sturges$breaks, 1))
[1] -1.8 -1.2 -0.6  0.0  0.6  1.2  1.8
> print(h.sturges$counts)
[1] 3 4 6 4 6 2
> print(cwidth)
[1] 0.605878
```

The bin width according to Sturges' rule is 0.605878, compared to the bin width 0.5 applied by `hist` by default. Note that the function

```
> nclass.Sturges
function (x) ceiling(log2(length(x)) + 1)
```

computes the number of classes according to Sturges' rule.

The density estimate for a point x in interval i is given by the height of the histogram on the ith bin. In this example we have the following estimates for the density at the point $x = 0.1$.

```
> print(h.default$density[5])
[1] 0.16
> print(h.sturges$density[4])
[1] 0.2640796
```

For the second estimate, the formula (10.1) is applied with $\nu_k = 4$ and $h = 0.605878$. (The standard normal density at $x = 0.1$ is 0.397.)

For larger samples of normal data, the default behavior of `hist` produces approximately the same density estimate as Sturges' Rule, as shown in Figure 10.1(b) for sample size $n = 1000$. ◇

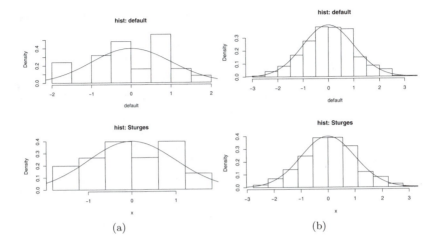

(a)   (b)

FIGURE 10.1: Histogram estimates of normal density in Example 10.1 for samples of size (a) 25 and (b) 1000 with standard normal density curve.

Example 10.2 (Density estimates from a histogram)

In general, to recover density estimates $\hat{f}(x)$ from a histogram, it is necessary to locate the bin containing the point x, then compute the relative frequency (10.1) for that bin. In the previous example with $n = 1000$, corresponding to Figure 10.1(b), we have the following estimates.

```
x0 <- .1
b <- which.min(h.default$breaks <= x0) - 1
print(c(b, h.default$density[b]))
b <- which.min(h.sturges$breaks <= x0) - 1
print(c(b, h.sturges$density[b]))

[1] 7.00 0.38
[1] 6.0000000 0.3889306
```

In the default histogram $\hat{f}_1$, the point $x_0 = 0.1$ is in bin 7, and $\hat{f}_1(0.1) = 0.38$. In $\hat{f}_2$ with breaks specified, x_0 is in bin 6 and $\hat{f}_2(0.1) = 0.3889306$. Alternately, the density estimate is the relative frequency weighted by bin width.

```
h.default$counts[7] / (n * 0.5)
h.sturges$counts[6] / (n * cwidth)
```

```
[1] 0.38
[1] 0.3889306
```

Both estimates are quite close to the value of the standard normal density $\phi(0.1) = 0.3969525$.   ◇

Sturges' Rule is motivated by the normal distribution, which is symmetric. To obtain better density estimates for skewed distributions, Doane [73] suggested a modification based on the sample skewness coefficient $\sqrt{b_1}$ (6.2). The suggested correction is to add

$$K_e = \log_2\left(1 + \frac{|\sqrt{b_1}|}{\sigma(\sqrt{b_1})}\right),\qquad(10.2)$$

classes, where

$$\sigma(\sqrt{b_1}) = \sqrt{\frac{6(n-2)}{(n+1)(n+3)}}$$

is the standard deviation of the sample skewness coefficient for normal data.

Scott's Normal Reference Rule

To select an optimal (or good) smoothing parameter for density estimation, one needs to establish a criterion for comparing smoothing parameters. One approach aims to minimize the squared error in the estimate. Following Scott's approach [244], we briefly summarize some of the main ideas on L_2 criteria. The mean squared error (MSE) of a density estimator $\hat{f}(x)$ at x is

$$MSE(\hat{f}(x)) = E(\hat{f}(x) - f(x))^2 = Var(\hat{f}(x)) + bias^2(\hat{f}(x)).$$

The MSE measures pointwise error. Consider the integrated squared error (ISE), which is the L_2 norm

$$ISE(\hat{f}(x)) = \int (\hat{f}(x) - f(x))^2 dx.$$

It is simpler to consider the statistic, mean integrated squared error (MISE), given by

$$MISE = E[ISE] = E\left[\int (\hat{f}(x) - f(x))^2 dx\right] = \int E[(\hat{f}(x) - f(x))^2] dx$$

$$= \int MSE(\hat{f}(x)) := IMSE$$

(the integrated mean squared error) by Fubini's Theorem. Under some regularity conditions on f, Scott [241] shows that

$$MISE = \frac{1}{nh} + \frac{h^2}{12} \int f'(x)^2 dx + O\left(\frac{1}{n} + h^3\right),$$

and the optimal choice of bin width is

$$h_n^* = \left(\frac{6n}{\int f'(x)^2 dx}\right)^{1/3} \tag{10.3}$$

with asymptotic MISE

$$AMISE^* = \left(\frac{9}{16} \int f'(x)^2 dx\right)^{1/3} n^{-2/3}. \tag{10.4}$$

In density estimation f is unknown, so the optimal h cannot be computed exactly, but the asymptotically optimal h depends on the unknown density only through its first derivative.

Scott's Normal Reference Rule [241], which is calibrated to a normal distribution with variance σ^2, specifies a bin width

$$\hat{h} \doteq 3.49\hat{\sigma}n^{-1/3},$$

where $\hat{\sigma}$ is an estimate of the population standard deviation σ. For normal distributions with variance σ^2, the optimal bin width is $h_n^* = 2(3^{1/3})\pi^{1/6}\sigma n^{-1/3}$. Substituting the sample estimate of standard deviation gives the *normal reference rule* for optimal bin width

$$\hat{h} = 3.490830212\,\hat{\sigma}n^{-1/3} \doteq 3.49\,\hat{\sigma}n^{-1/3}, \tag{10.5}$$

where $\hat{\sigma}^2$ is the sample variance S^2. There remains the choice of the location of the interval boundaries (bin origins or midpoints). On this subject see Scott [241] and the ASH density estimates in section 10.1.3 below.

R note 10.1 *The* `truehist` *(MASS) function [278] uses Scott's Rule by default. In* `hist` *and* `truehist` *the number of classes for Scott's Rule is computed by the function* `nclass.scott` *as*

```
h <- 3.5 * sqrt(stats::var(x)) * length(x)^(-1/3)
ceiling(diff(range(x))/h)
```

(If the vector **breaks** *of breakpoints is not specified, the number of classes is adjusted by the* **pretty** *function to obtain 'nice' breakpoints.)*

Example 10.3 (Density estimation for Old Faithful)

This example illustrates Scott's Normal Reference Rule to determine bin width for a histogram of data on the eruptions of the Old Faithful geyser. One version of the data is `faithful` in the base distribution of R. Another version [15], geyser (MASS), is analyzed by Venables and Ripley [278]. Here the geyser data set is analyzed. There are 299 observations on 2 variables, duration and waiting time. A density estimate for the time between eruptions (waiting) using Scott's Rule is computed below. For comparison, density estimation is repeated using breaks = "scott" in the hist function, and truehist (MASS) with breaks = "Scott".

Scott's Rule gives the estimate for bin width $\hat{h} = 3.5(13.89032 \cdot 0.1495465) = 7.27037$, and $\lceil (108 - 43)/7.27037 \rceil = 9$ bins.

```
library(MASS)  #for geyser and truehist
waiting <- geyser$waiting
n <- length(waiting)
# rounding the constant in Scott's rule
# and using sample standard deviation to estimate sigma
h <- 3.5 * sd(waiting) * n^(-1/3)

# number of classes is determined by the range and h
m <- min(waiting)
M <- max(waiting)
nclass <- ceiling((M - m) / h)
breaks <- m + h * 0:nclass

h.scott <- hist(waiting, breaks = breaks, freq = FALSE,
    main = "")
truehist(waiting, nbins = "Scott", x0 = 0, prob=TRUE,
    col = 0)
hist(waiting, breaks = "scott", prob=TRUE, density=5,
    add=TRUE)
```

The histograms from h.scott1 and h.scott2 are shown in Figures 10.2(a) and 10.2(b). The histograms suggest that the data are not normally distributed and that there are possibly two modes at about 55 and 75. ◇

Freedman-Diaconis Rule

Scott's normal reference rule above is a member of a class of rules that select the optimal bin width according to a formula $\hat{h} = Tn^{-1/3}$, where T is a statistic. These $n^{-1/3}$ rules are related to the fact that the optimal rate of decay of bin width with respect to L_p norms is $n^{-1/3}$ (see e.g. [288]). The Freedman-Diaconis Rule [99] is another member of this class. For the FD rule, the statistic T is twice the sample interquartile range. That is,

$$\hat{h} = 2(IQR)n^{-1/3},$$

where IQR denotes the sample interquartile range. Here the estimator $\hat{\sigma}$ is proportional to the IQR. The IQR is less sensitive than sample standard

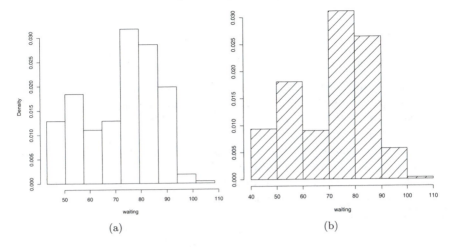

FIGURE 10.2: Histogram estimate of Old Faithful waiting time density in Example 10.3. (a) Scott's Rule suggests 9 bins. (b) `hist` with `breaks =` `"scott"` uses only 7 bins, after function `pretty` is applied to the breaks.

deviation to outliers in the data. The number of classes is the sample range divided by the bin width.

Table 10.1 summarizes results of a simulation experiment comparing Sturges' Rule, Scott's Normal Reference Rule, and the Freedman-Diaconis Rule. Each entry in the table represents a single standard normal or standard exponential sample. These distributions have equal variance, but each rule produces different optimal numbers of bins, particularly when the sample size is large. It appears that even for normal data, Sturges' Rule is oversmoothing the data.

TABLE 10.1:   Estimated Best Number of Class Intervals for Simulated Data According to Three Rules for Histograms

(a) Standard Normal				(b) Standard Exponential			
n	Sturges	Scott	FD	n	Sturges	Scott	FD
10	5	2	3	10	5	2	2
20	6	3	5	20	6	3	3
30	6	4	4	30	6	4	4
50	7	5	7	50	7	6	9
100	8	7	9	100	8	6	7
200	9	9	11	200	9	9	14
500	10	14	20	500	10	16	25
1000	11	19	25	1000	11	23	39
5000	14	40	52	5000	14	37	58
10000	15	46	60	10000	15	54	82

10.1.2   Frequency Polygon Density Estimate

All histogram density estimates are piecewise continuous but not continuous over the entire range of the data. A frequency polygon provides a continuous density estimate from the same frequency distribution used to produce the histogram. The frequency polygon is constructed by computing the density estimate at the midpoint of each class interval, and using linear interpolation for the estimates between consecutive midpoints.

Scott [243] derives the bin width for constructing the optimal frequency polygon by asymptotically minimizing the IMSE. The optimal frequency polygon bin width is

$$h_n^{fp} = 2 \left[\frac{49}{15} \int f''(x)^2 dx \right]^{-1/5} n^{-1/5} \qquad (10.6)$$

with

$$IMSE^{fp} = \frac{5}{12} \left[\frac{49}{15} \int f''(x)^2 dx \right]^{1/5} n^{-4/5} + O(n^{-1}).$$

Notice that in general (10.6) cannot be computed without the knowledge of the underlying distribution. In practice, f'' is estimated (e.g. a difference method is often used). For normal densities, $\int f''(x)^2 dx = 3/(8\sqrt{\pi}\sigma^5)$ and the optimal frequency polygon bin width is

$$h_n^{fp} = 2.15\sigma n^{-1/5}. \qquad (10.7)$$

The normal distribution as a reference distribution will not be optimal if the distribution is not symmetric. For data that is clearly skewed, a more appropriate reference distribution can be selected, such as a lognormal distribution. A skewness adjustment (Scott [244]) derived using a lognormal distribution as the reference distribution, is the factor

$$\frac{12^{1/5}\sigma}{e^{7\sigma^2/4}(e^{\sigma^2} - 1)^{1/2}(9\sigma^4 + 20\sigma^2 + 12)^{1/5}}. \qquad (10.8)$$

The adjustment factor should be multiplied times the bin width to obtain the appropriate smaller bin width. Similarly, if the distribution has heavier tails than the normal distribution, a kurtosis adjustment can be derived with reference to a t distribution.

Example 10.4 (Frequency polygon density estimate)

Construct a frequency polygon density estimate of the geyser (MASS) data. Determine the frequency polygon bin width by the normal reference rule,

$\hat{h}_n^{fp} = 2.15Sn^{-1/5}$, substituting the sample standard deviation S for σ. The calculations are straightforward using the returned value from hist. The vertices of the polygon are the sequence of points ($mids, $density) of the returned hist object. Then the histogram with frequency polygon density estimate is easily constructed by adding lines to the plot connecting these points. There are a few more steps involved, to close the polygon at the ends where the density estimate is zero. To draw the polygon there are several options, such as segments or polygon.

```
waiting <- geyser$waiting    #in MASS
n <- length(waiting)
# freq poly bin width using normal ref rule
h <- 2.15 * sqrt(var(waiting)) * n^(-1/5)

# calculate the sequence of breaks and histogram
br <- pretty(waiting, diff(range(waiting)) / h)
brplus <- c(min(br)-h, max(br+h))
histg <- hist(waiting, breaks = br, freq = FALSE,
    main = "", xlim = brplus)

vx <- histg$mids      #density est at vertices of polygon
vy <- histg$density
delta <- diff(vx)[1] # h after pretty is applied
k <- length(vx)
vx <- vx + delta      # the bins on the ends
vx <- c(vx[1] - 2 * delta, vx[1] - delta, vx)
vy <- c(0, vy, 0)
# add the polygon to the histogram
polygon(vx, vy)
```

The bin width is $h = 9.55029$. The frequency polygon is shown in Figure 10.3. If the density estimates are required for arbitrary points, approxfun can be applied for the linear interpolation. As a check on the estimate, verify that $\int_{\infty}^{\infty} \hat{f}(x)dx = 1$.

```
# check estimates by numerical integration
fpoly <- approxfun(vx, vy)
print(integrate(fpoly, lower=min(vx), upper=max(vx)))
1 with absolute error < 1.1e-14
```

◇

10.1.3   The Averaged Shifted Histogram

In the preceding sections we have considered several rules for determining the best number of classes or best class interval width. The optimal bin width

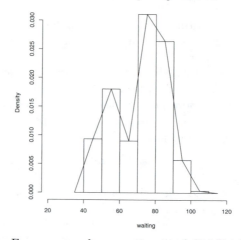

FIGURE 10.3:   Frequency polygon estimate of Old Faithful waiting time density in Example 10.4.

does not determine the location of the center or endpoints of the bin, however. For example, using truehist (MASS), we can easily shift the bins from left to right using the argument x0, while keeping the bin width constant. Shifting the class boundaries changes the density estimates, so several different density estimates are possible using the same bin width. Figure 10.4 on page 294 illustrates four histogram density estimates of a standard normal sample using the same number of bins, with bin origins offset by 0.25 from each other.

The Average Shifted Histogram (ASH) proposed by Scott [242] averages the density estimates. That is, the ASH estimate of density is

$$\hat{f}_{ASH}(x) = \frac{1}{m} \sum_{j=1}^{m} \hat{f}_j(x),$$

where the class boundaries for estimate $\hat{f}_{j+1}(x)$ are shifted by h/m from the boundaries for $\hat{f}_j(x)$. Here we are viewing the estimates as m histograms with class width h. Alternately we can view the ASH estimate as a histogram with widths h/m. The optimal bin width (see [244, Sec. 5.2]) for the naive ASH estimate of a Normal(μ, σ^2) density is

$$h^* = 2.576 \sigma n^{-1/5}. \tag{10.9}$$

Example 10.5 (Calculations for ASH estimate)

This numerical example illustrates the method of computing the ASH estimates. Four histogram estimates, each with bin width 1, are computed for a sample size $n = 100$. The bin origins for each of the densities are at 0, 0.25, 0.5, and 0.75 respectively. The bin counts and breaks are shown below.

```
breaks -4 -3 -2 -1  0  1  2  3  4
counts  0  2 11 27 38 16  6  0
```

```
breaks -3.75 -2.75 -1.75 -0.75  0.25  1.25  2.25  3.25  4.25
counts  0  4 17 23 38 16  2  0
```

```
breaks -3.5 -2.5 -1.5 -0.5  0.5  1.5  2.5  3.5  4.5
counts  0  7 21 23 34 15  0  0
```

```
breaks -3.25 -2.25 -1.25 -0.25  0.75  1.75  2.75  3.75  4.75
counts  2  9 26 30 21 12  0  0
```

To compute an ASH density estimate at the point $x = 0.2$, say, locate the intervals containing $x = 0.2$ and average these density estimates. The estimate is

$$\hat{f}_{ASH}(0.2) = \frac{1}{4} \sum_{k=1}^{4} \hat{f}_k(0.2) = \frac{1}{4} \times \frac{38 + 23 + 23 + 30}{100(1)} = \frac{114}{400} = 0.285.$$

Alternately, we can compute this estimate by considering the mesh over the subintervals with width $\delta = h/m = 0.25$. There are now 36 breakpoints at $-4 + 0.25i$, $i = 0, 1, \ldots, 35$, and 35 bin counts, $\nu_1, \ldots, \nu_{35}$. The point $x = 0.2$ is in the intervals $(-.75, .25], (-.5, .5], (-.25, .75]$, and $(0, 1]$ corresponding to the 14th through 20th subintervals. The bin counts are

```
[1:12]  0  0  0  0  0  0  2  0  2  3  4  2
[13:24] 8  7  9  3  4  7 16 11  4  3  3  6
[25:35] 4  2  0  0  0  0  0  0  0  0  0
```

and the estimate can be computed by rearranging the terms as

```
7  +  9  +  3  +  4                          =   23
      9  +  3  +  4  +  7                     =   23
            3  +  4  +  7  +  16             =   30
                  4  +  7  +  16  + 11  =        38
```

$$= 7 + 2(9) + 3(3) + 4(4) + 3(7) + 2(16) + 11 = 114$$

or

$$\hat{f}_{ASH}(0.2) = \frac{\nu_{14} + 2\nu_{15} + 3\nu_{16} + 4\nu_{17} + 3\nu_{18} + 2\nu_{19} + \nu_{20}}{mnh}.$$

◇

In general, if $t_k = \max\{t_j : t_j < x \le t_{j+1}\}$, we have

$$\hat{f}_{ASH}(x) = \frac{\nu_{k+1-m} + 2\nu_{k+2-m} + \cdots + m\nu_k + \cdots + 2\nu_{k+m-2} + \nu_{k+m-1}}{mnh}$$

$$= \frac{1}{nh} \sum_{j=1-m}^{m-1} \left(1 - \frac{|j|}{m}\right) \nu_{k+j}. \qquad (10.10)$$

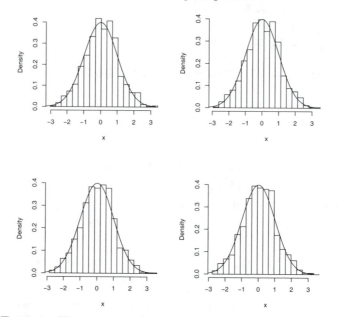

FIGURE 10.4: Histogram estimates of a normal sample with equal bin width but different bin origins, and standard normal density curve.

This computing formula requires that there are $m - 1$ empty bins on the left and the right. Equation (10.10) provides a formula for computing an ASH density estimate and shows that this estimate is a weighted average of the bin counts on the finer mesh. The weights $(1 - |j|/m)$ correspond to a discrete triangular distribution on $[-1, 1]$, which approaches the triangular density on $[-1, 1]$ as $m \to \infty$.

The ASH estimates can be generalized by replacing the weights $(1 - |j|/m)$ in (10.10) with a weight function $w(j) = w(j, m)$ corresponding to a symmetric density supported on $[-1, 1]$. The triangular kernel is used in (10.10), which is

$$K(t) = 1 - |t|, \qquad |t| < 1,$$

and $K(t) = 0$ otherwise. For other kernels see e.g. [244, 252] or the examples of density, and Section 10.2.

Example 10.6 (ASH density estimate)

Construct an ASH density estimate of the Old Faithful waiting time data in geyser$waiting (MASS) based on 20 histograms. For comparison with the naive histogram density estimate of this data in Example 10.3, the bin width is set to $h = 7.27037$. (The normal reference rule for ASH estimates in (10.9) gives $h = 11.44258$.)

```
library(MASS)
waiting <- geyser$waiting
n <- length(waiting)
m <- 20
a <- min(waiting) - .5
b <- max(waiting) + .5
h <- 7.27037
delta <- h / m

#get the bin counts on the delta-width mesh.
br <- seq(a - delta*m, b + 2*delta*m, delta)
histg <- hist(waiting, breaks = br, plot = FALSE)
nk <- histg$counts
K <- abs((1-m):(m-1))

fhat <- function(x) {
    # locate the leftmost interval containing x
    i <- max(which(x > br))
    k <- (i - m + 1):(i + m - 1)
    # get the 2m-1 bin counts centered at x
    vk <- nk[k]
    sum((1 - K / m) * vk) / (n * h)    #f.hat
    }

# density can be computed at any points in range of data
z <- as.matrix(seq(a, b + h, .1))
f.ash <- apply(z, 1, fhat)    #density estimates at midpts

# plot ASH density estimate over histogram
br2 <- seq(a, b + h, h)
hist(waiting, breaks = br2, freq = FALSE, main = "",
    ylim = c(0, max(f.ash)))
lines(z, f.ash, xlab = "waiting")
```

Compare the ASH estimate in Figure 10.5 with the histogram estimate in Figure 10.2(b) and the frequency polygon density estimate in Figure 10.3. ⋄

See the ash package [245] for an implementation of Scott's univariate and bivariate ASH routines.

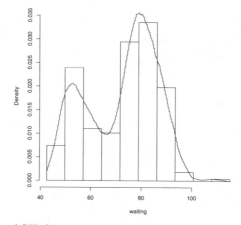

FIGURE 10.5: ASH density estimate of Old Faithful waiting times in Example 10.6.

10.2 Kernel Density Estimation

Kernel density estimation generalizes the idea of a histogram density estimate. If a histogram with bin width h is constructed from a sample $X_1, \ldots, X_n$, then a density estimate for a point x within the range of the data is

$$\hat{f}(x) = \frac{1}{2hn} \times k,$$

where k is the number of sample points in the interval $(x - h, x + h)$. This estimator can be written

$$\hat{f}(x) = \frac{1}{n} \sum_{i=1}^{n} \frac{1}{h} w\left(\frac{x - X_i}{h}\right), \tag{10.11}$$

where $w(t) = \frac{1}{2}I(|t| < 1)$ is a weight function. The density estimator $\hat{f}(x)$ in (10.11) with $w(t) = \frac{1}{2}I(|t| < 1)$ is called the *naive density estimator*. This weight function has the property that $\int_{-1}^{1} w(t)dt = 1$, and $w(t) \geq 0$, so $w(t)$ is a probability density supported on the interval $[-1, 1]$.

Kernel density estimation replaces the weight function $w(t)$ in the naive estimator with a function $K(\cdot)$ called a kernel function, such that

$$\int_{-\infty}^{\infty} K(t)dt = 1.$$

In probability density estimation, $K(\cdot)$ is usually a symmetric probability density function. The weight function $w(t) = \frac{1}{2}I(|t| < 1)$ is called the *rectangular kernel.* The rectangular kernel is a symmetric probability density centered at the origin, and

$$\frac{1}{nh} w \left(\frac{x - X_i}{h} \right),$$

corresponds to a rectangle of area $1/n$ centered at X_i. The density estimate at x is the sum of rectangles located within h units from x.

In this book, we restrict attention to symmetric positive kernel density estimators. Suppose that $K(\cdot)$ is another symmetric probability density centered at the origin, and define

$$\hat{f}_K(x) = \frac{1}{n} \sum_{i=1}^{n} \frac{1}{h} K \left(\frac{x - X_i}{h} \right). \tag{10.12}$$

Then $\hat{f}$ is a probability density function. For example, $K(x)$ may be the triangular density on $[-1, 1]$ (the *triangular kernel*) or the standard normal density (the *Gaussian kernel*). In section 10.1.3 we have seen that the ASH density estimate converges to a triangular kernel density estimate (see equation (10.10) for the kernel) as $n \to \infty$. The triangular kernel estimator corresponds to the sum of areas of triangles instead of rectangles. The Gaussian kernel estimator centers a normal density at each data point, as illustrated in Figure 10.6.

From the definition of the kernel density estimator in (10.12) it follows that certain continuity and differentiability properties of $K(x)$ also hold for $\hat{f}_K(x)$. If $K(x)$ is a probability density, then $\hat{f}_K(x)$ is continuous at x if $K(x)$ is continuous at x, and $\hat{f}_K(x)$ has an rth order derivative at x if $K^{(r)}(x)$ exists. In particular, if $K(x)$ is the Gaussian kernel, then $\hat{f}$ is continuous and has derivatives of all orders.

The histogram density estimator corresponds to the rectangular kernel density estimator. The bin width h is a smoothing parameter; small values of h reveal local features of the density, while large values of h produce a smoother density estimate. In kernel density estimation h is called the *bandwidth, smoothing parameter* or *window width*.

The effect of varying the bandwidth is illustrated in Figure 10.6. The $n = 10$ sample points in Figure 10.6,

```
-0.77 -0.60 -0.25   0.14   0.45   0.64   0.65   1.19   1.71   1.74
```

were generated from the standard normal distribution. As the window width h decreases, the density estimate becomes rougher, and larger h corresponds to smoother density estimates. (This example is presented simply to graphically illustrate the kernel method; density estimation is not very useful for such a small sample.)

Table 10.2 gives some kernel functions that are commonly applied in density estimation, which are also shown in Figure 10.7. The Epanechnikov kernel was first suggested for kernel density estimation by Epanechnikov [85]. The efficiency of a kernel is defined by Silverman [252, p. 42]. The rescaled

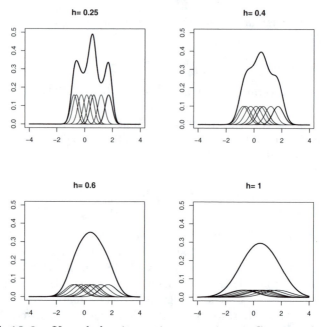

FIGURE 10.6: Kernel density estimates using a Gaussian kernel with bandwidth h.

Epanechnikov kernel has efficiency 1, which is an optimal kernel in the sense of MISE (Scott [244, pp. 138–140]). The asymptotic relative efficiencies given in Table 10.2 in fact show that there is not much difference among the kernels if the mean integrated squared error criterion is used (see [252, p. 43]). See the examples of `density` for a method of calculating the efficiencies (actually the reciprocal of efficiency in Table 10.2).

For a Gaussian kernel, the bandwidth h that optimizes IMSE is

$$h = (4/3)^{1/5}\sigma n^{-1/5} = 1.06\sigma n^{-1/5}. \tag{10.13}$$

This choice of bandwidth is an optimal (IMSE) choice when the distribution is normal. If the true density is not unimodal, however, (10.13) will tend to oversmooth. Alternately, one can use a more robust estimate of dispersion in (10.13), setting

$$\hat{\sigma} = \min(S, IQR/1.34),$$

where S is the standard deviation of the sample. Silverman [252, p. 48] indicates that an even better choice for a Gaussian kernel is the reduced width

$$h = 0.9\hat{\sigma}n^{-1/5} = 0.9\min(S, IQR/1.34)n^{-1/5}, \tag{10.14}$$

which is a good starting point appropriate for a wide range of distributions that are not necessarily normal, unimodal, or symmetric.

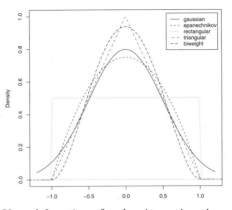

FIGURE 10.7: Kernel functions for density estimation.

The R reference manual [217] topic for bandwidth (`?bw.nrd`) refers to the rule in (10.14) as Silverman's "rule-of-thumb," which is applied unless the quartiles coincide. Various choices for bandwidth selection are illustrated in Examples 10.7 and 10.8 below.

TABLE 10.2: Kernel Functions for Density Estimation

Kernel	$K(t)$	Support	σ_K^2	Efficiency
Gaussian	$\frac{1}{\sqrt{2\pi}} \exp(-\frac{1}{2}t^2)$	$\mathbb{R}$	1	1.0513
Epanechnikov	$\frac{3}{4}(1 - t^2)$	$|t| < 1$	1/5	1
Rectangular	$\frac{1}{2}$	$|t| < 1$	1/3	1.0758
Triangular	$1 - |t|$	$|t| < 1$	1/6	1.0143
Biweight	$\frac{15}{16}(1 - t^2)^2$	$|t| < 1$	1/7	1.0061
Cosine	$\frac{\pi}{4} \cos \frac{\pi}{2} t$	$\mathbb{R}$	$1 - 8/\pi^2$	1.0005

For equivalent kernel rescaling, the bandwidth h_1 can be rescaled by setting

$$h_2 \approx \frac{\sigma_{K_1}}{\sigma_{K_2}} h_1.$$

Factors for equivalent smoothing are given by Scott [244, p. 142]. A kernel can also be scaled to "canonical" form such that the bandwidth is equivalent to the Gaussian kernel.

The `density` function in R computes kernel density estimates for seven kernels. The smoothing parameter is `bw` (bandwidth), but the kernels are scaled so that `bw` is the standard deviation of the kernel. The "canonical bandwidth" can be obtained using `density` with the option `give.Rkern = TRUE`. Choices for the kernel are `gaussian`, `epanechnikov`, `rectangular`, `triangular`, `biweight`, `cosine`, or `optcosine`. Run `example(density)` to see several plots of the corresponding density estimates. The cosine kernel given in Table 10.2 corresponds to the `optcosine` choice. The bandwidth adjustment for equivalent kernels in `density` is approximately 1, so the kernels are approximately equivalent.

Example 10.7 (Kernel density estimate of Old Faithful waiting time)

In this example we look at the result obtained by the default arguments to `density`. The default method applies the Gaussian kernel. For details on the default bandwidth selection see the help topics for `bandwidth` or `bw.nrd0`.

```
library(MASS)
waiting <- geyser$waiting
n <- length(waiting)
h1 <- 1.06 * sd(waiting) * n^(-1/5)
h2 <- .9 * min(c(IQR(waiting)/1.34, sd(waiting))) * n^(-1/5)
plot(density(waiting))

> print(density(waiting))
Call:
        density.default(x = waiting)

Data: waiting (299 obs.);        Bandwidth 'bw' = 3.998

      x                   y
Min.   : 31.01    Min.   :3.762e-06
1st Qu.: 53.25    1st Qu.:4.399e-04
Median : 75.50    Median :1.121e-02
Mean   : 75.50    Mean   :1.123e-02
3rd Qu.: 97.75    3rd Qu.:1.816e-02
Max.   :119.99    Max.   :3.342e-02

sdK <- density(kernel = "gaussian", give.Rkern = TRUE)
> print(c(sdK, sdK * sd(waiting)))
[1] 0.2820948 3.9183881
> print(c(sd(waiting), IQR(waiting)))
[1] 13.89032 24.00000
> print(c(h1, h2))
[1] 4.708515 3.997796
```

The default density estimate applied the Gaussian kernel with the bandwidth $h = 3.998$ corresponding to equation (10.14). The default density plot with bandwidth 3.998 is shown in Figure 10.8. Other choices of bandwidth are also shown for comparison. ◇

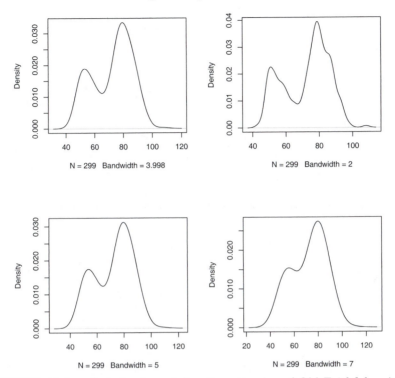

FIGURE 10.8: Gaussian kernel density estimates of Old Faithful waiting time in Example 10.7 using `density` with different bandwidths.

Example 10.8 (Kernel density estimate of precipitation data)

The dataset `precip` in R is the average amount of precipitation for 70 United States cities and Puerto Rico (see [217] for the source). We use the `density` function to construct kernel density estimates using the default and other choices for bandwidth.

```
n <- length(precip)
h1 <- 1.06 * sd(precip) * n^(-1/5)
h2 <- .9 * min(c(IQR(precip)/1.34, sd(precip))) * n^(-1/5)
h0 <- bw.nrd0(precip)

par(mfrow = c(2, 2))
plot(density(precip))                    #default Gaussian (h0)
plot(density(precip, bw = h1))  #Gaussian, bandwidth h1
plot(density(precip, bw = h2))  #Gaussian, bandwidth h2
plot(density(precip, kernel = "cosine"))
par(mfrow = c(1,1))
```

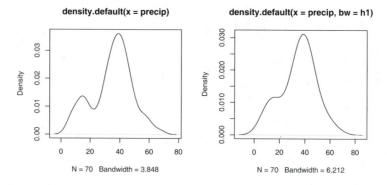

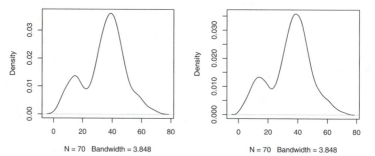

FIGURE 10.9: Kernel density estimates of precipitation data in Example 10.8 using `density` with different bandwidths.

The three values for bandwidth computed are

```
> print(c(h0, h1, h2))
[1] 3.847892 6.211802 3.847892
```

and the plots are shown in Figure 10.9. The default density plot applied the Gaussian kernel with the bandwidth $h = 3.848$ corresponding to equation (10.14) and the result of `bw.nrd0`.   ◇

Example 10.9 (Computing $\hat{f}(x)$ for arbitrary x)

To estimate the density for new points, use `approx`.

```
d <- density(precip)
xnew <- seq(0, 70, 10)
approx(d$x, d$y, xout = xnew)
```

The code above produces the estimates:

```
$x
[1]   0 10 20 30 40 50 60 70
$y
[1] 0.000952360 0.010971583 0.010036739
[4] 0.021100536 0.035776120 0.014421428
[7] 0.005478733 0.001172337
```

For certain applications it is helpful to create a function to return the estimates, which can be accomplished easily with `approxfun`. Below `fhat` is a
function returned by `approxfun`.

```
> fhat <- approxfun(d$x, d$y)
> fhat(xnew)
[1] 0.000952360 0.010971583 0.010036739
[4] 0.021100536 0.035776120 0.014421428
[7] 0.005478733 0.001172337
```

◇

Boundary kernels

Near the boundaries of the support set of a density, or discontinuity points,
kernel density estimates have larger errors. Kernel density estimates tend to
smooth the probability mass over the discontinuity points or boundary points.
For example, see the kernel density estimates of the precipitation data shown
in Figure 10.9. Note that the density estimates suggest that negative inches
of precipitation are possible.

In the next example, we illustrate the boundary problem with an exponential density, and compare the kernel estimate with the true density.

Example 10.10 (Exponential density)

A Gaussian kernel density estimate of an Exponential(1) density is shown in
Figure 10.10. The true exponential density is shown with a dashed line.

```
x <- rexp(1000, 1)
plot(density(x), xlim = c(-1, 6), ylim = c(0, 1), main="")
abline(v = 0)

# add the true density to compare
y <- seq(.001, 6, .01)
lines(y, dexp(y, 1), lty = 2)
```

Note that the smoothness of the kernel estimate does not fit the discontinuity
of the density at $x = 0$. ◇

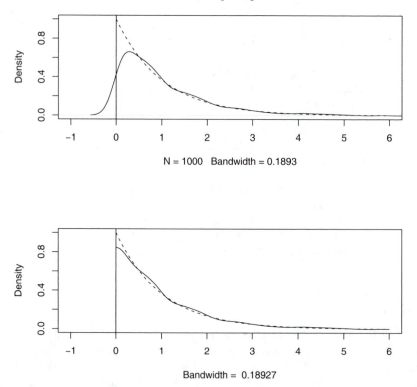

FIGURE 10.10:   Gaussian kernel density estimate (solid line) of an exponential density in Example 10.10, with true density (dashed line).   In the
second plot, the reflection boundary technique is applied on the same data.

Scott [244] discusses boundary kernels, which are finite support kernels
that are applied to obtain the density estimate in the boundary region. A
simple fix is to use a *reflection boundary technique* if the discontinuity occurs
at the origin. First add the reflection of the entire sample; that is, append
$-x_1, \ldots, -x_n$ to the data. Then estimate a density g using the $2n$ points,
but use n to determine the smoothness parameter. Then $\hat{f}(x) = 2\hat{g}(x)$. This
method is applied below.

Example 10.11 (Reflection boundary technique)

The reflection boundary technique can be applied when the density has a
discontinuity at 0, such as in Example 10.10.

```
xx <- c(x, -x)
g <- density(xx, bw = bw.nrd0(x))
a <- seq(0, 6, .01)
```

```
ghat <- approx(g$x, g$y, xout = a)
fhat <- 2 * ghat$y        # density estimate along a

bw <- paste("Bandwidth = ", round(g$bw, 5))
plot(a, fhat, type="l", xlim=c(-1, 6), ylim=c(0, 1),
     main = "", xlab = bw, ylab = "Density")
abline(v = 0)

# add the true density to compare
y <- seq(.001, 6, .01)
lines(y, dexp(y, 1), lty = 2)
```

The plot of the density estimate with reflection boundary is shown in Figure 10.10. ◇

See Scott [244] or Wand and Jones [289] for further discussion of methods for kernel density estimation near boundaries.

10.3 Bivariate and Multivariate Density Estimation

In this section examples are presented that illustrate some of the basic methods for bivariate and multivariate density estimation. Scott [244] is a comprehensive reference on multivariate density estimation. Also see Silverman [252, Ch. 4].

10.3.1 Bivariate Frequency Polygon

To construct a bivariate density histogram (polygon), it is necessary to define two-dimensional bins and count the number of observations in each bin. The bin2d function in the following example computes the two dimensional frequency table.

Example 10.12 (Bivariate frequency table: bin2d)

The function bin2d bins a bivariate data matrix, based on the univariate histogram hist in R. See the documentation for hist for an explanation of how the breakpoints are determined.

The frequencies are computed by constructing a two dimensional contingency table with the marginal breakpoints as the cut points. The return value of bin2d is a list including the table of bin frequencies, vectors of breakpoints, and vectors of midpoints.

```
bin2d <-
    function(x, breaks1 = "Sturges", breaks2 = "Sturges"){
    # Data matrix x is n by 2
    # breaks1, breaks2: any valid breaks for hist function
    # using same defaults as hist
    histg1 <- hist(x[,1], breaks = breaks1, plot = FALSE)
    histg2 <- hist(x[,2], breaks = breaks2, plot = FALSE)
    brx <- histg1$breaks
    bry <- histg2$breaks

    # bin frequencies
    freq <- table(cut(x[,1], brx),   cut(x[,2], bry))

    return(list(call = match.call(), freq = freq,
            breaks1 = brx, breaks2 = bry,
            mids1 = histg1$mids, mids2 = histg2$mids))
    }
```

To show the details of the bin2d function, it is applied to bin the bivariate sepal length and sepal width distribution of iris setosa data. Then in Example 10.13 bin2d is used to bin data for constructing a bivariate frequency polygon.

```
> bin2d(iris[1:50,1:2])
$call bin2d(x = iris[1:50, 1:2])

$freq
            (2,2.5] (2.5,3] (3,3.5] (3.5,4] (4,4.5]
(4.2,4.4]      0      3      1      0      0
(4.4,4.6]      1      0      3      1      0
(4.6,4.8]      0      2      5      0      0
(4.8,5]        0      2      8      2      0
(5,5.2]        0      0      6      4      1
(5.2,5.4]      0      0      2      4      0
(5.4,5.6]      0      0      1      0      1
(5.6,5.8]      0      0      0      2      1

$breaks1
[1] 4.2 4.4 4.6 4.8 5.0 5.2 5.4 5.6 5.8
$breaks2
[1] 2.0 2.5 3.0 3.5 4.0 4.5
$mids1
[1] 4.3 4.5 4.7 4.9 5.1 5.3 5.5 5.7
$mids2
[1] 2.25 2.75 3.25 3.75 4.25
```

◇

Example 10.13 (Bivariate density polygon)

Bivariate data is displayed in a 3D density polygon, using the `bin2d` function in Example 10.12 to compute the bivariate frequency table. After binning the bivariate data, the `persp` function plots the density polygon.

```
#generate standard bivariate normal random sample
n <- 2000;    d <- 2
x <- matrix(rnorm(n*d), n, d)

# compute the frequency table and density estimates
b <- bin2d(x)
h1 <- diff(b$breaks1)
h2 <- diff(b$breaks2)

# matrix h contains the areas of the bins in b
h <- outer(h1, h2, "*")

Z <- b$freq / (n * h)   # the density estimate

persp(x=b$mids1, y=b$mids2, z=Z, shade=TRUE,
      xlab="X", ylab="Y", main="",
      theta=45, phi=30, ltheta=60)
```

The perspective plot, a three dimensional density polygon, is shown in Figure 10.11. Also see Figure 4.7 on page 109 for another view of bivariate normal data, in a "flat" hexagonal histogram.   ◇

See the `persp` examples for more options, including color. Also see the `wireframe` function in the `lattice` [239] package. Other functions that bin bivariate data are e.g. `bin2` (`ash`) [245] and `hist2d` (`gplots`) [290].

3D Histogram

A 3D histogram can be displayed by functions in the `rgl` [2] package, an interactive 3D graphics package. To see a demo, type

```
library(rgl)
demo(hist3d)
```

After running the demo, the source code for two functions named `hist3d` and `binplot.3d` that are used in the demo should have appeared in the console window (scroll up to see it). To apply the `rgl` demo histogram to this example, copy the two functions `hist3d` and `binplot.3d` into a source file. These functions are in the file *hist3d.r* located in the *demo* directory of *library/rgl*.

```
library(rgl)
#run demo(hist3d) or
#source binplot.3d and hist3d functions
n <- 1000
d <- 2
x <- matrix(rnorm(n*d), n, d)
rgl.clear()
hist3d(x[,1], x[,2])
```

As Silverman [252, p. 78] points out, there are serious presentational difficulties with a 3D histogram. The surface and wireframe plots of bivariate densities are better, particularly when they are generated from a continuous density estimator.

10.3.2   Bivariate ASH

The average shifted histogram estimator of density can be extended to multivariate density estimation. Suppose that bivariate data $\{(x, y)\}$, have been sorted into an $nbin_1$ by $nbin_2$ array of bins with frequencies $\nu = (\nu_{ij})$ and bin widths $h = (h_1, h_2)$ (see e.g. the $\texttt{bin2d}$ function in Example 10.12). The parameter $m = (m_1, m_2)$ is the number of shifted histograms on each axis used in the estimate. The histograms are shifted in two directions, so that there are $m_1 m_2$ histogram density estimates to be averaged.

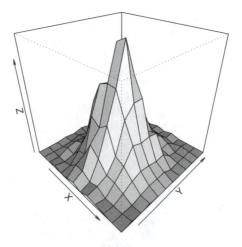

FIGURE 10.11: Density polygon of bivariate normal data in Example 10.13, using normal reference rule (Sturges' Rule) to determine bin widths.

The bivariate ASH estimate of the joint density $f(x, y)$ is

$$\hat{f}_{ASH}(x, y) = \frac{1}{m_1 m_2} \sum_{i=1}^{m_1} \sum_{j=1}^{m_2} \hat{f}_{ij}(x, y).$$

The bin weights are given by

$$w_{ij} = \left(1 - \frac{|i|}{m_1}\right)\left(1 - \frac{|j|}{m_2}\right), \quad i = 1-m_1, \ldots, m_1-1, \, j = 1-m_2, \ldots, m_2-1.$$

$$(10.15)$$

One can apply a similar algorithm for computing the individual estimates $\hat{f}_{ij}(x, y)$ as in the univariate ASH. See Scott [244, Sec. 5.2] for a bivariate ASH algorithm. The ASH estimates can be generalized by replacing the weights $(1 - |i|/m_1)$ and $(1 - |j|/m_2)$ in (10.15) with other kernels. The triangular kernel is applied in (10.15). Also note that the bivariate ASH methods can be generalized to dimension $d \geq 2$.

Example 10.14 (Bivariate ASH density estimate)

This example computes a bivariate ASH estimate of a bivariate normal sample, using Scott's routines in the **ash** package [245]. The function **ash2** returns a list containing (among other things) the coordinates of the bin centers and the density estimates, labeled x, y, z. The generator **rmvn.eigen** is given in Example 3.16 on page 71. Alternately, samples can be generated using e.g. **mvrnorm (MASS)**.

```
library(ash)  # for bivariate ASH density est.
# generate N_2(0,Sigma) data
n <- 2000
d <- 2
nbin <- c(30, 30)          # number of bins
m <- c(5, 5)               # smoothing parameters

# First example with positive correlation
Sigma <- matrix(c(1, .9, .9, 1), 2, 2)
set.seed(345)
x <- rmvn.eigen(n, c(0, 0), Sigma=Sigma)
b <- bin2(x, nbin = nbin)
# kopt is the kernel type, here triangular
est <- ash2(b, m = m, kopt = c(1,0))

persp(x = est$x, y = est$y, z = est$z, shade=TRUE,
        xlab = "X", ylab = "Y", zlab = "", main="",
        theta = 30, phi = 75, ltheta = 30, box = FALSE)
contour(x = est$x, y = est$y, z = est$z, main="")
```

The perspective and contour plots from the ASH estimates are shown in Figures 10.12(a) and 10.12(c). The variables in the first example have positive correlation $\rho = 0.9$. In the second example, the variables have negative correlation $\rho = -0.9$.

```
# Second example with negative correlation
Sigma <- matrix(c(1, -.9, -.9, 1), 2, 2)
set.seed(345)
x <- rmvn.eigen(n, c(0, 0), Sigma=Sigma)
b <- bin2(x, nbin = nbin)
est <- ash2(b, m = m, kopt = c(1,0))

persp(x = est$x, y = est$y, z = est$z, shade=TRUE,
        xlab = "X", ylab = "Y", zlab = "", main="",
        theta = 30, phi = 75, ltheta = 30, box = FALSE)
contour(x = est$x, y = est$y, z = est$z, main="")
par(ask = FALSE)
```

The perspective plots and contour plots from the ASH estimates of the densities in the second case are shown in Figures 10.12(b) and 10.12(d). ◇

10.3.3 Multidimensional kernel methods

Suppose $X = (X_1, \ldots, X_d)$ is a random vector in $\mathbb{R}^d$, and $K(X) : \mathbb{R}^d \to \mathbb{R}$ is a kernel function, such that $K(X)$ is a density function on $\mathbb{R}^d$. Let the $n \times d$ matrix (x_{ij}) be an observed sample from the distribution of X. The smoothing parameter is a d-dimensional vector h. If the bandwidth is equal in all dimensions, the multivariate kernel density estimator of $f(X)$ with smoothing parameter h_1 is

$$\hat{f}_K(X) = \frac{1}{nh_1^d} \sum_{i=1}^{n} K\left(\frac{X - x_{i\cdot}}{h_1}\right), \tag{10.16}$$

where $x_{i\cdot}$ is the ith row of (x_{ij}). Usually $K(X)$ will be a symmetric and unimodal density on $\mathbb{R}^d$, such as a standard multivariate normal density. The Gaussian kernels have unbounded support. An example of a kernel with bounded support is the multivariate version of the Epanechnikov kernel, defined

$$K(X) = \frac{1}{2c_d}(d+2)(1 - X^T X)\, I(X^T X < 1),$$

where $c_d = 2\pi^{d/2}/(d\Gamma(d/2))$ is the volume of the d-dimensional unit sphere. When $d = 1$ the constant is $c_1 = 2$ and $K(x) = (3/4)(1-x^2)\, I(|x| < 1)$, which is the univariate Epanechnikov kernel given in Table 10.2.

In the bivariate case, choosing equal bandwidths $h_1 = h_2$ and the standard Gaussian kernel corresponds to centering identical weight functions like

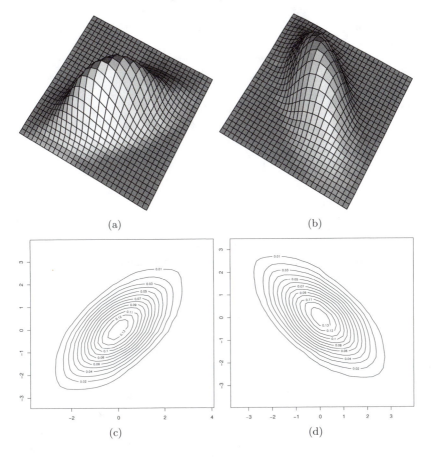

FIGURE 10.12: Bivariate ASH density estimates of bivariate normal data in Example 10.14.

smooth bumps at each sample point and summing the heights of these surfaces to obtain the density estimate at a given point. For the bivariate Gaussian kernel, in a graphical representation corresponding to Figure 10.6 the small bumps will be surfaces (bivariate normal densities) rather than curves.

The product kernel density estimate of $f(X)$ with smoothing parameter $h = (h_1, \ldots, h_d)$ is

$$\hat{f}(X) = \frac{1}{nh_1 \cdots h_d} \sum_{i=1}^{n} \prod_{j=1}^{d} K\left(\frac{X_i - x_{ij}}{h_j}\right). \qquad (10.17)$$

For this estimator and the multivariate frequency polygon, the optimal smoothing parameter has

$$h_j^* = O(n^{-1/(4+d)}), \qquad AMISE^* = O(n^{-4/(4+d)}),$$

and for uncorrelated multivariate normal data the optimal bandwidths are

$$h_j^* = \left(\frac{4}{d+2}\right)^{1/(d+4)} \times \sigma_i \, n^{-1/(d+4)}.$$

The constant $(4/(d+2))^{1/(d+4)}$ is close to 1 and converges to 1 as $d \to \infty$, thus Scott's multivariate normal reference rule [244] for d-dimensional data is

$$\hat{h}_i = \hat{\sigma}_i n^{-1/(d+4)}.$$

Example 10.15 (Product kernel estimate of a bivariate normal mixture)

This example plots the density estimate for a bivariate normal location mixture using kde2d (MASS). The mixture has three components with different mean vectors and identical variance $\Sigma = I_2$. The mean vectors are

$$\mu_1 = \begin{bmatrix} 0 \\ 1 \end{bmatrix}, \qquad \mu_2 = \begin{bmatrix} 4 \\ 0 \end{bmatrix}, \qquad \mu_3 = \begin{bmatrix} 3 \\ -1 \end{bmatrix},$$

and the mixing probabilities are $p = (0.2, 0.3, 0.5)$. The code to generate the mixture data and plots in Figure 10.13 follows.

```
library(MASS)  #for mvrnorm and kde2d
#generate the normal mixture data
n <- 2000
p <- c(.2, .3, .5)
mu <- matrix(c(0, 1, 4, 0, 3, -1), 3, 2)
Sigma <- diag(2)
i <- sample(1:3, replace = TRUE, prob = p, size = n)
k <- table(i)

x1 <- mvrnorm(k[1], mu = mu[1,], Sigma)
x2 <- mvrnorm(k[2], mu = mu[2,], Sigma)
x3 <- mvrnorm(k[3], mu = mu[3,], Sigma)
X <-  rbind(x1, x2, x3)   #the mixture data
x <- X[,1]
y <- X[,2]
> print(c(bandwidth.nrd(x), bandwidth.nrd(y)))
[1] 1.876510 1.840368

# accepting the default normal reference bandwidth
fhat <- kde2d(x, y)
contour(fhat)
persp(fhat, phi = 30, theta = 20, d = 5, xlab = "x")
```

```
# select bandwidth by unbiased cross-validation
h = c(ucv(x), ucv(y))
fhat <- kde2d(x, y, h = h)
contour(fhat)
persp(fhat, phi = 30, theta = 20, d = 5, xlab = "x")
```

The bandwidth by normal reference is $h \doteq (1.877, 1.840)$, and by cross-validation $h \doteq (0.556, 1.132)$. The first choice results in a smoother estimate. Although in Figure 10.13 three modes are evident for both estimates, it appears that the density estimate corresponding to unbiased cross-validation may be too rough in this example. ◇

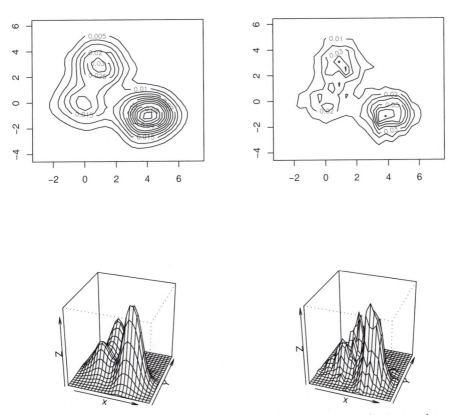

FIGURE 10.13: Product kernel estimates of bivariate normal mixture data in Example 10.15 (normal reference rule at left.)

For kernel density estimates for multivariate data also see kde (ks) [76] and KernSnooth [286]. Readers are referred to the examples of kde2d (MASS) for a Gaussian kernel density estimate of the bivariate geyser (MASS) data with default normal reference bandwidth (also see [278, 5.6]).

10.4 Other Methods of Density Estimation

Orthogonal systems provide an alternate approach to density estimation [244, 252, 285]. Suppose that the random variable X is supported on the interval $[0, 1]$. Then one approach to estimation of the density f of X is to represent f by its Fourier expansion and estimate the Fourier coefficients from the observed random sample $X_1, \ldots, X_n$. Although intuitively appealing, the resulting estimator is not useful because it will tend to a sum of delta functions that place probability mass at the individual observations. See [244], [252], or [285] for an explanation of how this problem is resolved by smoothing to obtain a more useful density estimator, and how it is generalized to densities with unbounded support. Scott [244, p. 129] shows that the resulting estimator is in the form of a fixed kernel estimator. Walter and Shen [285, Sec. 13.3] show that an estimator based on the Haar wavelets is the traditional histogram estimator of a density.

Scott [244] and Silverman [252] discuss several other approaches to density estimation including adaptive kernel methods and cross-validation, near neighbor estimates, and penalized likelihood methods. An L_1 approach to density estimation is covered by Devroye and Györfi [71]. Many other criteria have been applied, such as the Kullback-Liebler distance, Hellinger distance, AIC, etc. Other approaches focus on regression and smoothing [79, 128, 129, 130, 203], splines [86, 284], or generalized additive models [135, 137]. Some related R packages are ash [245], gam [134], gss [123], KernSmooth [286], ks [76], locfit [180], MASS [278], sm [29], and splines.

Exercises

10.1 Construct a histogram estimate of density for a random sample of standard lognormal data using Sturges' Rule, for sample size $n = 100$. Repeat the estimate *for the same sample* using the correction for skewness proposed by Doane [73] in equation (10.2). Compare the number of bins and break points using both methods. Compare the density estimates at the deciles of the lognormal distribution with the lognormal density at the same points. Does the suggested correction give better density estimates in this example?

10.2 Estimate the IMSE for three histogram density estimates of standard normal data, from a sample size $n = 500$. Use Sturges' Rule, Scott's Normal Reference Rule, and the FD Rule.

10.3 Construct a frequency polygon density estimate for the `precip` dataset in R. Verify that the estimate satisfies $\int_{-\infty}^{\infty} \hat{f}(x)dx \doteq 1$ by numerical integration of the density estimate.

10.4 Construct a frequency polygon density estimate for the `precip` dataset, using a bin width determined by substituting

$$\hat{\sigma} = IQR/1.348$$

for standard deviation in the usual Normal Reference Rule for a frequency polygon.

10.5 Construct a frequency polygon density estimate for the `precip` dataset, using a bin width determined by the Normal Reference Rule for a frequency polygon adjusted for skewness. The skewness adjustment factor is given in 10.8.

10.6 Construct an ASH density estimate for the `faithful$eruptions` dataset in R, using width h determined by the normal reference rule. Use a weight function corresponding to the biweight kernel,

$$K(t) = \frac{15}{16}(1 - t^2)^2 \text{ if } |t| < 1, \qquad K(t) = 0 \text{ otherwise.}$$

10.7 Construct an ASH density estimate for the `precip` dataset in R. Choose the best value for width h^* empirically by computing the estimates over a range of possible values of h and comparing the plots of the densities. Does the optimal value h_n^{fp} correspond to the optimal value h^* suggested by comparing the density plots?

10.8 The `buffalo` dataset in the `gss` [123] package contains annual snowfall accumulations in Buffalo, New York from 1910 to 1973. The 64 observations are

126.4	82.4	78.1	51.1	90.9	76.2	104.5	87.4	110.5	25.0	69.3	53.5	39.8
63.6	46.7	72.9	79.6	83.6	80.7	60.3	79.0	74.4	49.6	54.7	71.8	49.1
103.9	51.6	82.4	83.6	77.8	79.3	89.6	85.5	58.0	120.7	110.5	65.4	39.9
40.1	88.7	71.4	83.0	55.9	89.9	84.8	105.2	113.7	124.7	114.5	115.6	102.4
101.4	89.8	71.5	70.9	98.3	55.5	66.1	78.4	120.5	97.0	110.0		

This data was analyzed by Scott [242]. Construct kernel density estimates of the data using Gaussian and biweight kernels. Compare the estimates for different choices of bandwidth. Is the estimate more influenced by the type of kernel or the bandwidth?

10.9 Construct a kernel density estimate for simulated data from the normal location mixture $\frac{1}{2}N(0,1) + \frac{1}{2}N(3,1)$. Compare several choices of bandwidth, including (10.13) and (10.14). Plot the true density of the mixture over the density estimate, for comparison. Which choice of smoothing parameter appears to be best?

10.10 Apply the reflection boundary technique to obtain a better kernel density estimate for the precipitation data in Example 10.8. Compare the estimates in Example 10.8 and the improved estimates in a single graph. Also try setting `from = 0` or `cut = 0` in the `density` function.

10.11 Write a bivariate density polygon plotting function based on Examples 10.12 and 10.13. Use Example 10.13 to check the results, and then apply your function to display the bivariate `faithful` data (Old Faithful geyser).

10.12 Plot a bivariate ASH density estimate of the `geyser(MASS)` data.

10.13 Generalize the bivariate ASH algorithm to compute an ASH density estimate for a d-dimensional multivariate density, $d \geq 2$.

10.14 Write a function to bin three-dimensional data into a three-way contingency table, following the method in the `bin2d` function of Example 10.12. Check the result on simulated $N_3(0, I)$ data. Compare the marginal frequencies returned by your function to the expected frequencies from a standard univariate normal distribution.

R Code

Code to generate data as shown in Table 10.1 on page 289.

```
N <- c(10, 20, 30, 50, 100, 200, 500, 1000, 5000, 10000)
m <- length(N)
out <- matrix(0, nrow = m, ncol = 8)
out[ ,1] <- N
out[ ,5] <- N
for (i in 1:m) {
    x <- rnorm(N[i])
    out[i, 2:4] <- c(nclass.Sturges(x),
        nclass.scott(x), nclass.FD(x))
    x <- rexp(N[i])
    out[i, 6:8] <- c(nclass.Sturges(x),
        nclass.scott(x), nclass.FD(x))
}
print(out)
```

Code to plot the histograms in Figure 10.4 on page 294.

```
library(MASS)  #for truehist
par(mfrow = c(2, 2))
x <- sort(rnorm(1000))
y <- dnorm(x)
o <- (1:4) / 4
h <- .35
for (i in 1:4) {
    truehist(x, prob = TRUE, h = .35, x0 = o[i],
        xlim = c(-3.5, 3.5), ylim = c(0, 0.45),
        ylab = "Density", main = "")
    lines(x, y)
}
par(mfrow = c(1, 1))
```

Code to plot Figure 10.6 on page 298.

To display the type of plot in Figure 10.6, first open a new plot to set up the plotting window, but use `type="n"` in the `plot` command so that nothing is drawn in the graph window yet. Then add the density curves for each point inside the loop using `lines`. Finally, add the density estimate using `lines` again.

```
for (h in c(.25, .4, .6, 1)) {
    x <- seq(-4, 4, .01)
    fhat <- rep(0, length(x))
    # set up the plot window first
    plot(x, fhat, type="n", xlab="", ylab="",
        main=paste("h=",h), xlim=c(-4,4), ylim=c(0, .5))
    for (i in 1:n) {
        # plot a normal density at each sample pt
        z <- (x - y[i]) / h
        f <- dnorm(z)
        lines(x, f / (n * h))
        # sum the densities to get the estimates
        fhat <- fhat + f / (n * h)
    }
    lines(x, fhat, lwd=2) # add density estimate to plot
}
```

Use `par(mfrow = c(2, 2))` to display four plots in one screen.

Code to plot kernels in Figure 10.7 on page 299.

```
#see examples for density, kernels in S parametrization
(kernels <- eval(formals(density.default)$kernel))

plot(density(0, from=-1.2, to=1.2, width=2,
    kern="gaussian"), type="l", ylim=c(0, 1),
    xlab="", main="")
for(i in 2:5)
    lines(density(0, width=2, kern=kernels[i]), lty=i)
legend("topright", legend=kernels[1:5],
    lty=1:5, inset=.02)
```

Chapter 11

Numerical Methods in R

11.1 Introduction

This chapter begins with a review of some concepts that should be understood by any statistician who will apply numerical methods that are implemented in statistical packages such as R. Following this introduction, a selection of examples are presented that illustrate the application of numerical methods using functions provided in R. Readers should refer to one or more of the relevant references for a thorough and rigorous presentation of the underlying principles.

Many excellent references are available on numerical methods. Two recent texts written to address the problems of statistical computing in particular are Monahan [202] and Lange [168]. The Monahan text is an excellent resource for statisticians with a limited background in numerical analysis. Nocedal and Wright [206] is a graduate level text on optimization. Lange [169] is another graduate level optimization text that features statistical applications. Thisted [269] covers numerical computation for statistics, including numerical analysis, numerical integration, and smoothing.

Computer representation of real numbers

A positive decimal number x is represented by the ordered coefficients $\{d_j\}$ in the series

$$d_n 10^n + d_{n-1} 10^{n-1} + \cdots + d_1 10^1 + d_0 + d_{-1} 10^{-1} + d_{-2} 10^{-2} + \dots$$

and decimal point separating d_0 and d_{-1}, where d_j are integers in $\{0, 1, \dots, 9\}$. The same number can be represented in base 2 using the binary digits $\{0, 1\}$ by $a_k a_{k-1} \dots a_1 a_0 . a_{-1} a_{-2} \dots$, where

$$x = a_k 2^k + a_{k-1} 2^{k-1} + \cdots + a_1 2 + a_0 + a_{-1} 2^{-1} + a_{-2} 2^{-2} + \dots,$$

$a_j \in \{0, 1\}$. The point separating a_0 and a_{-1} is called the radix point. Similarly, x can be represented in any integer base $b > 1$ by expanding in powers of b.

R note 11.1 *The function* `digitsBase` *in the package* `sfsmisc` *[183] returns the vector of digits that represent a given integer in another base.*

Whenever a computer is involved in mathematical calculations, it is very likely to involve the conversions "from" and "to" decimal, because machines and humans represent numbers in different bases. Both types of conversions introduce errors that could be significant in certain cases.

At the lowest level, the computer recognizes exactly two states, like a switch that is on or off, or a circuit that is open or closed. Therefore, at some level, the base 2 representation is used in computer arithmetic. Other powers of 2 such as 8 (octal) or 16 (hexadecimal) are also more natural for low level routines than base 10.

Positive integers can always be represented by a finite sequence of digits, ending with an implicit radix point. For this reason, integers are called *fixed point* numbers. Numbers that require an explicit radix point in the sequence of digits may be fixed point or floating point (generally treated as floating point in calculations). *Floating point* numbers are represented by a sign, a finite sequence of digits, and an exponent, similar to the representation of real numbers in scientific notation. In general, this representation of a real number is approximate, not exact.

Even though the internal representation of numbers is usually transparent to the user, who conveniently interacts with the software in the decimal system, it is important in statistical computing to understand that there are fundamental differences between mathematical calculations and computer calculations. Mathematical ideas such as *limit*, *supremum*, *infimum*, etc. cannot be exactly reproduced in the computer. No computer has infinite storage capacity, so only finitely many numbers can be represented in the computer; there is a smallest and a largest positive number. See Monahan [202, Ch. 2] for a discussion of fixed point and floating point arithmetic, and inaccuracies that can occur in algorithms as simple as calculation of sample variance.

R note 11.2 *The R variable* .Machine *holds machine specific constants with information on the largest integer, smallest number, etc. For example, in R-2.5.0 for Windows, the largest integer (*.Machine\$integer.max*) is $2^{31} - 1 = 2147483647$. Type* .Machine *at the command prompt for the complete list. For portability and reusability of code, tolerances or convergence criteria should be given in terms of machine constants. For example, the* uniroot *function, which seeks a root of a univariate function, has a default tolerance of* .Machine\$double.eps^0.25.

Occasionally users are surprised to find that some mathematical identities appear to be contradicted by the software. A typical example is

```
> (.3 - .1)
[1] 0.2
> (.3 - .1) == .2
[1] FALSE
> .2 - (.3 - .1)
[1] 2.775558e-17
```

The base 2 representation of 0.2 is an infinite series of digits $0.00110011\ldots$, which cannot be represented exactly in the computer. Notice that although the result above is not exactly equal to 0.2, the error is negligible. Good programming practice avoids testing the equality of two floating point numbers.

Example 11.1 (Identical and nearly equal)

R provides the function `all.equal` to check for near equality of two R objects. In a logical expression, use `isTRUE` to obtain a logical value.

```
> isTRUE(all.equal(.2, .3 - .1))
[1] TRUE
> all.equal(.2, .3)              #not a logical value
[1] "Mean relative  difference: 0.5"
> isTRUE(all.equal(.2, .3))  #always a logical value
[1] FALSE
```

The `isTRUE` function is applied in Example 11.9. The `identical` function is available for testing whether two objects are identical. The help topic for `identical` gives very clear and explicit advice to programmers: "A call to `identical` is the way to test exact equality in `if` and `while` statements, as well as in logical expressions that use `&&` or `||`. In all these applications you need to be assured of getting a single logical value." Also see the examples below.

```
> x <- 1:4
> y <- 2
> y == 2
[1] TRUE
> x == y #not necessarily a single logical value
[1] FALSE   TRUE FALSE FALSE
> identical(x, y)  #always a single logical value
[1] FALSE
> identical(y, 2)
[1] TRUE
```

◇

Overflow occurs when the result of an arithmetic operation exceeds the maximum floating point number that can be represented. Underflow occurs when the result is smaller than the minimum floating point number. In the case of underflow, the result might unexpectedly be returned as zero. This could lead to division by zero or other problems that produce unexpected and possibly inaccurate results – without warning. Overflow is usually more obvious, but should be avoided. Good algorithms should set underflows to zero and give a warning if this may produce unexpected results. Programmers can avoid many of these problems, however, by carefully coding arithmetic expressions with the limitations of the machine in mind.

Often the expression to be evaluated is not impossible to compute, but one needs to be careful about the order of operations. One of the most common, and easily avoided problems occurs when we need to compute a ratio of two very large or very small numbers. For example, $n!/(n-2)! = n(n-1)$, but we could easily have trouble computing the numerator or denominator if n is large. A good approach for this type of problem is to take the logarithm of the quotient and exponentiate the result. A typical example that arises in statistical applications is the following.

Example 11.2 (Ratio of two large numbers)

Evaluate

$$\frac{\Gamma((n-1)/2)}{\Gamma(1/2)\Gamma((n-2)/2)}.$$

This could be coded using the gamma function in R, but $\Gamma(n) = (n-1)!$, so when n is large, gamma may return Inf and the arithmetic operations could return NaN. On the other hand, although numerator and denominator are both large, the ratio is much smaller. Compute the ratio $\Gamma((n-1)/2)/\Gamma((n-2)/2)$ using the logarithm of the gamma function lgamma. That is, $\Gamma(n)/\Gamma(m) = $ exp(lgamma(n) - lgamma(m)). Also, recall that $\Gamma(1/2) = \sqrt{\pi}$.

```
> n <- 400
> (gamma((n-1)/2) / (sqrt(pi) * gamma((n-2)/2)))
[1] NaN
> exp(lgamma((n-1)/2) - lgamma((n-2)/2)) / sqrt(pi)
[1] 7.953876
```

◇

A thorough discussion of computer arithmetic is beyond the scope of this text. Among the references, Monahan [202] or Thisted [269] are good starting points on this topic for statistical computing; on computer arithmetic and algorithms see e.g. Higham [142] or Knuth [164].

Evaluating Functions

The power series expansion of a function is commonly applied. If $f(x)$ is analytic, then $f(x)$ can be evaluated in a neighborhood of the point x_0 by a power series

$$f(x) = \sum_{k=0}^{\infty} a_k(x - x_0)^k.$$

The Taylor series representation of $f(x)$ in a neighborhood of x_0 is

$$f(x) = \sum_{k=0}^{\infty} \frac{f^{(k)}(x_0)}{k!}(x - x_0)^k,$$

also called a Maclaurin series when $x_0 = 0$. The infinite series must be truncated in order to obtain a numerical approximation. The power series approximation is thus a (high degree) polynomial approximation. If f has continuous derivatives up to order $(n + 1)$ in a neighborhood of 0, then the finite Taylor expansion of $f(x)$ at $x_0 = 0$ is

$$\lim_{x \to 0} f(x) = \sum_{k=0}^{n} \frac{f^{(k)}(0)}{k!} x^k + R_n(x),$$

where $R_n(x) = O(x^{n+1})$.

Recall that "O" (big oh) and "o" (little oh) describe the order of convergence of functions. Let f and g be defined on a common interval (a, b) and let $a \le x_0 \le b$. Suppose that $g(x) \ne 0$ for all $x \ne x_0$ in a neighborhood of x_0. Then $f(x) = O(g(x))$ if there exists a constant M such that $|f(x)| \le M|g(x)|$ as $x \to x_0$. If $\lim_{x \to x_0} f(x)/g(x) = 0$ then $f(x) = o(g(x))$.

If the finite Taylor expansion is computed in a language such as C or fortran, a method is used that avoids repeated multiplications. That is, if $y_k = x^k/k!$ then

$$\frac{1}{k!} f^{(k)}(0)x^k = y_k f^{(k)}(0) = y_{k-1}(x/k)f^{(k)}(0),$$

saving many multiplications. Computing in R, however, it will usually be faster to take advantage of the vectorized operations, provided it is known how many terms are required.

Example 11.3 (Taylor expansion)

Consider the finite Taylor expansion for the sine function,

$$\sin x = \sum_{k=0}^{n} \frac{(-1)^k}{(2k + 1)!} x^{2k+1}.$$

For example, evaluate $\sin(\pi/6)$ from the Taylor polynomial.

The remainder term $R_n(x) = O(x^{n+1})$ can be used to determine the approximate number of terms required in the finite expansion. Suppose that a 24th degree polynomial is sufficiently accurate at $x = \pi/6$. Two methods of computing the Taylor polynomial are compared below.

The following method of calculation is efficient in C or fortran code, but not in R. The timer measures 1000 calculations of the Taylor polynomial.

```
system.time({
    for (i in 1:1000) {
        a <- rep(0, 24)
        a0 <- pi / 6
        a2 <- a0 * a0
        a[1] <- -a0^3 / 6
        for (i in 2:24)
            a[i] <- - a2 * a[i-1] / ((2*i+1)*(2*i))
        a0 + sum(a)}
})
[1] 0.36 0.01 0.49   NA   NA
```

Compare the version above to the vectorized version below. The vectorized version appears to be about 5 times faster than the method above. In R code, vectorized operations like the code below are usually more efficient than loops.

```
system.time({
    for (i in 1:1000) {
        K <- 2 * (0:24) + 1
        i <- rep(c(1, -1), length=25)
        sum(i * (pi/6)^K / factorial(K))}
})
[1] 0.07 0.01 0.08   NA   NA
```

◇

Power series expansions are also useful for numerical evaluation of derivatives. Within the common region of the radius of convergence of the power series and its derivative, one can differentiate the finite expansion term by term. The next example illustrates this method with a useful function for the derivative of the zeta function.

Example 11.4 (Derivative of zeta function)

The Riemann zeta function is defined by

$$\zeta(a) = \sum_{i=1}^{\infty} \frac{1}{i^a},$$

which converges for all $a > 1$. Write a function to evaluate the first derivative of the zeta function.

It can be shown that

$$\zeta(a) = \frac{1}{z-1} + \sum_{n=0}^{\infty} \frac{(-1)^n}{n!} \gamma_n (z-1)^n,$$

where

$$\gamma_n = \zeta^{(n)}(z) - \frac{(-1)^n n!}{(z-1)^{n+1}}\bigg|_{z=1} = \lim_{m \to \infty} \left[\sum_{k=1}^{m} \frac{(\log k)^n}{k} - \frac{(\log m)^{n+1}}{n+1} \right]$$

are the Stieltjes constants. Differentiating $\zeta(a)$ gives

$$\zeta'(a) = -\frac{1}{(z-1)^2} - \gamma_1 + \gamma_2(z-1) - \frac{1}{2}\gamma_3(z-1)^2 + \dots, \qquad a > 1.$$

The Stieltjes constants can be evaluated numerically, and tables of the Stieltjes constants are available [1]. More terms can be added if greater accuracy is needed, but to conserve space, only five of the constants are used in the version below. This "light" version of the zeta derivative gives remarkably good results over the interval $(1, 2)$ (see the next example).

```
zeta.deriv <- function(a) {
    z <- a - 1
    # Stieltjes constants gamma_k for k=1:5
    g <- c(
            -.7281584548367672e-1,
            -.9690363192872318e-2,
            .2053834420303346e-2,
            .2325370065467300e-2,
            .7933238173010627e-3)
    i <- c(-1, 1, -1, 1, -1)
    n <- 0:4
    -1/z^2 + sum(i * g * z^n / factorial(n))
}
```

◇

Another approach to numerical evaluation of the derivative of a function applies the following central difference formula

$$f'(x) \approx \frac{f(x+h) - f(x-h)}{2h},$$

for a small value of h. According to [213], h should be chosen so that x and $x + h$ differ by an exactly representable number.

Example 11.5 (Derivative of zeta function, cont.)

Compare the finite series approximation of the numerical derivative in Example 11.4 with the central difference formula. That is, for small h, compare

$$\frac{\zeta(a+h) - \zeta(a-h)}{2h}$$

with the value returned by `zeta.deriv(a)` in Example 11.4. The $\zeta(\cdot)$ function is implemented in the GNU scientific library, available in the `gsl` package [127].

```
library(gsl)    #for zeta function
z <- c(1.001, 1.01, 1.5, 2, 3, 5)
h <- .Machine$double.eps^0.5
dz <- dq <- rep(0, length(z))
for (i in 1:length(z)) {
    v <- z[i] + h
    h <- v - z[i]
    a0 <- z[i] - h
    if (a0 < 1) a0 <- (1 + z[i])/2
    a1 <- z[i] + h
    dq[i] <- (zeta(a1) - zeta(a0)) / (a1 - a0)
    dz[i] <- zeta.deriv(z[i])
}

h
[1] 1.490116e-08

cbind(z, dz, dq)
        z            dz            dq
[1,] 1.001 -9.999999e+05 -9.999999e+05
[2,] 1.010 -9.999927e+03 -9.999927e+03
[3,] 1.500 -3.932240e+00 -3.932240e+00
[4,] 2.000 -9.375469e-01 -9.375482e-01
[5,] 3.000 -1.981009e-01 -1.981262e-01
[6,] 5.000 -2.853446e-02 -2.857378e-02
```

Values of z are given in the first column, values of $\zeta'(z)$ computed by the finite series approximation are given in column `dz`, and values of $\zeta'(z)$ computed by the central difference formula are given in column `dq`. Although the two estimates are quite close, it appears that the difference in the two estimates is increasing with z. The finite series approximation can be improved by adding more terms. ◇

11.2 Root-finding in One Dimension

This section will briefly summarize the main ideas behind the Brent minimization algorithm [32], on which the R root-finding function `uniroot` is based, and illustrate its application with examples. Refer to [32, 169, 213, 206] for more details. The source code of the fortran implementation "zeroin.f" in the GNU Scientific Library (GSL) can be found at the web site `http://www.gnu.org/software/gsl/`.

Let $f(x)$ be a continuous function $f : \mathbb{R}^1 \to \mathbb{R}^1$. A root (or zero) of the equation $f(x) = c$ is a number x such that $g(x) = f(x) - c = 0$. Thus, we can restrict attention to solving $f(x) = 0$.

One can choose from numerical methods that require evaluation of the first derivative of $f(x)$, and algorithms that do not require the first derivative. Newtons's method or Newton-Raphson method are examples of the first type, while Brent's algorithm is an example of the second type of method. In either case, one must bracket the root between two endpoints where $f(\cdot)$ has opposite signs.

Bisection method

If $f(x)$ is continuous on $[a, b]$, and $f(a), f(b)$ have opposite signs, then by the intermediate value theorem it follows that $f(c) = 0$ for some $a < c < b$. The bisection method simply checks the sign of $f(x)$ at the midpoint $x = (a+b)/2$ of the interval at each iteration. If $f(a), f(x)$ have opposite signs, then the interval is replaced by $[a, x]$ and otherwise it is replaced by $[x, b]$. At each iteration, the length of the interval containing the root decreases by half. The method cannot fail, and the number of iterations needed to achieve a specified tolerance is known in advance. If the initial interval $[a, b]$ contains more than one root, then bisection will find one of the roots. The rate of convergence of the bisection algorithm is linear.

Example 11.6 (Solving $f(x) = 0$)

Solve

$$a^2 + y^2 + \frac{2ay}{n - 1} = n - 2,$$

where a is a specified constant and $n > 2$ is an integer. Of course, this equation can be solved directly by elementary algebra, to obtain the exact solution:

$$y = \frac{-a}{n - 1} \pm \sqrt{n - 2 + a^2 + \left(\frac{a}{n - 1}\right)^2}.$$

Let us compare the exact solution with a numerical solution. Apply the bisection method to seek a positive solution. If we restate the problem as: find the solutions of

$$f(y) = a^2 + y^2 + \frac{2ay}{n - 1} - (n - 2) = 0,$$

the first step is to code the function f. The next step is to determine an interval such that $f(y)$ has opposite signs at the endpoints. For example, if $a = 1/2$ and $n = 20$, there will be a positive and a negative root. In the following, the positive root is found, starting from the interval $(0, 5n)$.

```
f <- function(y, a, n) {
    a^2 + y^2 + 2*a*y/(n-1) - (n-2)
}

a <- 0.5
n <- 20
b0 <- 0
b1 <- 5*n

#solve using bisection
it <- 0
eps <- .Machine$double.eps^0.25
r <- seq(b0, b1, length=3)
y <- c(f(r[1], a, n), f(r[2], a, n), f(r[3], a, n))
if (y[1] * y[3] > 0)
    stop("f does not have opposite sign at endpoints")

while(it < 1000 && abs(y[2]) > eps) {
    it <- it + 1
    if (y[1]*y[2] < 0) {
        r[3] <- r[2]
        y[3] <- y[2]
    } else {
        r[1] <- r[2]
        y[1] <- y[2]
    }
    r[2] <- (r[1] + r[3]) / 2
    y[2] <- f(r[2], a=a, n=n)
    print(c(r[1], y[1], y[3]-y[2]))
}
```

The estimate of the root when the stopping condition is satisfied is the value in r[2] and the value of the function at r[2] is in y[2].

```
>      r[2]
[1] 4.186845
>      y[2]
[1] 2.984885e-05
>      it
[1] 21
```

Our exact formula gives the roots $y = 4.186841, -4.239473$. (Most problems, including this one, can be solved more efficiently using the uniroot function, which is shown in Example 11.7 below.)   ◇

Other methods, such as the secant method, may (formally) converge faster than the bisection method, but the root may not remain bracketed. These superlinear methods may be faster for many problems, but may fail to converge to a root. The secant method assumes that $f(x)$ is approximately linear on the interval bracketing the root. Inverse quadratic interpolation approximates $f(x)$ with a quadratic function fitted to the three prior points.

Brent's method

Brent's method combines the root bracketing and bisection with inverse quadratic interpolation. It fits x as a quadratic function of y. If the three points are $(a, f(a)), (b, f(b)), (c, f(c))$, with b as the current best estimate, the next estimate for the root is found by interpolation, setting $y = 0$ in the Lagrange interpolation polynomial

$$x = \frac{[y - f(a)][y - f(b)]c}{[f(c) - f(a)][f(c) - f(b)]}$$
$$+ \frac{[y - f(b)][y - f(c)]a}{[f(a) - f(b)][f(a) - f(c)]} + \frac{[y - f(c)][y - f(a)]b}{[f(b) - f(c)][f(b) - f(a)]}.$$

If this estimate is outside of the interval known to bracket the root, bisection is used at this step. (For details see [32] or[213] or the zeroin.f fortran code.) Brent's method is generally faster than bisection, and it has the sure convergence of the bisection method.

Brent's method is implemented in the R function `uniroot`, which searches for a zero of a univariate function between two points where the function has opposite signs.

Example 11.7 (Solving $f(x) = 0$ with Brent's method: `uniroot`)

Solve

$$a^2 + y^2 + \frac{2ay}{n - 1} = n - 2,$$

with $a = 0.5, n = 20$ as in Example 11.6. The first step is to code the function f. This function is not complicated, so we code this function inline in the `uniroot` statement. The next step is to determine an interval such that $f(y)$ has opposite signs at the endpoints. The call to `uniroot` and result are shown below.

```
a <- 0.5
n <- 20
out <- uniroot(function(y) {
            a^2 + y^2 + 2*a*y/(n-1) - (n-2) },
            lower = 0, upper = n*5)
```

```
> unlist(out)
        root          f.root          iter    estim.prec
4.186870e+00 2.381408e-04 1.400000e+01 6.103516e-05
```

In the call to `uniroot`, we can optionally specify the maximum number of iterations (default 1000) or the tolerance (default `.Machine$double.eps^0.25`). The positive solution to $f(y) = 0$ is (approximately) $y = 4.186870$. To seek the negative root, we can apply `uniroot` again. The interval can be specified as above, or as shown below.

```
uniroot(function(y) {a^2 + y^2 + 2*a*y/(n-1) - (n-2)},
        interval = c(-n*5, 0))$root
[1]  -4.239501
```

Our exact formula (see Example 11.6) gives $y = 4.186841, -4.239473$.   ◇

R note 11.3 *Also see the* `polyroot` *function, to find zeroes of a polynomial with real or complex coefficients. See the help topic* `Complex` *for description of functions in R that support complex arithmetic.*

11.3   Numerical Integration

Basic numerical integration using the `integrate` function is illustrated in the following examples, where useful functions are developed for the density and cdf of the sample correlation statistic.

Numerical integration methods can be adaptive or non-adaptive. Non-adaptive methods apply the same rules over the entire range of integration. The integrand is evaluated at a finite set of points and a weighted sum of these function values is used to obtain the estimate. The numerical estimate of $\int_a^b f(x)$ is of the form $\sum_{i=0}^n f(x_i)w_i$, where $\{x_i\}$ are points in an interval containing $[a, b]$ and $\{w_i\}$ are suitable weights.

For example, the trapezoidal rule divides $[a, b]$ into n equal length subintervals length $h = (b - a)/n$, with endpoints $x_0, x_1, \ldots, x_n$, and uses the area of the trapezoid to estimate the integral over each subinterval. The estimate on (x_i, x_{i+1}) is $(f(x_i) + f(x_{i+1}))(h/2)$. The numerical estimate of $\int_a^b f(x)dx$ is

$$\frac{h}{2}f(a) + h\sum_{i=1}^{n-1} f(x_i) + \frac{h}{2}f(b).$$

If $f(x)$ is twice continuously differentiable, the error is $O(f''(x^*)/n^2)$, where $x^* \in (a, b)$. This is an example of a closed Newton-Cotes integration formula. See e.g. [269, Ch. 5] for more examples.

Quadrature methods evaluate the integrand at a finite set of points (nodes), but these nodes need not be evenly spaced. Suppose that w is a non-negative function such that $\int_a^b x^k w(x)dx < \infty$, for all $k \geq 0$. Then the integrand $f(x)$ can be expressed as $g(x)w(x)$.

Note that we have assumed that $w(x)/\int_a^b w(x)dx$ is a density function with finite positive moments. For example, we can take $w(x) = \exp(-x^2/2)$, called Gauss-Hermite quadrature. In this case, $\int_{-\infty}^\infty x^k w(x)dx < \infty$, for all $k \geq 0$, which applies to arbitrary intervals (a, b) on the real line. In Gaussian quadrature, the nodes $\{x_i\}$ selected are the roots of a set of orthogonal polynomials with respect to w. The normalized orthogonal polynomials also determine weights $\{w_i\}$.

The Gaussian Quadrature Theorem implies that if $g(x)$ is $2n$ times continuously differentiable, then the error in the numerical estimate $\sum_{i=1}^n w_i g(x_i)$ is

$$\int_a^b g(x)w(x)dx - \sum_{i=1}^n w_i g(x_i) = \frac{g^{(2n)}(x^*)}{(2n)!k_n^2},$$

where k_n is the leading coefficient of the nth polynomial and $x^* \in (a, b)$. Quadrature and other approaches to numerical integration are discussed in more detail in [121, 168, 269].

When an integrand behaves well in one part of the range of integration, but not so well in another part, it helps to treat each part differently. Adaptive methods choose the subintervals based on the local behavior of the integrand.

The `integrate` function provided in R uses an adaptive quadrature method to find the approximate value of the integral of a one variable function. The limits of integration can be infinite. The maximum number of subintervals, the relative error and the absolute error can be specified, but have reasonable default values for many problems.

Example 11.8 (Numerical integration with `integrate`)

Compute

$$\int_0^\infty \frac{dy}{(\cosh y - \rho r)^{n-1}}, \tag{11.1}$$

where $-1 < \rho < 1$, $-1 < r < 1$ are constants and $n \geq 2$ is an integer. The graph of the integrand is shown in Figure 11.1(a). We apply adaptive quadrature implemented by the `integrate` function provided in R.

First write a function that returns the value of the integrand. This function should take as its first argument a vector containing the nodes, and return a vector of the same length. Additional arguments can also be supplied. This function or the name of this function is the first argument to `integrate`.

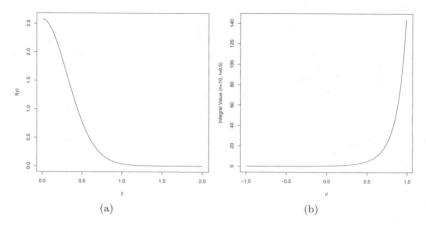

FIGURE 11.1:   Example 11.8 ($n = 10$, $r = 0.5$, $\rho = 0.2$) (a) Integrand,
(b) Value of the integral as a function of ρ.

  A simple way to compute the integral for fixed parameters, say ($n = 10$,
$r = 0.5$, $\rho = 0.2$) is

```
> integrate(function(y){(cosh(y) - 0.1)^(-9)}, 0, Inf)
1.053305 with absolute error < 2.3e-05
```

The integral for arbitrary (n, r, ρ) is needed, so write a more general integrand
function with these arguments, and supply the extra arguments in the call to
integrate.

```
f <- function(y, N, r, rho) {
    (cosh(y) - rho * r)^(1 - N)
}
integrate(f, lower=0, upper=Inf,
            rel.tol=.Machine$double.eps^0.25,
            N=10, r=0.5, rho=0.2)
```

This version produces the same estimate as above.
  To see how the result depends on ρ, fix $n = 10$ and $r = 0.5$ and plot the
value of the integral as a function of ρ. The plot is shown in Figure 11.1(b),
as produced by the following code.

```
ro <- seq(-.99, .99, .01)
v <- rep(0, length(ro))
for (i in 1:length(ro)) {
    v[i] <- integrate(f, lower=0, upper=Inf,
                rel.tol=.Machine$double.eps^0.25,
                N=10, r=0.5, rho=ro[i])$value
    }
```

```
plot(ro, v, type="l", xlab=expression(rho),
     ylab="Integral Value (n=10, r=0.5)")
```

◇

R note 11.4 *Sometimes there is a conflict between named arguments and optional user-supplied arguments. To avoid the conflict, either choose another name for the optional argument, or supply both arguments. For example, the following produces an error, because of apparent ambiguity between argument* rel.tol *and* r.

```
> integrate(f, lower=0, upper=Inf, n=10, r=0.5, rho=0.2)
Error in f(x, ...) : argument "r" is missing, with no default
```

The integral (11.1) appears in a density function in the following example.

Example 11.9 (Density of sample correlation coefficient)

The sample product-moment correlation coefficient measures linear association between two variables. The population correlation coefficient of (X, Y) is

$$\rho = \frac{E\left[(X - E(X))(Y - E(Y))\right]}{\sqrt{Var(X)Var(Y)}}.$$

If $\{(X_j, Y_j), \ j = 1, \ldots, n\}$ are paired sample observations, the sample correlation coefficient is

$$R = \frac{\sum_{j=1}^{n}(X_j - \overline{X})(Y_j - \overline{Y})}{\left[\sum_{j=1}^{n}(X_j - \overline{X})^2 \sum_{j=1}^{n}(Y_j - \overline{Y})^2\right]^{1/2}}.$$

Assume that $\{(X_j, Y_j), \ j = 1, \ldots, n\}$ are iid with $BVN(\mu_1, \mu_2, \sigma_1, \sigma_2, \rho)$ (bivariate normal) distribution. If $\rho = 0$, the density function of R (see e.g. [157, Ch. 32]) is given by

$$f(r) = \frac{\Gamma((n-1)/2)}{\Gamma(1/2)\Gamma((n-2)/2)}(1 - r^2)^{(n-4)/2}, \qquad -1 < r < 1. \qquad (11.2)$$

For $0 < |\rho| < 1$, the density function is more complicated. Several forms of the density function are given in [157, p. 549], including:

$$f(r) = \frac{(n-2)(1-\rho^2)^{(n-1)/2}(1-r^2)^{(n-4)/2}}{\pi} \int_0^\infty \frac{dw}{(\cosh w - \rho r)^{n-1}}, \qquad (11.3)$$

for $-1 < r < 1$.

To evaluate the density function (11.3), the integral must be evaluated. This is covered in Example 11.8. The case $\rho = 0$ can be handled separately using the simpler formula (11.2). The method for evaluating the constant $\Gamma((n - 1)/2)/\Gamma((n - 2)/2)$ was discussed in Example 11.2. The following function combines these results to evaluate the density of the correlation statistic.

```
.dcorr <- function(r, N, rho=0) {
    # compute the density function of sample correlation
    if (abs(r) > 1 || abs(rho) > 1) return (0)
    if (N < 4) return (NA)

    if (isTRUE(all.equal(rho, 0.0))) {
        a <- exp(lgamma((N - 1)/2) - lgamma((N - 2)/2)) /
                sqrt(pi)
        return (a * (1 - r^2)^((N - 4)/2))
    }

    # if rho not 0, need to integrate
    f <- function(w, R, N, rho)
        (cosh(w) - rho * R)^(1 - N)

    #need to insert some error checking here
    i <- integrate(f, lower=0, upper=Inf,
            R=r, N=N, rho=rho)$value
    c1 <- (N - 2) * (1 - rho^2)^((N - 1)/2)
    c2 <- (1 - r^2)^((N - 4) / 2) / pi
    return(c1 * c2 * i)
}
```

Some error checking should be added to this function in case the numerical integration fails.

As an informal check on the density calculations, plot the density curve. For $\rho = 0$ the density curve should be symmetric about 0 and the shape should resemble a symmetric beta density. The plot is shown in Figure 11.2.

```
r <- as.matrix(seq(-1, 1, .01))
d1 <- apply(r, 1, .dcorr, N=10, rho=.0)
d2 <- apply(r, 1, .dcorr, N=10, rho=.5)
d3 <- apply(r, 1, .dcorr, N=10, rho=-.5)
plot(r, d2, type="l", lty=2, lwd=2, ylab="density")
lines(r, d1, lwd=2)
lines(r, d3, lty=4, lwd=2)
legend("top", inset=.02, c("rho = 0", "rho = 0.5",
    "rho = -0.5"), lty=c(1,2,4), lwd=2)
```

◇

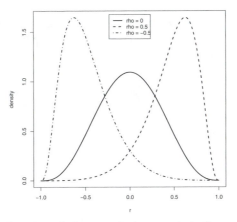

FIGURE 11.2: Density of the correlation statistic for sample size 10.

R note 11.5 *Density functions in R are vectorized, but the function* `.dcorr` *of Example 11.3 is really expecting a single number* r, *rather than a vector. Later this function can be extended to a general version* `dcorr`, *like the density functions* `dnorm`, `dgamma`, *etc. in R that accept vector arguments.*

11.4 Maximum Likelihood Problems

Maximum likelihood is a method of estimation of parameters of a distribution. The abbreviation MLE may refer to maximum likelihood estimation (the method), to the estimate, or to the estimator. The method finds a value of the parameter that maximizes the likelihood function. Thus, an important class of optimization problems in statistics are maximum likelihood problems.

Suppose that $X_1, \ldots, X_n$ are random variables with parameter $\theta \in \Theta$ (θ may be a vector). The *likelihood function* $L(\theta)$ of random variables $X_1, \ldots, X_n$ evaluated at $x_1, \ldots, x_n$ is defined as the joint density

$$L(\theta) = f(x_1, \ldots, x_n; \theta).$$

If $X_1, \ldots, X_n$ are a random sample (so $X_1, \ldots, X_n$ are iid) with density $f(x; \theta)$, then

$$L(\theta) = \prod_{i=1}^{n} f(x_i; \theta).$$

A maximum likelihood estimate of θ is a value $\hat{\theta}$ that maximizes $L(\theta)$. That is, $\hat{\theta}$ is a solution (not necessarily unique) to

$$L(\hat{\theta}) = f(x_1, \ldots, x_n; \hat{\theta}) = \max_{\theta \in \Theta} f(x_1, \ldots, x_n; \theta). \tag{11.4}$$

If $\hat{\theta}$ is unique, $\hat{\theta}$ is the *maximum likelihood estimator* (MLE) of θ.

If θ is a scalar, the parameter space Θ is an open interval, and $L(\theta)$ is differentiable and assumes a maximum on Θ, then $\hat{\theta}$ is a solution of

$$\frac{d}{d\theta}L(\theta) = 0. \tag{11.5}$$

The solutions to (11.5) are solutions to

$$\frac{d}{d\theta}\ell(\theta) = 0, \tag{11.6}$$

where $\ell(\theta) = \log L(\theta)$ is the *log-likelihood*. In the case where $X_1, \ldots, X_n$ are a random sample, we have

$$\ell(\theta) = \log \prod_{i=1}^{n} f(x_i; \theta) = \sum_{i=1}^{n} \log f(x_i; \theta),$$

so (11.6) is often easier to solve than (11.5).

Example 11.10 (MLE using `mle`)

Suppose Y_1, Y_2 are iid with density $f(y) = \theta e^{-\theta y}$, $y > 0$. Find the MLE of θ.

By independence,

$$L(\theta) = (\theta e^{-\theta y_1})(\theta e^{-\theta y_2}) = \theta^2 e^{-\theta(y_1 + y_2)}.$$

Thus $\ell(\theta) = 2\log\theta - \theta(y_1 + y_2)$ and the log-likelihood equation to be solved is

$$\frac{d}{d\theta}\ell(\theta) = \frac{2}{\theta} - (y_1 + y_2) = 0, \qquad \theta > 0.$$

The unique solution is $\hat{\theta} = 2/(y_1 + y_2)$, which maximizes $L(\theta)$. Therefore the MLE is the reciprocal of the sample mean in this example.

Although we have the analytical solution, let us see how the problem can be solved numerically using the `mle` (`stats4`) function. The `mle` function takes as its first argument the function that evaluates $-\ell(\theta) = -\log(L(\theta))$. The negative log-likelihood is minimized by a call to `optim`, an optimization routine.

```
#the observed sample
y <- c(0.04304550, 0.50263474)

mlogL <- function(theta=1) {
    #minus log-likelihood of exp. density, rate 1/theta
    return( - (length(y) * log(theta) - theta * sum(y)))
}
```

```
library(stats4)
fit <- mle(mlogL)
summary(fit)

Maximum likelihood estimation

Call: mle(minuslogl = mlogL)

Coefficients:
      Estimate Std. Error
theta  3.66515   2.591652

-2 log L: -1.195477
```

Alternately, the initial value for the optimizer could be supplied in the call to mle; two examples are

```
mle(mlogL, start=list(theta=1))
mle(mlogL, start=list(theta=mean(y)))
```

In this example, the maximum likelihood estimate is $\hat{\theta} = 1/\overline{Y} = 3.66515$. The maximum log-likelihood is $\ell(\hat{\theta}) = 2\log(1/\bar{y}) - (1/\bar{y})(y_1 + y_2) = 0.5977386$, or $-2\log(L) = -1.195477$. The same result was obtained by mle. ◇

Suppose $\hat{\theta}$ satisfies (11.6). Then $\hat{\theta}$ may be a relative maximum, relative minimum, or an inflection point of $\ell(\theta)$. If $\ell''(\hat{\theta}) < 0$, then $\hat{\theta}$ is a local maximum of $\log \ell(\theta)$.

The second derivative of the log-likelihood also contains information about the variance of $\hat{\theta}$. The Fisher information (see e.g. [39, 231]) on X at θ is defined

$$\mathcal{I}(\theta) = [-E_\theta \ell''(\theta)]_{|\theta}.$$

The Fisher information gives a bound on the variance of unbiased estimators of θ. The larger the information $\mathcal{I}(\theta)$, the more information the sample contains about the value of θ, and the smaller the variance of the best unbiased estimator.

If θ is a vector in $\mathbb{R}^d$, Θ is an open subset of $\mathbb{R}^d$, and the first order partial derivatives of $L(\theta)$ exist in all coordinates of θ, then $\hat{\theta}$ must satisfy simultaneously the d equations

$$\frac{\partial}{\partial \theta_j} L(\hat{\theta}) = 0, \qquad j = 1, \ldots, d, \tag{11.7}$$

or the d corresponding log-likelihood equations.

If the log-likelihood is not quadratic, the solution of the likelihood equations (11.11) is a nonlinear system of d equations in d variables. Thus, maximum likelihood estimation and maximum likelihood based inference often require nonlinear numerical methods.

Note that there are several potential problems to finding a solution: the derivatives of the likelihood function may not exist, or may not exist on all of Θ; the optimal θ may not be an interior point of Θ; or the likelihood equation (11.5) or (11.7) may be difficult to solve. In this case, numerical methods of optimization may succeed in finding optimal solutions $\hat{\theta}$ satisfying (11.4).

11.5   One-dimensional Optimization

There are several approaches to one-dimensional optimization implemented in R. Many types of problems can be restated so that the root-finding function `uniroot` can be applied. The `nlm` function implements nonlinear minimization with a Newton-type algorithm. The documentation for the `optimize` function indicates that it is C translation of Fortran code based on the Algol 60 procedure "localmin" given in [32], which implements a combination of golden section search and successive parabolic interpolation.

Example 11.11 (One-dimensional optimization with `optimize`)

Maximize the function

$$f(x) = \frac{\log(1 + \log(x))}{\log(1 + x)}$$

with respect to x. The graph of $f(x)$ in Figure 11.3 shows that the maximum occurs between 4 and 8.

```
x <- seq(2, 8, .001)
y <- log(x + log(x))/(log(1+x))
plot(x, y, type = "l")
```

Apply `optimize` on the interval (4, 8). The default is to minimize the function. To maximize $f(x)$, set `maximum = TRUE`. The default tolerance is `.Machine$double.eps^0.25`.

```
f <- function(x)
    log(x + log(x))/log(1+x)

> optimize(f, lower = 4, upper = 8, maximum = TRUE)
$maximum
[1] 5.792299
$objective
[1] 1.055122
```

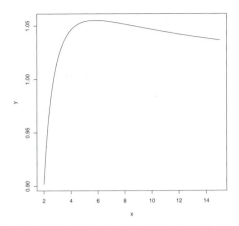

FIGURE 11.3: The function $f(x)$ in Example 11.11.

◇

Example 11.12 (MLE: Gamma distribution)

Let $x_1, \ldots, x_n$ be a random sample from a Gamma(r, λ) distribution (r is the shape parameter and λ is the rate parameter). In this example, $\theta = (r, \lambda) \in \mathbb{R}^2$ and $\Theta = \mathbb{R}^+ \times \mathbb{R}^+$. Find the maximum likelihood estimator of $\theta = (r, \lambda)$.

The likelihood function is

$$L(r, \lambda) = \frac{\lambda^{nr}}{\Gamma(r)^n} \prod_{i=1}^{n} x_i^{r-1} \exp\left(-\lambda \sum_{i=1}^{n} x_i\right), \qquad x_i \geq 0,$$

and the log-likelihood function is

$$\ell(r, \lambda) = nr \log \lambda - n \log \Gamma(r) + (r - 1) \sum_{i=1}^{n} \log x_i - \lambda \sum_{i=1}^{n} x_i. \qquad (11.8)$$

The problem is to maximize (11.8) with respect to r and λ. In this form it is a two-dimensional optimization problem. This problem can be reduced to a one-dimensional root-finding problem. Find the simultaneous solution(s) (r, λ) to

$$\frac{\partial}{\partial \lambda} \ell(r, \lambda) = \frac{nr}{\lambda} - \sum_{i=1}^{n} x_i = 0; \qquad (11.9)$$

$$\frac{\partial}{\partial r} \ell(r, \lambda) = n \log \lambda - n \frac{\Gamma'(r)}{\Gamma(r)} + \sum_{i=1}^{n} \log x_i = 0. \qquad (11.10)$$

Equation (11.9) implies $\hat{\lambda} = \hat{r}/\bar{x}$. Substituting $\hat{\lambda}$ for λ in (11.10) reduces the problem to solving

$$n \log \frac{\hat{r}}{\bar{x}} + \sum_{i=1}^{n} \log x_i - n\frac{\Gamma'(\hat{r})}{\Gamma(\hat{r})} = 0 \qquad (11.11)$$

for $\hat{r}$. Thus, the MLE $(\hat{r}, \hat{\lambda})$ is the simultaneous solution (r, λ) of

$$\log \lambda + \frac{1}{n} \sum_{i=1}^{n} \log x_i = \psi(\lambda\bar{x}); \qquad \bar{x} = \frac{r}{\lambda},$$

where $\psi(t) = \frac{d}{dt} \log \Gamma(t) = \Gamma'(t)/\Gamma(t)$ (the digamma function in R). A numerical solution is easily obtained using the uniroot function.

In the following simulation experiment, random samples of size $n = 200$ are generated from a Gamma($r = 5, \lambda = 2$) distribution, and the parameters are estimated by optimizing the likelihood equations using uniroot. The sampling and estimation is repeated 20000 times. Below is a summary of the estimates obtained by this method.

```
m <- 20000
est <- matrix(0, m, 2)
n <- 200
r <- 5
lambda <- 2

obj <- function(lambda, xbar, logx.bar) {
    digamma(lambda * xbar) - logx.bar - log(lambda)
    }

for (i in 1:m) {
    x <- rgamma(n, shape=r, rate=lambda)
    xbar <- mean(x)
    u <- uniroot(obj, lower = .001, upper = 10e5,
            xbar = xbar, logx.bar = mean(log(x)))
    lambda.hat <- u$root
    r.hat <- xbar * lambda.hat
    est[i, ] <- c(r.hat, lambda.hat)
}

ML <- colMeans(est)
[1] 5.068116 2.029766
```

The average estimate for the shape parameter r was 5.068116 and the average estimate for λ was 2.029766. The estimates are positively biased, but close to the target parameters ($r = 5, \lambda = 2$).

Recall that a maximum likelihood estimator is asymptotically normal. For large n, $\hat{\lambda} \sim N(\lambda, \sigma_1^2)$ and $\hat{r} \sim N(r, \sigma_2^2)$ where σ_1^2 and σ_2^2 are the Cramér-Rao lower bounds of λ and r, respectively. The histogram of replicates $\hat{\lambda}$ is shown in Figure 11.4(a), and the histogram of replicates $\hat{r}$ is shown in Figure 11.4(b). Here $n = 200$ is not very large, and the histogram of replicates in both cases is slightly skewed but close to normal.

```
hist(est[, 1], breaks="scott", freq=FALSE,
    xlab="r", main="")
points(ML[1], 0, cex=1.5, pch=20)
hist(est[, 2], breaks="scott", freq=FALSE,
    xlab=bquote(lambda), main="")
points(ML[2], 0, cex=1.5, pch=20)
```

◇

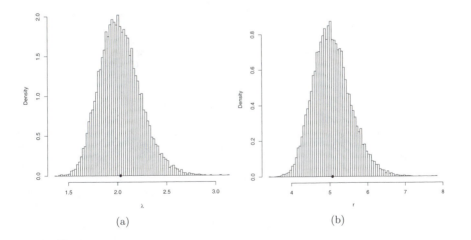

(a)   (b)

FIGURE 11.4:   Replicates of maximum likelihood estimates by numerical optimization of the likelihood of a Gamma($r = 5, \lambda = 2$) random variable in Example 11.12.

11.6 Two-dimensional Optimization

In the gamma MLE problem we seek the maximum of a two parameter likelihood function. Although it is possible to simplify the problem and solve it as in Example 11.12, it serves as a simple example to illustrate the `optim` general purpose optimization function in R. It implements Nelder-Mead [205], quasi-Newton, and conjugate-gradient algorithms [96], and also methods for box-constrained optimization and simulated annealing. See Nocedal and Wright [206] and the R manual [217] for reference on these methods and their implementation. The syntax for `optim` is

```
optim(par, fn, gr = NULL, method =
    c("Nelder-Mead", "BFGS", "CG", "L-BFGS-B", "SANN"),
    lower = -Inf, upper = Inf,
    control = list(), hessian = FALSE, ...)
```

The default method is Nelder-Mead. The first argument `par` is a vector of initial values of the target parameters, and `fn` is the objective function. The first argument to `fn` is the vector of target parameters and its return value should be a scalar.

Example 11.13 (Two-dimensional optimization with `optim`)

The objective function to be maximized is the log-likelihood function

$$\log L(\theta|x) = nr \log \lambda + (r - 1) \sum_{i=1}^{n} \log x_i - \lambda \sum_{i=1}^{n} x_i - n \log \Gamma(r),$$

and the parameters are $\theta = (r, \lambda)$. The log-likelihood function is implemented as

```
LL <- function(theta, sx, slogx, n) {
    r <- theta[1]
    lambda <- theta[2]
    loglik <- n * r * log(lambda) + (r - 1) * slogx -
        lambda * sx - n * log(gamma(r))
    - loglik
    }
```

which avoids some repeated calculation of the sums $\mathtt{sx} = \sum_{i=1}^{n} x_i$ and `slogx` $= \sum_{i=1}^{n} \log x_i$. As `optim` performs minimization by default, the return value is $-\log L(\theta)$. Initial values for the estimates must be chosen carefully. For this problem, the method of moments estimators could be given for the initial values of the parameters, but for simplicity $r = 1$ and $\lambda = 1$ are used here as the initial values. If `x` is the random sample of size `n`, the `optim` call is

```
optim(c(1,1), LL, sx=sum(x), slogx=sum(log(x)), n=n)
```

The return object includes an error code $convergence, which is 0 for success and otherwise indicates a problem. The MLE is computed for one sample below.

```
n <- 200
r <- 5;      lambda <- 2
x <- rgamma(n, shape=r, rate=lambda)

optim(c(1,1), LL, sx=sum(x), slogx=sum(log(x)), n=n)

# results from optim
par1                    5.278565
par2                    2.142059
value                 284.550086
counts.function        73.000000
counts.gradient               NA
convergence             0.000000
```

This result indicates that the Nelder-Mead (default) method successfully converged to $\hat{r} = 5.278565$ and $\hat{\lambda} = 2.142059$. The precision can be adjusted by reltol. The algorithm stops if it is unable to reduce the value by a factor of reltol, which defaults to sqrt(.Machine$double.eps) = 1.490116e-08 in this computation.

The simulation experiment below repeats the estimation procedure for comparison with the results in Example 11.12.

```
mlests <- replicate(20000, expr = {
    x <- rgamma(200, shape = 5, rate = 2)
    optim(c(1,1), LL, sx=sum(x), slogx=sum(log(x)), n=n)$par
    })
colMeans(t(mlests))
[1] 5.068109 2.029763
```

The estimates obtained by the two-dimensional optimization of (11.8) have approximately the same average value as the estimates obtained by the one-dimensional root-finding approach in Example 11.12. ◇

R note 11.6 *When replicating a vector, note that* **replicate** *fills a matrix in column major order. In the example above, the vector in each replicate is length 2, so the matrix has 2 rows and 20000 columns. The transpose of this result is the two dimensional sample of replicates.*

Example 11.14 (MLE for a quadratic form)

Consider the problem of estimating the parameters of a quadratic form of centered Gaussian random variables given by

$$Y = \lambda_1 X_1^2 + \lambda_2 X_2^2 + \cdots + \lambda_k X_k^2,$$

where X_j are iid standard normal random variables, $j = 1, \ldots, k$, and $\lambda_1 > \cdots > \lambda_k > 0$. By elementary transformations, each $Y_j = \lambda_j X_j^2$ has a gamma distribution with shape parameter $1/2$ and rate parameter $1/(2\lambda_j)$, $j = 1, \ldots, k$. Hence Y can be represented as the mixture of the k independent gamma variables,

$$Y \overset{D}{=} \frac{1}{k} G\left(\frac{1}{2}, \frac{1}{2\lambda_1}\right) + \cdots + \frac{1}{k} G\left(\frac{1}{2}, \frac{1}{2\lambda_k}\right).$$

The notation above means that Y can be generated from a two-stage experiment. First a random integer J is observed, where J is uniformly distributed on the integers 1 to k. Then a random variate Y from the distribution of $Y_J \sim \text{Gamma}(\frac{1}{2}, \frac{1}{2\lambda_J})$ is observed.

Assume that $\sum_{j=1}^{k} \lambda_j = 1$. Suppose a random sample $y_1, \ldots, y_m$ is observed from the distribution of Y, and $k = 3$. Find the maximum likelihood estimate of the parameters λ_j, $j = 1, 2, 3$.

This problem can be approached by numerical optimization of the log-likelihood function with two unknown parameters λ_1 and λ_2. The density of the mixture is

$$f(y|\lambda) = \sum_{j=1}^{3} f_j(y|\lambda),$$

where $f_j(y|\lambda)$ is the gamma density with shape parameter $1/2$ and rate parameter $1/(2\lambda_j)$. The log-likelihood can be written in terms of two unknown parameters λ_1 and λ_2, with $\lambda_3 = 1 - \lambda_1 - \lambda_2$.

```
LL <- function(lambda, y) {
    lambda3 <- 1 - sum(lambda)
    f1 <- dgamma(y, shape=1/2, rate=1/(2*lambda[1]))
    f2 <- dgamma(y, shape=1/2, rate=1/(2*lambda[2]))
    f3 <- dgamma(y, shape=1/2, rate=1/(2*lambda3))
    f <- f1/3 + f2/3 + f3/3    #density of mixture
    #returning -loglikelihood
    return( -sum(log(f)))
    }
```

The sample data in this example is generated from the quadratic form with $\lambda = (0.60, 0.25, 0.15)$. Then the optim function is applied to search for the minimum of LL, starting with initial estimates $\lambda = (0.5, 0.3, 0.2)$.

```
set.seed(543)
m <- 2000
lambda <- c(.6, .25, .15)   #rate is 1/(2 lambda)
lam <- sample(lambda, size = 2000, replace = TRUE)
y <- rgamma(m, shape = .5, rate = 1/(2*lam))

opt <- optim(c(.5,.3), LL, y=y)
theta <- c(opt$par, 1 - sum(opt$par))
```

Results are shown below. The return code in opt$convergence is 0, indicating successful convergence. The optimal value obtained for the log-likelihood was 736.325 at the point $(\lambda_1, \lambda_2) = (0.5922404, 0.2414725)$.

```
> as.data.frame(unlist(opt))
                   unlist(opt)
par1                 0.5922404
par2                 0.2414725
value             -736.3250225
counts.function     43.0000000
counts.gradient             NA
convergence          0.0000000

> theta
[1]  0.5922404 0.2414725 0.1662871
```

The maximum likelihood estimate is $\hat{\lambda} \doteq (0.592, 0.241, 0.166)$. The data was generated with parameter values $(0.60, 0.25, 0.15)$. For another approach to estimating λ see Example 11.15.   ◇

Remark 11.1 *The problem of approximating the distribution of quadratic forms has received much attention in the literature over the years. Many theoretical results and numerical methods have been developed for this important class of distributions. On numerical approximations for the distribution of quadratic forms of normal variables see Imhof [150, 151] and Kuonen [166].*

11.7   The EM Algorithm

The EM (Expectation–Maximization) algorithm is a general optimization method that is often applied to find maximum likelihood estimates when data are incomplete. Following the seminal paper of Dempster, Laird and Rubin [67] in 1977, the method has been widely applied and extended to solve many other types of statistical problems. For a recent review of EM methods and extensions see [178, 194, 292].

Incompleteness of data may arise from missing data as is often the case with multivariate samples, or from other types of data such as samples from censored or truncated distributions, or latent variables. Latent variables are unobservable variables that are introduced in order to simplify the analysis in some way.

The main idea of the EM algorithm is simple, and although it may be slow to converge relative to other available methods, it is reliable at finding a global maximum. Start with an initial estimate of the target parameter, and then alternate the E (expectation) step and M (maximization) step. In the E step compute the conditional expectation of the objective function (usually a log-likelihood function) given the observed data and current parameter estimates. In the M step, the conditional expectation is maximized with respect to the target parameter. Update the estimates and iteratively repeat the E and M steps until the algorithm converges according to some criterion. Although the main idea of EM is simple, for some problems computing the conditional expectation in the E step can be complicated. For incomplete data, the E step requires computing the conditional expectation of a function of the complete data, given the missing data.

Example 11.15 (EM algorithm for a mixture model)

In this example the EM algorithm is applied to estimate the parameters of the quadratic form introduced in Example 11.14. Recall that the problem can be formulated as estimation of the rate parameters of a mixture of gamma random variables. Although the EM algorithm is not the best approach for this problem, as an exercise we repeat the estimation for $k = 3$ components (two unknown parameters) as outlined in Example 11.14.

The EM algorithm first updates the posterior probability p_{ij} that the ith sample observation y_i was generated from the jth component. At the tth step,

$$p_{ij}^{(t)} = \frac{\frac{1}{k} f_j(y_i | y, \lambda^{(t)})}{\sum_{j=1}^{k} \frac{1}{k} f_j(y_j | y, \lambda^{(t)})},$$

where $\lambda^{(t)}$ is the current estimate of the parameters $\{\lambda_j\}$, and $f_j(y_i | y, \lambda^{(t)})$ is the Gamma$(1/2, 1/(2\lambda_j^{(t)}))$ density evaluated at y_i. Note that the mean of the jth component is λ_j so the updating equation is

$$\mu_j^{(t+1)} = \frac{\sum_{i=1}^{m} p_{ij}^{(t)} y_i}{\sum p_{ij}^{(t)}}.$$

In order to compare the estimates, we generate the data from the mixture Y using the same random number seed as in Example 11.14.

```
set.seed(543)
lambda <- c(.6, .25, .15)  #rate is 1/(2lambda)
lam <- sample(lambda, size = 2000, replace = TRUE)
y <- rgamma(m, shape = .5, rate = 1/(2*lam))

N <- 10000               #max. number of iterations
L <- c(.5, .4, .1)       #initial est. for lambdas
tol <- .Machine$double.eps^0.5
L.old <- L + 1

for (j in 1:N) {
    f1 <- dgamma(y, shape=1/2, rate=1/(2*L[1]))
    f2 <- dgamma(y, shape=1/2, rate=1/(2*L[2]))
    f3 <- dgamma(y, shape=1/2, rate=1/(2*L[3]))
    py <- f1 / (f1 + f2 + f3) #posterior prob y from 1
    qy <- f2 / (f1 + f2 + f3) #posterior prob y from 2
    ry <- f3 / (f1 + f2 + f3) #posterior prob y from 3

    mu1 <- sum(y * py) / sum(py) #update means
    mu2 <- sum(y * qy) / sum(qy)
    mu3 <- sum(y * ry) / sum(ry)
    L <- c(mu1, mu2, mu3)   #update lambdas
    L <- L / sum(L)

    if (sum(abs(L - L.old)/L.old) < tol) break
    L.old <- L
}
```

Results are shown below.

```
print(list(lambda = L/sum(L), iter = j, tol = tol))

$lambda [1] 0.5954759 0.2477745 0.1567496
$iter [1] 592
$tol [1] 1.490116e-08
```

Here the EM algorithm converged in 592 iterations (within $< 1.5e - 8$) to the estimate $\hat{\lambda} \doteq (0.595, .248, .157)$. The data was generated with parameters $(0.60, 0.25, 0.15)$. Compare this result with the maximum likelihood estimate obtained by two-dimensional numerical optimization of the log-likelihood function in Example 11.14. ◇

11.8   Linear Programming – The Simplex Method

The simplex method is a widely applied optimization method for a special class of constrained optimization problems with linear objective functions and linear constraints. The constraints usually include inequalities, and therefore the region over which the objective function is to be optimized (the feasible region) can be described by a simplex. Linear programming methods include the simplex method and interior point methods, but here we illustrate the simplex method only. See Nocedal and Wright [206, Ch. 13] for a summary of the simplex method.

Given m linear constraints in n variables, let A be the $m \times n$ matrix of coefficients, so that the constraints are given by $Ax \geq b$, where $b \in \mathbb{R}^m$. Here we suppose that $m < n$. An element $x \in \mathbb{R}^n$ of the feasible set satisfies the constraint $Ax \geq b$. The objective function is a linear function of n variables with coefficients given by vector c. Hence, the objective is to minimize $c^T x$ subject to the constraint $Ax \geq b$.

The problem as stated above is the *primal* problem. The *dual* problem is: maximize $b^T y$ subject to the constraint $A^T y \leq c$, where $y \in \mathbb{R}^n$. The duality theorem states that if either the primal or the dual problem has an optimal solution with a finite objective value, then the primal and the dual problems have the same optimal objective value.

The vertices of the simplex are called the basic feasible points of the feasible set. When the optimal value of the objective function exists, it will be achieved at one of the basic feasible points. The simplex algorithm evaluates the objective function at the basic feasible points, but selects the points at each iteration in such a way that an optimal solution is found in relatively few iterations. It can be shown (see e.g. [206, Thm. 13.4]) that if the linear program is bounded and not degenerate, the simplex algorithm will terminate after finitely many iterations at one of the basic feasible points.

The simplex method is implemented by the `simplex` function in the `boot` package [34]. The `simplex` function will maximize or minimize the linear function ax subject to the constraints $A_1 x \leq b_1$, $A_2 x \geq b_2$, $A_3 x = b_3$, and $x \geq 0$. Either the primal or dual problem is easily handled by the `simplex` function.

Example 11.16 (Simplex algorithm)

Use the simplex algorithm to solve the following problem.
Maximize $2x + 2y + 3z$ subject to

$$-2x + y + z \leq 1$$
$$4x - y + 3z \leq 3$$
$$x \geq 0,\ y \geq 0,\ z \geq 0.$$

For such a small problem, it would not be too difficult to solve it directly, because the theory implies that if there is an optimal solution, it will be achieved at one of the vertices of the feasible set. Hence, we need only evaluate the objective function at each of the finitely many vertices. The vertices are determined by the intersection of the linear constraints. The simplex method also evaluates the objective function as it moves from one vertex to another, usually changing the coordinates in one vertex only at each step. The trick is to decide which vertex to check next by moving in the direction of greatest increase/decrease in the objective function. Eventually, for bounded, nondegenerate problems, the value of the objective function cannot be improved and the algorithm terminates with the solution. The `simplex` function implements the algorithm.

```
library(boot)    #for simplex function
A1 <- rbind(c(-2, 1, 1), c(4, -1, 3))
b1 <- c(1, 3)
a <- c(2, 2, 3)
simplex(a = a, A1 = A1, b1 = b1, maxi = TRUE)

Optimal solution has the following values
x1 x2 x3
 2  5  0
The optimal value of the objective  function is 14.
```

◇

11.9   Application: Game Theory

In the linear program of Example 11.16, the constraints are inequalities. Equality constraints are also possible. Equality constraints might arise if, for example, the sum of the variables is fixed. If the variables represent a discrete probability mass function, the sum of the probabilities must equal one. We solve for a probability mass function in the next problem. It is a classical problem in game theory.

Example 11.17 (Solving the Morra game)

One of the world's oldest known games of strategy is the Morra game. In the 3-finger Morra game, each player shows 1, 2, or 3 fingers, and simultaneously each calls his guess of the number of fingers his opponent will show. If both players guess correctly, the game is a draw. If exactly one player guesses correctly, he wins an amount equal to the sum of the fingers shown by both players. This example appears in Dresher [74] and in Székely and Rizzo [264]. For more details on methods of solving games, see Owen [208].

The strategies for each player are pairs (d, g), where d is the number of fingers and g is the guess. Thus, each player has nine pure strategies, $(1,1)$, $(1,2), \ldots, (3,3)$. This is a zero-sum game: the gain of the first player is the loss of the second player. Player 1 seeks to maximize his winnings, and Player 2 seeks to minimize his losses. The game can be represented by the payoff matrix in Table 11.1.

TABLE 11.1: Payoff Matrix of the Game of Morra

Strategy	1	2	3	4	5	6	7	8	9
1	0	2	2	-3	0	0	-4	0	0
2	-2	0	0	0	3	3	-4	0	0
3	-2	0	0	-3	0	0	0	4	4
4	3	0	3	0	-4	0	0	-5	0
5	0	-3	0	4	0	4	0	-5	0
6	0	-3	0	0	-4	0	5	0	5
7	4	4	0	0	0	-5	0	0	-6
8	0	0	-4	5	5	0	0	0	-6
9	0	0	-4	0	0	-5	6	6	0

Denote the payoff matrix by $A = (a_{ij})$. By von Neumann's minimax theorem [282], the optimal strategies of both players in this game are mixed strategies because $\min_i \max_j a_{ij} > \max_j \min_i a_{ij}$. A *mixed strategy* is simply a probability distribution $(x_1, \ldots, x_9)$ on the set of strategies, where strategy j is chosen with probability x_j.

The minimax theorem implies that if both players apply optimal strategies x^* and y^* respectively, then each player has expected payoff $v = x^{*T} A y^*$, the *value* of the game. If the first player applies an optimal strategy x^* against any strategy y of the other player, his expected gain is at least v. Introduce the variable $x_{10} = v$, and let $x = (x_1, \ldots, x_9, x_{10})$.

Let A_1 be the matrix formed by augmenting A with a column of -1's. Then since $x^{*T} A y \geq v$ for every pure strategy $y_j = 1$, we have the system of constraints $A_1 x \leq 0$. The equality constraint is $\sum_{i=1}^{m} x_i = 1$. The simplex function automatically includes the constraints $x_i \geq 0$. (To be sure that $v \geq 0$, one can translate the payoff matrix by subtracting min(A) from each element. The set of optimal strategies does not change.)

Define the $1 \times (n+1)$ vector $A_3 = [1, 1, \ldots, 1, 0]$. Maximize $v = x_{10}$ subject to the constraints $A_1 x \geq 0$ and $A_3 x = 1$. Keep in mind that the optimal x returned by simplex will be $x^* = (x_1, \ldots, x_m)$ and $v = x_{m+1}$.

Note that we are interested in optimal solutions of both the primal and the dual problem, with analogous constraints and objective for the second player. All two-player zero-sum games have similar representations as linear programs, so the solution can be obtained for general $m \times n$ two-player zero-sum games. Our function `solve.game` has the payoff matrix as its single argument, and returns in a list, the payoff matrix, optimal strategies, and the value of the game.

```
solve.game <- function(A) {
    #solve the two player zero-sum game by simplex method
    #optimize for player 1, then player 2
    #maximize v subject to ...
    #let x strategies 1:m, and put v as extra variable
    #A1, the <= constraints
    #
    min.A <- min(A)
    A <- A - min.A    #so that v >= 0
    max.A <- max(A)
    A <- A / max(A)
    m <- nrow(A)
    n <- ncol(A)
    it <- n^3
    a <- c(rep(0, m), 1) #objective function
    A1 <- -cbind(t(A), rep(-1, n)) #constraints <=
    b1 <- rep(0, n)
    A3 <- t(as.matrix(c(rep(1, m), 0))) #constraints sum(x)=1
    b3 <- 1
    sx <- simplex(a=a, A1=A1, b1=b1, A3=A3, b3=b3,
                  maxi=TRUE, n.iter=it)
    #the 'solution' is [x1,x2,...,xm | value of game]
    #
    #minimize v subject to ...
    #let y strategies 1:n, with v as extra variable
    a <- c(rep(0, n), 1) #objective function
    A1 <- cbind(A, rep(-1, m)) #constraints <=
    b1 <- rep(0, m)
    A3 <- t(as.matrix(c(rep(1, n), 0))) #constraints sum(y)=1
    b3 <- 1
    sy <- simplex(a=a, A1=A1, b1=b1, A3=A3, b3=b3,
                  maxi=FALSE, n.iter=it)

    soln <- list("A" = A * max.A + min.A,
                 "x" = sx$soln[1:m],
                 "y" = sy$soln[1:n],
                 "v" = sx$soln[m+1] * max.A + min.A)
    soln
    }
```

Although the function `solve.game` applies in principle to arbitrary $m \times n$ games, it is of course limited in practice to systems that are not too large for the `simplex (boot)` function to solve.

Now we apply the function `solve.game` to solve the Morra game. A list object is returned that contains optimal strategies for each player and the value of the game.

```
#enter the payoff matrix
A <- matrix(c(  0,-2,-2,3,0,0,4,0,0,
                2,0,0,0,-3,-3,4,0,0,
                2,0,0,3,0,0,0,-4,-4,
                -3,0,-3,0,4,0,0,5,0,
                0,3,0,-4,0,-4,0,5,0,
                0,3,0,0,4,0,-5,0,-5,
                -4,-4,0,0,0,5,0,0,6,
                0,0,4,-5,-5,0,0,0,6,
                0,0,4,0,0,5,-6,-6,0), 9, 9)

library(boot)   #needed for simplex function

s <- solve.game(A)
```

The optimal strategies returned by `solve.game` are the same for both players (the game is symmetric).

```
> round(cbind(s$x, s$y), 7)
          [,1]        [,2]
x1 0.0000000 0.0000000
x2 0.0000000 0.0000000
x3 0.4098361 0.4098361
x4 0.0000000 0.0000000
x5 0.3278689 0.3278689
x6 0.0000000 0.0000000
x7 0.2622951 0.2622951
x8 0.0000000 0.0000000
x9 0.0000000 0.0000000
```

Each player should randomize their strategies according to the probability distributions above.

It can be shown (see e.g. [74]) that the extreme points of the set of optimal strategies of either player for this Morra game are

$$(0, 0, 5/12, 0, 4/12, 0, 3/12, 0, 0), \tag{11.12}$$

$$(0, 0, 16/37, 0, 12/37, 0, 9/37, 0, 0), \tag{11.13}$$

$$(0, 0, 20/47, 0, 15/47, 0, 12/47, 0, 0), \tag{11.14}$$

$$(0, 0, 25/61, 0, 20/61, 0, 16/61, 0, 0). \tag{11.15}$$

Notice that the solutions obtained by the simplex method in this example correspond to the extreme point (11.15). ◇

For linear and integer programming, also see the `lp` function in the contributed package `lpSolve` [26].

Exercises

11.1 The natural logarithm and exponential functions are inverses of each other, so that mathematically $\log(\exp x) = \exp(\log x) = x$. Show by example that this property does not hold exactly in computer arithmetic. Does the identity hold with near equality? (See `all.equal`.)

11.2 Suppose that X and Y are independent random variables variables, $X \sim$ Beta(a, b) and $Y \sim$ Beta(r, s). Then it can be shown [7] that

$$P(X < Y) = \sum_{k=\max(r-b,0)}^{r-1} \frac{\binom{r+s-1}{k}\binom{a+b-1}{a+r-1-k}}{\binom{a+b+r+s-2}{a+r-1}}.$$

Write a function to compute $P(X < Y)$ for any $a, b, r, s > 0$. Compare your result with a Monte Carlo estimate of $P(X < Y)$ for $(a, b) = (10, 20)$ and $(r, s) = (5, 5)$.

11.3 (a) Write a function to compute the kth term in

$$\sum_{k=0}^{\infty} \frac{(-1)^k}{k!\,2^k} \frac{\|a\|^{2k+2}}{(2k+1)(2k+2)} \frac{\Gamma\left(\frac{d+1}{2}\right)\Gamma\left(k+\frac{3}{2}\right)}{\Gamma\left(k+\frac{d}{2}+1\right)},$$

where $d \geq 1$ is an integer, a is a vector in $\mathbb{R}^d$, and $\|\cdot\|$ denotes the Euclidean norm. Perform the arithmetic so that the coefficients can be computed for (almost) arbitrarily large k and d. (This sum converges for all $a \in \mathbb{R}^d$.)
(b) Modify the function so that it computes and returns the sum.
(c) Evaluate the sum when $a = (1, 2)^T$.

11.4 Find the intersection points $A(k)$ in $(0, \sqrt{k})$ of the curves

$$S_{k-1}(a) = P\left(t(k-1) > \sqrt{\frac{a^2(k-1)}{k-a^2}}\right)$$

and

$$S_k(a) = P\left(t(k) > \sqrt{\frac{a^2 k}{k+1-a^2}}\right),$$

for $k = 4 : 25, 100, 500, 1000$, where $t(k)$ is a Student t random variable with k degrees of freedom. (These intersection points determine the critical values for a t-test for scale-mixture errors proposed by Székely [260].)

11.5   Write a function to solve the equation

$$\frac{2\Gamma(\frac{k}{2})}{\sqrt{\pi(k-1)}\Gamma(\frac{k-1}{2})} \int_0^{c_{k-1}} \left(1 + \frac{u^2}{k-1}\right)^{-k/2} du$$

$$= \frac{2\Gamma(\frac{k+1}{2})}{\sqrt{\pi k}\Gamma(\frac{k}{2})} \int_0^{c_k} \left(1 + \frac{u^2}{k}\right)^{-(k+1)/2} du$$

for a, where

$$c_k = \sqrt{\frac{a^2 k}{k+1-a^2}}.$$

Compare the solutions with the points $A(k)$ in Exercise 11.4.

11.6   Write a function to compute the cdf of the Cauchy distribution, which has density

$$\frac{1}{\theta\pi(1 + [(x-\eta)/\theta]^2)}, \qquad -\infty < x < \infty,$$

where $\theta > 0$. Compare your results to the results from the R function pcauchy. (Also see the source code in pcauchy.c.)

11.7   Use the simplex algorithm to solve the following problem.
Minimize $4x + 2y + 9z$ subject to

$$2x + y + z \le 2$$
$$x - y + 3z \le 3$$
$$x \ge 0, \ y \ge 0, \ z \ge 0.$$

11.8   In the Morra game, the set of optimal strategies are not changed if a constant is subtracted from every entry of the payoff matrix, or a positive constant is multiplied times every entry of the payoff matrix. However, the simplex algorithm may terminate at a different basic feasible point (also optimal). Compute B <- A + 2, find the solution of game B, and verify that it is one of the extreme points (11.12)–(11.15) of the original game A. Also find the value of game A and game B.

Appendix A

Notation

Selected notation and abbreviations used throughout the text are summarized here. Notation that is specific to a particular chapter is not included.

Symbol Description

$E[X]$	Expected value of the random variable X	Φ^{-1}	Inverse cdf of the standard normal distribution: $\Phi^{-1}(\alpha) = z \Rightarrow \Phi(z) = \alpha$
$I(A)$	Indicator function on the set A: $I(x) = 1$ if $x \in A$ and $I(x) = 0$ if $x \notin A$	$\overset{D}{=}$	equal in distribution
I_d	The $d \times d$ identity matrix	$\doteq$	is approximately equal to
$\log x$	Natural logarithm of x	$X \sim$	X has the distribution named on right of $\sim$.
$\mathbb{P}$	Transition matrix of a Markov chain	$\overset{iid}{\sim}$	Variables on the left are iid from distribution named on the right.
$\mathbb{R}$	The one dimensional field of real numbers		
$\mathbb{R}^d$	The d-dimensional real coordinate space	$\|x\|$	Euclidean norm of x
$\Gamma(\cdot)$	Complete gamma function	$|A|$	Determinant of matrix A
$\Phi(\cdot)$	cdf of the standard normal distribution	A^T	Transpose of A
		$\overline{X}$	Sample mean or vector of sample means

Abbreviations

ASL	achieved significance level
ASH	average shifted histogram (density estimate)
BVN	bivariate normal
cdf	cumulative distribution function
dCor	distance correlation
dCov	distance covariance
ecdf, edf	empirical cumulative distribution function
GUI	graphical user interface
iid	independent and identically distributed
IMSE	integrated mean squared error
LRT	likelihood ratio test
M-H	Metropolis-Hastings
MC	Monte Carlo
MCMC	Markov Chain Monte Carlo
MISE	mean integrated squared error
MLE	maximum likelihood estimator or estimate
MSE	mean squared error
MVN	multivariate normal
$N(\mu, \sigma^2)$	Normal distribution with mean μ and variance σ^2
$N_d(\mu, \Sigma)$	d-dimensional multivariate normal distribution with mean vector μ and variance-covariance matrix Σ
$\chi^2(\nu)$	Chi-squared distribution with ν degrees of freedom
$W_d(\Sigma, n)$	Wishart distribution with parameters (Σ, n, d)
se	standard error
svd	singular value decomposition

Appendix B

Working with Data Frames and Arrays

B.1 Resampling and Data Partitioning

B.1.1 Using the `boot` function

Bootstrap is implemented in the `boot` function (`boot` package [34]), which provides functions and arguments for the book [63]. In ordinary bootstrap, the samples are selected with replacement. The basic syntax for ordinary bootstrap is

```
boot(data, statistic, R)
```

where `data` is the observed sample and `R` is the number of bootstrap replicates. The default is `sim = "ordinary"`, the ordinary bootstrap (sampling with replacement).

The second argument (`statistic`) is a function, or the name of a function, which calculates the statistic to be replicated. Suppose we call this function f. The `boot` function generates the random indices $i = (i_1, \ldots, i_n)$ for each bootstrap replicate, and passes to the function f a copy of the `data` and the index vector i. The function f then computes the statistic $\hat{\theta}^{(b)}$ corresponding to the resampled observations. Example B.1 discusses how to extract the samples for the calculations inside f.

Example B.1 (Extracting a bootstrap sample using an index vector)

We have seen that the `sample` function can be used to sample from a vector with replacement. Equivalently, if x is a vector of length n, we can sample with replacement from the vector of indices `1:n`, and use the resulting value to extract the elements of x. Notice that the two methods below generate the same samples.

```
>      set.seed(123)
>      sample(letters[1:10], size = 10, replace = TRUE)
[1] "c" "h" "e" "i" "j" "a" "f" "i" "f" "e"

>      set.seed(123)
>      i <- sample(1:10, size = 10, replace = TRUE)
>      letters[i]
[1] "c" "h" "e" "i" "j" "a" "f" "i" "f" "e"
```

Similarly, the [ ] operator can be used to extract bootstrap samples from data frames and matrices using x[i,].

```
>      x
       [,1] [,2] [,3] [,4]
[1,]    16   14   17   12
[2,]    14   13   16   14
[3,]    13   13   14   11
[4,]    19   11   15   11
[5,]    14   10    8   11

>      i
[1] 1 3 3 2 1

>      x[i, ]
       [,1] [,2] [,3] [,4]
[1,]    16   14   17   12
[2,]    13   13   14   11
[3,]    13   13   14   11
[4,]    14   13   16   14
[5,]    16   14   17   12
```

The boot function will pass a copy of the observed sample x and the bth index vector i; the user's function f (statistic) should compute the test statistic on x[i,] or x[i]. For example, if x is a bivariate sample, and the statistic to replicate is correlation, then the function f can be written as follows.

```
f <- function(x, i) {
    cor(x[i, 1], x[i, 2])
}
```

For a resampling experiment, it is helpful to code the calculations for the statistic in a function like f above, whether or not the boot function will be used to run the bootstrap.   ◊

B.1.2   Sampling without replacement

The boot function can also be applied in situations where the resampling should be without replacement. For example, in permutation tests, the method of resampling should be sim = "permutation".

If `boot` is not used, then it is necessary to generate for each replicate a permutation of the sample observations. To obtain a permutation of the sample observations in a data frame or matrix x, use `x[i, ]`, where i is a permutation of the indices of the sample elements. A permutation of the integers `1:n` is generated by `sample(1:n)`.

In situations like the jackknife and cross-validation, it is more convenient to specify what should not be extracted. To specify which elements to exclude, use the `[ ]` operator with a negative argument. For example, to extract all but row i of a matrix A, use `A[-i, ]`. In general, i can be a vector and `A[-i, ]` extracts a submatrix from A that excludes the rows indexed by i.

Example B.2 (Extracting rows from a matrix)

```
> A <- matrix(1:25, 5, 5)
> A[-(2:3), ]

      [,1] [,2] [,3] [,4] [,5]
[1,]    1    6   11   16   21
[2,]    4    9   14   19   24
[3,]    5   10   15   20   25

> A[-(2:3), 4]
[1] 16 19 20
```

In the last line, notice that the result has been converted to a vector. To extract the 3×1 matrix use `as.matrix(A[-(2:3), 4])`.   ◇

A random sample of size k or $n - k$ can be selected without replacement from a sample x of size n by

```
i <- sample(1:n, size = k)}
x1 <- x[i, ]}
x2 <- x[-i, ]}
```

Then { x1, x2 } form a partition of the original sample x.

Some exact tests require that all permutations of a sample be generated. The `permutations` function in package e1071 [72] generates a matrix containing all $n!$ permutations of an index set `1:n`. Each row of the returned matrix is a permutation of `1:n`.

To generate random two-way contingency tables with given marginals see the function `r2dtable`.

B.2   Subsetting and Reshaping Data

When working with real data, it is often the case that the format or layout of the data does not match what is required by the methods one would like to apply, there are missing values, or other issues. R provides several utilities for reshaping a dataset. The following simple examples illustrate some of the operations that are possible, such as merging, subsetting or reshaping data. These operations can be very complicated and difficult in practice. Refer to the documentation for each of the individual topics for more detailed explanations and examples.

The examples that follow are provided for convenient reference on a few special topics only, and readers should refer to one of the references for a good introduction to data analysis using R, such as Dalgaard [62] or Verzani [280].

B.2.1   Subsetting Data

Subsets of data frames can be extracted using the operators $, [[ ]], and array indexing [ ], as shown above. The subset function provides another approach to subsetting data. The subset function expects the name of the data set, the condition satisfied (subset) by the desired subset, and/or a list of variables (select).

Example B.3 (Subsetting data frames)

Means and summary statistics computed for the iris data in Examples 1.1 and 1.4 can also be computed as follows. The first subset uses the condition that the species is versicolor and selects the variable petal length. The second subset selects sepal length and width without restricting species.

```
# versicolor petal length
y <- subset(iris, Species == "versicolor",
            select = Petal.Length)

summary(y)

Petal.Length
Min.   :3.00
1st Qu.:4.00
Median :4.35
Mean   :4.26
3rd Qu.:4.60
Max.   :5.10
```

```
# sepal width, all species
y <- subset(iris, select = c(Sepal.Length, Sepal.Width))
mean(y)

Sepal.Length  Sepal.Width
    5.843333     3.057333
```

◇

B.2.2 Stacking/Unstacking Data

A data frame or list can be stacked or unstacked using the stack (unstack) function.

Example B.4 (Unstacking data)

The InsectSprays data frame contains two variables, count (an integer) and spray (a factor). The format is stacked. The first few observations are shown below.

```
> attach(InsectSprays)
> InsectSprays
   count spray
1     10     A
2      7     A
3     20     A
4     14     A
5     14     A
6     12     A
. . .
```

The data can be unstacked by the default formula unstack(InsectSprays), or by explicitly specifying the formula as shown below.

```
> unstack(count, count ~ spray)
    A  B C  D E  F
1  10 11 0  3 3 11
2   7 17 1  5 5  9
3  20 21 7 12 3 15
4  14 11 2  6 5 22
5  14 16 3  4 3 15
6  12 14 1  3 6 16
7  10 17 2  5 1 13
8  23 17 1  5 1 10
9  17 19 3  5 3 26
10 20 21 0  5 2 26
11 14  7 1  2 6 24
12 13 13 4  4 4 13
```

If the result is stored in an object u, then the unstacking could be reversed
by stack(u). In the result of stack(u), the counts would then be labeled
"values" and the spray (indices) will be labeled "ind".   ◇

R note B.1 *The formula* count ~ spray *represents the linear model where
the response is* count *and the single predictor is the factor* spray. *An intercept
term is included by default. The default model formula associated with a data
frame is supplied by* formula. *For example, the default formula associated
with the* iris *data is the following one, which might not be what is expected.*

```
> formula(iris)
Sepal.Length ~ Sepal.Width + Petal.Length + Petal.Width + Species
```

B.2.3   Merging Data Frames

Two data frames can be merged by common variable (column) names or
common row names, using the merge function.

Example B.5 (Merge by ID)

In this example, we have created two sets of scores, data1 and data2. The
common variable is the ID number in the first column. This example is typical
of repeated measurement data. We wish to merge the two scores into a single
data frame, by ID. The ID is the first variable in data1 and the first variable
in data2, so by=c(1,1) specifies that the merge will match by ID. In the first
version below, only the observations with common ID numbers, labeled "V1"
will be retained in the new data set. This corresponds to a listwise deletion
of any subjects with missing values.

```
data1
      [,1] [,2]
[1,]    1    9
[2,]    2   12
[3,]    3    9
[4,]    4   13
[5,]    5   13

data2
      [,1] [,2]
[1,]    3    6
[2,]    4   10
[3,]    5   13
[4,]    6   10
[5,]    7   10
```

Now merge the data sets. By default, only the complete cases are included in the result. In the second version below, all observations are retained in the new data set. Missing scores are assigned the missing value NA.

The syntax is

```
merge(x, y)  #default

merge(x, y, by = intersect(names(x), names(y)),
        by.x = by, by.y = by, all = FALSE, ...)
```

where ... indicates more arguments (see the help topic).

```
# keep only the common ID's
merge(data1, data2, by=c(1,1))
   V1 V2.x V2.y
1  3   9    6
2  4   13   10
3  5   13   13

#keep all observations
merge(data1, data2, by=c(1,1), all=TRUE)
   V1 V2.x V2.y
1  1   9    NA
2  2   12   NA
3  3   9    6
4  4   13   10
5  5   13   13
6  6   NA   10
7  7   NA   10
```

◇

B.2.4   Reshaping Data

Suppose we need to reshape Example B.5 data into a "long" format, introducing a time variable. The reshape function is provided to convert between the "wide" and "long" formats. The syntax is

```
reshape(data, varying, v.names, timevar, idvar, ids,
        times, drop, direction, new.row.names,
        split, include))
```

and all of the parameters except data and direction have default values. To keep all observations use all=TRUE. The repeated measurements or time-varying measurements are specified by varying. The direction is "wide" or "long."

Example B.6 (Reshape)

Convert Example B.5 data from "wide" to "long" format.

```
#keep all observations
a <- merge(data1, data2, by=c(1,1), all=TRUE)
reshape(a, idvar="ID", varying=c(2,3),
    direction="long", v.names="Scores")
```

	V1	time	Scores	ID
1.1	1	1	9	1
2.1	2	1	12	2
3.1	3	1	9	3
4.1	4	1	13	4
5.1	5	1	13	5
6.1	6	1	NA	6
7.1	7	1	NA	7
1.2	1	2	NA	1
2.2	2	2	NA	2
3.2	3	2	6	3
4.2	4	2	10	4
5.2	5	2	13	5
6.2	6	2	10	6
7.2	7	2	10	7

◇

B.3   Data Entry and Data Analysis

B.3.1   Manual Data Entry

A spreadsheet-like interface to create a data frame is provided in the `edit` function.

```
mydata <- edit(data.frame())
```

This command opens a spreadsheet-like editor for data entry. When the editor is closed, a data frame `mydata` is created. Then `mydata` can be edited by `edit(mydata)`. It is probably easier to enter a large data set in a spreadsheet and read it into a data frame via `read.table`, described below.

B.3.2   Recoding Missing Values

The first step in recoding missing values is to find the missing values. The function `is.na` tests for missing values, returning logical values. The `which` function returns the indices of a logical vector that are TRUE. Applying which to the result of is.na gives a vector containing the indices of the missing values. Then if i contains the indices of the missing data of a vector x, recoding NA to 0, for example, is as simple as x[i] <- 0.

Example B.7 (Recode)

With the repeated measures data in Example B.6, recode the missing scores to 0. The function is.na tests for missing values. Extract the row indices of the missing scores using the which function. Below, which returns the indices 6,7,8,9, indicating that scores with those subscripts are missing.

```
#store the previous result into b
b <- reshape(a, idvar="ID", varying=c(2,3),
                direction="long", v.names="Scores")
i <- which(is.na(b$Scores))    #these are missing
```

Now the indices stored in i are 6, 7, 8, 9, and we replace the corresponding NA's with 0.

```
b$Scores[i] <- 0    #replace NA with 0
b
      V1 time Scores ID
1.1  1    1       9  1
2.1  2    1      12  2
3.1  3    1       9  3
4.1  4    1      13  4
5.1  5    1      13  5
6.1  6    1       0  6
7.1  7    1       0  7
1.2  1    2       0  1
2.2  2    2       0  2
3.2  3    2       6  3
4.2  4    2      10  4
5.2  5    2      13  5
6.2  6    2      10  6
7.2  7    2      10  7
```

The which function can also be used to extract array indices, by setting arr.ind=TRUE. From the result of the second version of the merge operation in Example B.5, we can extract the array indices of the missing values as follows.

```
m <- merge(data1, data2, by=c(1,1), all=TRUE)
i <- which(is.na(m), arr.ind=TRUE)    #these are missing
>i
       row col
[1,]    6   2
[2,]    7   2
[3,]    1   3
[4,]    2   3
```

◇

B.3.3   Reading and Converting Dates

A time series for financial data usually has a calendar date corresponding to each observation. In this section we discuss some basic methods for importing files with dates, converting dates to useful formats, and extracting the day, month, and year. Date arithmetic and formatting is a complicated subject, however, and depends in part on the locale. Refer to the R manual [217] for thorough documentation.

Our first example illustrates how to convert a string format date from "mm/dd/yyyy" format into "yyyymmdd". See the help topics for as.Date, format.Date, and strptime for more details and other examples.

Example B.8 (Date formats)

Convert the string representation of a date into a date object, and display the result in several formats. The default format is "yyyy-mm-dd". The date is printed in four different formats below.

```
d <- "3/27/1995"
thedate <- as.Date(d, "%m/%d/%Y")
print(thedate)
[1] "1995-03-27"
print(format(thedate, "%Y%m%d"))
[1] "19950327"
print(format(thedate, "%B %d, %Y"))
[1] "March 27, 1995"
print(format(thedate, "%y-%b-%d"))
[1] "95-Mar-27"
```

◇

To extract year, month, day, or other components from the date or time, we can use the POSIXlt date-time class (?DateTimeClasses).

Example B.9 (Date-time class)

Continuing with the previous example, use the POSIXlt date-time class to extract the year, month, and day from the date 1995-03-27. The commands and results are below. Notice that the months Jan., ..., Dec. are numbered $0, 1, \ldots, 11$, and year is years since 1900.

```
> pdate <- as.POSIXlt(thedate)
> print(pdate$year)
[1] 95
> print(pdate$mon)
```

```
[1] 2
> print(pdate$mday)
[1] 27
```

Type `?DateTimeClasses` to see the documentation on the date-time objects `POSIXlt` and `POSIXct`.   ◇

B.3.4 Importing/exporting .csv files

Data is often supplied in comma-separated-values (.csv) format, which is a text file that separates data with special text characters called delimiters. Files in .csv format can be opened in most spreadsheet applications. Spreadsheet data should be saved in .csv format before importing into R. In a .csv file, the dates are likely to be given as strings, delimited by double quotation marks.

Example B.10 (Importing/exporting .csv files)

This example illustrates how to export the contents of a data frame to a .csv file, and how to import the data from a .csv file into an R data frame.

```
#create a data frame
dates <- c("3/27/1995", "4/3/1995",
           "4/10/1995", "4/18/1995")
prices <- c(11.1, 7.9, 1.9, 7.3)
d <- data.frame(dates=dates, prices=prices)

#create the .csv file
filename <- "/Rfiles/temp.csv"
write.table(d, file = filename, sep = ",",
            row.names = FALSE)
```

The new file "temp.csv" can be opened in most spreadsheets. When displayed in a text editor (not a spreadsheet), the file "temp.csv" contains the following lines (without the leading spaces).

```
"dates","prices"
"3/27/1995",11.1
"4/3/1995",7.9
"4/10/1995",1.9
"4/18/1995",7.3
```

Most .csv format files can be read using `read.table`. In addition there are functions `read.csv` and `read.csv2` designed for .csv files.

```
#read the .csv file
read.table(file = filename, sep = ",", header = TRUE)
read.csv(file = filename) #same thing
```

```
        dates prices
1 3/27/1995   11.1
2  4/3/1995    7.9
3 4/10/1995    1.9
4 4/18/1995    7.3
```

See Example B.8 for converting the character representation of the dates to date objects.   ◇

B.3.5   Examples of data entry and analysis

Although it is not the subject of this text, users new to R generally need to know how to analyze typical textbook examples with small data sets. For Monte Carlo studies, one also may need to extract certain results from a fitted model. We conclude this section with a few simple examples of this type.

Stacked data entry

Example B.11 (One-way ANOVA)

Weight measurements are collected for two treatment groups of subjects and a control group. This is a completely randomized design, and we want to obtain the one-way Analysis of Variance (ANOVA). The layout of the data is the one-way layout, and for ANOVA we will need stacked data. The factor has three levels. Here we create a vector for the response variable (weight) and a vector for the group variable, encoding it as a `factor`. See Example B.13 for another approach to stacking the data for the one-way layout.

```
# One-way ANOVA example
# Completely randomized design
ctl <- c(4.17,5.58,5.18,6.11,4.50,4.61,5.17,4.53,5.33,5.14)
trt1 <- c(4.81,4.17,4.41,3.59,5.87,3.83,6.03,4.89,4.32,4.69)
trt2 <- c(5.19,3.33,3.20,3.13,6.46,5.36,6.95,4.19,3.16,4.95)

group <- factor(rep(1:3, each=10)) #factor
weight <- c(ctl, trt1, trt2)        #response
a <- lm(weight ~ group)
```

Note that encoding the group variable as a `factor` is important. If `group` is not a factor, but simply a vector of integers, then `lm` will fit a regression model. The output for `anova` is the ANOVA table. More detailed output is available with the `summary` method.

```
> anova(a)                              #brief summary
Analysis of Variance Table
Response: weight
          Df  Sum Sq Mean Sq F value Pr(>F)
group      2  1.1200  0.5600  0.5656 0.5746
Residuals 27 26.7344  0.9902

> summary(a)                            #more detailed summary
Call:
lm(formula = weight ~ group)
Residuals:
    Min      1Q  Median      3Q     Max
-1.4620 -0.5245  0.0685  0.5005  2.3580

Coefficients:
            Estimate Std. Error t value Pr(>|t|)
(Intercept)   5.0320     0.3147  15.991 2.71e-15 ***
group2       -0.3710     0.4450  -0.834    0.412
group3       -0.4400     0.4450  -0.989    0.332
---
Signif. codes:  0 '***' 0.001 '**' 0.01 '*' 0.05 '.' 0.1 ' ' 1
Residual standard error: 0.9951 on 27 degrees of freedom
Multiple R-Squared: 0.04021,    Adjusted R-squared: -0.03089
F-statistic: 0.5656 on 2 and 27 DF,  p-value: 0.5746
```

◇

Extracting statistics and estimates from fitted models

In Monte Carlo studies, we often want to extract the p-values, F statistics, or R-squared values from the analysis, rather than print a summary of it. The following example shows how to extract various results from an **anova** object or the summary.

Example B.12 (Extract p-values and statistics from ANOVA)

To extract p-values, F statistics and other information from the **anova** object or result of **summary**, we need the **names** of these values. Then the information can be extracted by name or by position using square brackets. (This example continues from the analysis in Example B.11.)

```
A <- anova(a)
names(A)
[1] "Df"       "Sum Sq"  "Mean Sq" "F value" "Pr(>F)"
```

Then, suppose we need the F statistic. It is a vector of length 2, corresponding to the two rows in the ANOVA table. The F statistic in each row corresponds to the factor in the same row.

```
> A$"F value"
[1] 0.5655666          NA
> A$"F value"[1]
[1] 0.5655666
```

Similarly, we can use **names** to find the names of the values in the object returned by the summary method.

```
B <- summary(a)
names(B)
[1] "call"      "terms"    "residuals"    "coefficients"  "aliased"
[6] "sigma"     "df"       "r.squared"    "adj.r.squared" "fstatistic"
[11] "cov.unscaled"
```

Now suppose that we want to extract the R-squared, the MSE, and the degrees of freedom for error from this model.

```
> B$sigma
[1] 0.9950695
> B$r.squared
[1] 0.0402093
> B$df[2]
[1] 27
```

◇

Create data frame in stacked layout

The next example shows an alternate method for entering data in the one-way layout. In this case, we create a data frame and use the **stack** function.

Example B.13 (Stacked data entry)

The small data set in this example is given in Case Study 12.3.1 of Larsen and Marx [170]. The factor (type of antibiotic) has five levels. The response variable measures the binding of the drug to serum proteins. The layout of the data frame must be stacked for the ANOVA.

```
P <- c(29.6, 24.3, 28.5, 32)
T <- c(27.3, 32.6, 30.8, 34.8)
S <- c(5.8,  6.2, 11, 8.3)
E <- c(21.6, 17.4, 18.3, 19)
C <- c(29.2, 32.8, 25,  24.2)

#glue the columns together in a data frame
x <- data.frame(P, T, S, E, C)

#now stack the data for ANOVA
y <- stack(x)
names(y) <- c("Binding", "Antibiotic")
```

The first few rows of the stacked data in y are

```
  Binding Antibiotic
1   29.6         P
2   24.3         P
3   28.5         P
4   32.0         P
5   27.3         T
6   32.6         T
  . . .
```

and this data is in the one-way layout for ANOVA. Now y is a data frame, so there is a default formula associated with it.

```
> #check the default formula
> print(formula(y))   #default formula is right one
Binding ~ Antibiotic
```

As the default formula is the same model that we want to fit, lm can be applied without specifying the formula.

```
> lm(y)
Call:
lm(formula = y)
Coefficients:
(Intercept)  AntibioticE  AntibioticP  AntibioticS  AntibioticT
     27.800       -8.725        0.800      -19.975        3.575

> anova(lm(y))
Analysis of Variance Table
Response: Binding
            Df  Sum Sq Mean Sq F value   Pr(>F)
Antibiotic   4 1480.82  370.21  40.885 6.74e-08 ***
Residuals   15  135.82    9.05
---
Signif. codes:  0 '***' 0.001 '**' 0.01 '*' 0.05 '.' 0.1 ' ' 1
```

Statistics, p-values, and estimates can be extracted from the fitted model in the same way as shown in Example B.12.   ◇

Example B.14 (Two-way ANOVA)

The leafshape (DAAG) [185] data is already in stacked format, with two factors location and leaf architecture arch.

```
> data(leafshape, package = "DAAG")
> anova(lm(petiole ~ location * arch))
```

```
Analysis of Variance Table

Response: petiole
                Df Sum Sq Mean Sq F value     Pr(>F)
location         5   209.9    42.0  1.8107     0.1108
arch             1  1098.5  1098.5 47.3786 3.983e-11 ***
location:arch    5   232.6    46.5  2.0066     0.0779 .
Residuals      274  6352.8    23.2
---
Signif. codes:  0 '***' 0.001 '**' 0.01 '*' 0.05 '.' 0.1 ' ' 1
```

Use the formula petiole~location+arch to fit the model without the inter-
action term.   ◇

Example B.15 (Multiple comparisons)

In Example B.13, one can follow up with a multiple comparison procedure to
decide which means are significantly different. One such method is Tukey's
procedure.   The critical value of the studentized range statistic at $\alpha = 0.05$
can be obtained by

```
qtukey(p = .95, nmeans = 5, df = 15)
[1] 4.366985
```

For TukeyHSD use aov to fit the model rather than lm.

```
#alternately: Tukey Honest Significant Difference
a <- aov(formula(y), data = y)
TukeyHSD(a, conf.level=.95)

  Tukey multiple comparisons of means
    95% family-wise confidence level
Fit: aov(formula = formula(y), data = y)

$Antibiotic
        diff         lwr        upr    p adj
E-C   -8.725 -15.295401  -2.154599 0.0071611
P-C    0.800  -5.770401   7.370401 0.9952758
S-C  -19.975 -26.545401 -13.404599 0.0000010
T-C    3.575  -2.995401  10.145401 0.4737713
P-E    9.525   2.954599  16.095401 0.0034588
S-E  -11.250 -17.820401  -4.679599 0.0007429
T-E   12.300   5.729599  18.870401 0.0003007
S-P  -20.775 -27.345401 -14.204599 0.0000006
T-P    2.775  -3.795401   9.345401 0.6928357
T-S   23.550  16.979599  30.120401 0.0000001
```

                                                                      ◇

Example B.16 (Regression)

Other examples of `formula` (see e.g. Example 7.17) for regression rather than ANOVA are the following.

```
library(DAAG)
attach(ironslag)

# simple linear regression model
lm(magnetic ~ chemical)

# quadratic regression model
lm(magnetic ~ chemical + I(chemical^2))

# exponential regression model
lm(log(magnetic) ~ chemical)

# log-log model
lm(log(magnetic) ~ log(chemical))

# cubic polynomial model
lm(magnetic ~ poly(chemical, degree = 3))

detach(ironslag)
detach(package:DAAG)
```

In the quadratic model, the "as is" operator `I( )` indicates that the exponentiation operator is an arithmetic operator, and should not be interpreted as a formula operator. Note that `poly` evaluates an orthogonal polynomial.

```
> cor(poly(chemical, 2))      #uncorrelated
              1              2
1  1.000000e+00 -4.956837e-18
2 -4.956837e-18  1.000000e+00

> cor(chemical, chemical^2) #correlated
[1] 0.9919215
```

◇

References

[1] M. Abramowitz and I. A. Stegun, editors. *Handbook of Mathematical Functions with Formulas, Graphs, and Mathematical Tables*. Dover, New York, 1972.

[2] D. Adler and D. Murdoch. *rgl: 3D visualization device system (OpenGL)*, 2007. R package version 0.74.

[3] J. H. Ahrens and U. Dieter. Computer methods for sampling from the exponential and normal distributions. *Comm. ACM*, 15:873–882, 1972.

[4] J. H. Ahrens and U. Dieter. Sampling from the binomial and Poisson distributions: A method with bounded computation times. *Computing*, 25:193–208, 1980.

[5] J. Albert. *Bayesian Computation with R*. Springer, New York, 2007.

[6] J. H. Albert. Teaching Bayesian statistics using sampling methods and MINITAB. *The American Statistician*, 47:182–191, 1993.

[7] P. M. E. Altham. Exact Bayesian analysis of a 2×2 contingency table and Fisher's "exact" significance test. *Journal of the Royal Statistical Society. Series B*, 31:261–269, 1969.

[8] T. W. Anderson. *An Introduction to Multivariate Statistical Analysis*. Wiley, New York, Second edition, 1984.

[9] T. W. Anderson and D. A. Darling. A test of goodness-of-fit. *Journal of the American Statistical Association*, 49:765–769, 1954.

[10] D. F. Andrews. Plots of high dimensional data. *Biometrics*, 28:125–136, 1972.

[11] F. J. Anscombe and W. J. Glynn. Distribution of the kurtosis statistic b_2 for normal statistics. *Biometrika*, 70:227–234, 1986.

[12] S. Arya and D. M. Mount. Approximate nearest neighbor searching. In *Proceedings of the fourth annual ACM-SIAM Symposium on Discrete Algorithms (SODA 1993)*, pages 271–280, 1993.

[13] S. Arya, D. M. Mount, N. S. Netanyahu, R. Silverman, and A. Y. Wu. An optimal algorithm for approximate nearest neighbor searching. *Journal of the ACM*, 45:891–923, 1998.

[14] D. Asimov. The grand tour: a tool for viewing multidimensional data. *SIAM Journal on Scientific and Statistical Computing*, 6(1):128–143, 1985.

[15] A. Azzalini and A. W. Bowman. A look at some data on the Old Faithful geyser. *Applied Statistics*, 39:357–365, 1990.

[16] L. J. Bain and M. Engelhardt. *Introduction to Probability and Mathematical Statistics*. Duxbury Classic Series. Brooks-Cole, Pacific Grove, CA, 1991.

[17] N. K. Bakirov, M. L. Rizzo, and G. J. Székely. A multivariate nonparametric test of independence. *Journal of Multivariate Analysis*, 93:1742–1756.

[18] J. Banks, J. Carson, B. L. Nelson, and D. Nicol. *Discrete-Event System Simulation*. Prentice-Hall, Upper Saddle River, NJ, fourth edition, 2004.

[19] P. Barbe and P. Bertail. *The Weighted Bootstrap*. Springer, New York, 1995.

[20] L. Baringhaus and C. Franz. On a new multivariate two-sample test. *Journal of Multivariate Analysis*, 88:190–206, 2004.

[21] M. S. Bartlett. On the theory of statistical regression. *Proceedings of the Royal Society of Edinburgh*, 53:260–283.

[22] K. E. Basford and G. J. McLachlan. Likelihood estimation with normal mixture models. *Journal of the Royal Statistical Society. Series C*, 34(3):282–289, 1985.

[23] M. A. Bean. *Probability: The Science of Uncertainty with Applications to Investments, Insurance, and Engineering*. Brooks-Cole, Pacific Grove, CA, 2001.

[24] R. A. Becker, J. M. Chambers, and A. R. Wilks. *The New S Language: A Programming Environment for Data Analysis and Graphics*. Wadsworth & Brooks/Cole, Pacific Grove, CA, 1988.

[25] J. L. Bentley. Multidimensional binary search trees used for associative searching. *Communications of the ACM*, 18(9):509–517, 1975.

[26] M. Berkelaar et al. *lpSolve: Interface to Lp solve v. 5.5 to solve linear/integer programs*, 2006. R package version 5.5.7.

[27] P. J. Bickel. A distribution free version of the Smirnov two-sample test in the multivariate case. *Annals of Mathematical Statistics*, 40:1–23.

[28] P. J. Bickel and L. Breiman. Sums of functions of nearest neighbor distances, moment bounds, limit theorems and a goodness of fit test. *Annals of Probability*, 11:185–214, 1983.

[29] A. W. Bowman and A. Azzalini. *sm: Smoothing methods for nonpara-metric regression and density estimation*, 2005. Ported to R by B. D. Ripley up to version 2.0 and later versions by Adrian W. Bowman and Adelchi Azzalini. R package version 2.1-0.

[30] A. W. Bowman, P. Hall, and D. M. Titterington. Cross-validation in nonparametric estimation of probabilities and probability densities. *Biometrika*, 71(2):341–351, 1984.

[31] G. E. P. Box and M. E. Müller. A note on the generation of random normal deviates. *The Annals of Mathematical Statistics*, 29:610–611, 1958.

[32] R. Brent. *Algorithms for Minimization without Derivatives*. Prentice-Hall, New Jersey, 1973.

[33] S. P. Brooks, P. Dellaportas, and G. O. Roberts. An approach to di-agnosing total variation convergence of MCMC algorithms. *Journal of Computational and Graphical Statistics*, 6(3):251–265, 1997.

[34] A. Canty and B. Ripley. *boot: Bootstrap R (S-Plus) Functions (Canty)*, 2006. S original by Angelo Canty, R port by Brian Ripley. R package version 1.2-28.

[35] O. Cappe and C. P. Robert. Markov Chain Monte Carlo: 10 years and still running! *Journal of the American Statistical Association*, 95(452):1282–1286, 2000.

[36] A. E. Carlin, B. P. Gelfand and A. F. M. Smith. Hierarchical Bayesian analysis of changepoint problems. *Applied Statistics*, 41:389–405, 1992.

[37] B. P. Carlin and T. A. Louis. *Bayes and Empirical Bayes Methods for Data Analysis*. Chapman and Hall/CRC, Boca Raton, FL, 2000.

[38] D. Carr. *hexbin: Hexagonal Binning Routines*, 2006. Ported by Nicholas Lewin-Koh and Martin Maechler. R package version 1.8.0.

[39] G. Casella and R. Berger. *Statistical Inference*. Duxbury Press, Bel-mont, California, 1990.

[40] G. Casella and E. E. George. Explaining the Gibbs sampler. *The American Statistician*, 46:167–174, 1992.

[41] J. M. Chambers. *Programming with Data: A Guide to the S Language*. Springer, New York, 1998.

[42] J. M. Chambers and T. J. Hastie. *Statistical Models in S*. Chapman & Hall, London, 1992.

[43] J. M. Chambers, C. L. Mallows, and B. W. Stuck. A method for sim-ulating stable random variables. *Journal of the American Statistical Association*, 71:304–344, 1976.

[44] M.-H. Chen, Q.-M. Shao, and J. G. Ibrahim. *Monte Carlo Methods in Bayesian Computation.* Springer, New York, 2000.

[45] M. A. Chernick. *Bootstrap Methods: A Practitioner's Guide.* Wiley, New York, 1999.

[46] H. Chernoff. The use of faces to represent points in k-dimensional space graphically. *Journal of the American Statistical Association,* 68:361–368, 1973.

[47] S. Chib and E. Greenberg. Understanding the Metropolis-Hastings algorithm. *The American Statistician,* 49:327–335, 1995.

[48] W. S. Cleveland. *Visualizing Data.* Summit Press, New Jersey, 1993.

[49] W. S. Cleveland. Coplots, nonparametric regression, and conditionally parametric fits. In *Multivariate Analysis and its Applications (Hong Kong, 1992),* volume 24 of *IMS Lecture Notes Monograph Series,* pages 21–36. Inst. Math. Statist., Hayward, CA, 1994.

[50] W. S. Cleveland and R. McGill. The many faces of a scatterplot. *Journal of the American Statistical Association,* 79(388):807–822, 1984.

[51] J.-F. Coeurjolly. Simulation and identification of the fractional Brownian motion: a bibliographical and comparative study. *Journal of Statistical Software,* 5, 2000.

[52] D. Cook and D. F. Swayne. *Interactive and Dynamic Graphics for Data Analysis: With R and GGobi.* Springer, New York, 2007.

[53] G. Cornuejols and R. Tütüncü. *Optimization Methods in Finance.* Cambridge University Press, Cambridge, 2007.

[54] M. K. Cowles and B. P. Carlin. Markov Chain Monte Carlo convergence diagnostics: A comparative review. *Journal of the American Statistical Association,* 91(434):883–904, 1996.

[55] D. R. Cox and N. J. H. Small. Testing multivariate normality. *Biometrika,* 65:263–272, 1978.

[56] H. Cramér. On the composition of elementary errors. II Statistical applications. *Skandinavisk Aktuarietidskrift,* 11:141–180, 1928.

[57] M. J. Crawley. *Statistical Computing: An Introduction to Data Analysis using S-Plus.* Wiley, New York, 2002.

[58] R. B. D'Agostino. Tests for the normal distribution. In R. B. D'Agostino and M. A. Stephens, editors, *Goodness-of-Fit Techniques,* pages 367–420. Marcel Dekker, New York, 1986.

[59] R. B. D'Agostino and E. S. Pearson. Tests for departure from normality. empirical results for the distributions of b_2 and $\sqrt{b_1}$. *Biometrika,* 60:613–622, 1973.

[60] R. B. D'Agostino and M. A. Stephens. *Goodness-of-Fit Techniques.* Marcel Dekker, New York, 1986.

[61] D. B. Dahl. *xtable: Export tables to LaTeX or HTML*, 2007. With contributions from many others. R package version 1.4-3.

[62] P. Dalgaard. *Introductory Statistics with R.* Springer, New York, 2002.

[63] A. C. Davison and D. V. Hinkley. *Bootstrap Methods and their Application.* Cambridge University Press, Oxford, 1997.

[64] M. H. DeGroot and M. J. Schervish. *Probability and Statistics.* Addison-Wesley, New York, third edition, 2002.

[65] S. Déjean and S. Cohen. FracSim: An R package to simulate multifractional Lévy motions. *Journal of Statistical Software*, 14, 2005.

[66] S. Déjean and S. Cohen. *FracSim: Simulation of Lévy motions*, 2005. R package version 0.2.

[67] A. P. Dempster, N. M. Laird, and D. B. Rubin. Maximum likelihood from incomplete data via the EM algorithm (with discussion). *Journal of the Royal Statistical Society. Series B. Methodological*, 39:1–38, 1977.

[68] L. Devroye. The computer generation of Poisson random variables. *Computing*, 26:197–207, 1981.

[69] L. Devroye. *Non-Uniform Random Variate Generation.* Springer, New York, 1986.

[70] L. Devroye. *A Course in Density Estimation.* Birkhäuser, Boston, 1987.

[71] L. Devroye and L. Györfi. *Nonparametric Density Estimation: The L_1 View.* John Wiley, New York, 1985.

[72] E. Dimitriadou, K. Hornik, F. Leisch, D. Meyer, and A. Weingessel. *e1071: Misc Functions of the Department of Statistics (e1071), TU Wien*, 2006. R package version 1.5-16.

[73] D. P. Doane. Aesthetic frequency classification. *The American Statistician*, 30:181–183, 1976.

[74] M. Dresher. *Games of Strategy: Theory and Application.* Dover, New York, 1981.

[75] R. O. Duda, P. E. Hart, and D. G. Stork. *Pattern Classification.* Wiley, New York, second edition, 2001.

[76] T. Duong. *ks: Kernel smoothing*, 2007. R package version 1.4.9.

[77] R. Durrett. *Probability: Theory and Examples.* Wadsworth Publishing (Duxbury Press), Belmont, CA, second edition, 1996.

[78] R. Eckhardt. Stan Ulam, John von Neumann, and the Monte Carlo method. *Los Alamos Science*, (15, Special Issue):131–137, 1987. With contributions by Tony Warnock, Gary D. Doolen, and John Hendricks, Stanislaw Ulam 1909–1984.

[79] S. Efromovich. Density estimation for the case of supersmooth measurement error. *Journal of the American Statistical Association*, 92(438):526–535, 1997.

[80] B. Efron. Bootstrap methods: another look at the jackknife. *Annals of Statistics*, 7:1–26, 1979.

[81] B. Efron. Nonparametric estimates of standard error: the jackknife, the bootstrap, and other methods. *Biometrika*, 68:589–599, 1981.

[82] B. Efron. Nonparametric standard errors and confidence intervals (with discussion). *Canadian Journal of Statistics*, 9:139–172, 1981.

[83] B. Efron. *The Jackknife, the Bootstrap and Other Resampling Plans.* Society for Industrial and Applied Mathematics, Philadelphia, 1982.

[84] B. Efron and R. J. Tibshirani. *An Introduction to the Bootstrap.* Chapman & Hall/CRC, Boca Raton, FL, 1993.

[85] V. K. Epanechnikov. Non-parametric estimation of a multivariate probability density. *Theory of Probability and its Applications*, 14:153–158, 1969.

[86] R. L. Eubank. *Spline Smoothing and Nonparametric Regression.* Marcel Dekker, New York, 1988.

[87] M. Evans and T. Schwartz. *Approximating Integrals via Monte Carlo and Deterministic Methods.* Oxford University Press, Oxford, 2000.

[88] B. Everitt and T. Hothorn. *A Handbook of Statistical Analyses Using R.* Chapman & Hall/CRC, Boca Raton, FL, 2006.

[89] B. S. Everitt and D. J. Hand. *Finite Mixture Distributions.* Chapman & Hall, London, 1981.

[90] J. J. Faraway. *Linear Models with R.* Chapman & Hall/CRC, Boca Raton, FL, 2004.

[91] J. J. Faraway. *Extending Linear Models with R: Generalized Linear, Mixed Effects and Nonparametric Regression Models.* Chapman & Hall/CRC, Boca Raton, FL, 2006.

[92] R. A. Fisher. On the 'probable error' of a coefficient of correlation deduced from a small sample. *Metron*, 1:3–32, 1921.

[93] R. A. Fisher. The moments of the distribution for normal samples of measures of departures from normality. *Proceedings of the Royal Society of London*, A, 130:16–28, 1930.

[94] G. S. Fishman. *Monte Carlo Concepts, Algorithms, and Applications.* Springer, New York, 1995.

[95] G. S. Fishman. *Discrete-Event Simulation.* Springer, New York, 2001.

[96] R. Fletcher and C. M. Reeves. Function minimization by conjugate gradients. *Computer Journal,* 7:148–154.

[97] J. Fox. *An R and S-Plus Companion to Applied Regression.* Sage Publications, Thousand Oaks, CA, 2002.

[98] J. N. Franklin. Numerical simulation of stationary and non-stationary Gaussian random processes. *SIAM Review,* 7:68–80.

[99] D. Freedman and P. Diaconis. On the histogram as a density estimator: L_2 theory. *Zeitschrift für Wahrscheinlichkeitstheorie und verwandte Gebiete,* 57:453–476.

[100] J. Friedman and J. Tukey. A projection pursuit algorithm for exploratory data analysis. *IEEE Transactions on Computers,* 23:881–889, 1975.

[101] J. H. Friedman and L. C. Rafsky. Multivariate generalizations of the Wald-Wolfowitz and Smirnov two-sample tests. *Annals of Statistics,* 7:697–717, 1979.

[102] M. Friendly. *Visualizing Categorical Data.* SAS Press, Cary, NC, 2000.

[103] D. Gamerman. *Markov Chain Monte Carlo. Stochastic simulation for Bayesian inference.* Chapman & Hall, London, 1997.

[104] A. E. Gelfand. Gibbs sampling. *Journal of the American Statistical Association,* 95(452):1300–1304, 2000.

[105] A. E. Gelfand, S. E. Hills, A. Racine-Poon, and A. F. M. Smith. Illustration of Bayesian inference in normal data models using Gibbs sampling. *Journal of the American Statistical Association,* 85:972–985, 1990.

[106] A. E. Gelfand and A. F. M. Smith. Sampling based approaches to calculating marginal densities. *Journal of the American Statistical Association,* 85:398–409, 1990.

[107] A. Gelman. Inference and monitoring convergence. In W. R. Gilks, S. Richardson, and D. J. Spiegelhalter, editors, *Markov Chain Monte Carlo in Practice,* pages 131–143. Chapman & Hall, Boca Raton, FL, 1996.

[108] A. Gelman, J. B. Carlin, H. S. Stern, and D. B. Rubin. *Bayesian Data Analysis.* Chapman and Hall, Boca Raton, second edition, 2004.

[109] A. Gelman and D. B. Rubin. Inference from iterative simulation using multiple sequences (with discussion). *Statistical Science,* 7:457–511, 1992.

[110] A. Gelman and D. B. Rubin. A single sequence from the Gibbs sampler gives a false sense of security. In J. M. Bernardo, J. O. Berger, O. P. Dawid, and A. F. M. Smith, editors, *Bayesian Statistics 4*, pages 625–631. Oxford University Press, Oxford, 1992.

[111] S. Geman and D. Geman. Stochastic relaxation, Gibbs distributions and the Bayesian restoration of images. *IEEE Transactions on Pattern Analysis and Machine Intelligence*, 6:721–741, 1984.

[112] J. E. Gentle. *Random Number Generation and Monte Carlo Methods*. Springer, New York, 1998.

[113] J. E. Gentle. *Elements of Computational Statistics*. Springer, New York, 2002.

[114] J. E. Gentle, W. Härdle, and Y. Mori, editors. *Handbook of Computational Statistics : Concepts and Methods*. Springer, New York, 2004.

[115] A. Genz and F. Bretz. *mvtnorm: Multivariate Normal and T Distribution*, 2007. R port by Torsten Hothorn. R package version 0.8-1.

[116] C. J. Geyer. Practical Markov Chain Monte Carlo (with discussion). *Statistical Science*, 7:473–511, 1992.

[117] C. J. Geyer. *mcmc: Markov Chain Monte Carlo*, 2005. R package version 0.5-1.

[118] S. Ghahramani. *Fundamentals of Probability*. Prentice-Hall, New Jersey, second edition, 2000.

[119] W. R. Gilks. Full conditional distributions. In W. R. Gilks, S. Richardson, and D. J. Spiegelhalter, editors, *Markov Chain Monte Carlo in Practice*, pages 75–88. Chapman & Hall, Boca Raton, FL, 1996.

[120] W. R. Gilks, S. Richardson, and D. J. Spiegelhalter. *Markov Chain Monte Carlo in Practice*. Chapman & Hall, Boca Raton, FL, 1996.

[121] G. H. Givens and J. A. Hoeting. *Computational Statistics*. Wiley, New Jersey, 2005.

[122] R. Gnanadesikan. *Methods for the Statistical Analysis of Multivariate Observations*. Wiley, New York, Second edition, 1997.

[123] C. Gu. *gss: General Smoothing Splines*. R package version 0.9-3.

[124] F. A. Haight. *Handbook of the Poisson Distribution*. Wiley, New York, 1967.

[125] P. Hall and J. S. Marron. Choice of kernel order in density estimation. *Annals of Statistics*, 16:161–173, 1987.

[126] D. J. Hand, F. Daly, A. D. Lunn, K. J. McConway, and E. Ostrokowski. *A Handbook of Small Data Sets*. Chapman & Hall, Boca Raton, FL, 1996.

[127] R. K. S. Hankin. *gsl: wrapper for the Gnu Scientific Library*, 2005. qrng functions by Duncan Murdoch. R package version 1.6-7.

[128] W. Härdle. *Applied Nonparametric Regression*, volume 19 of *Econometric Society Monographs*. Cambridge University Press, Cambridge, 1990.

[129] W. Härdle. *Smoothing Techniques*. Springer-Verlag, New York, 1991. With implementation in S.

[130] W. Härdle, M. Müller, S. Sperlich, and A. Werwatz. *Nonparametric and Semiparametric Models*. Springer-Verlag, New York, 2004.

[131] F. E. Harrell. *Regression Modeling Strategies, with Applications to Linear Models, Survival Analysis and Logistic Regression*. Springer, 2001.

[132] F. E. Harrell Jr. *Hmisc: Harrell Miscellaneous*, 2007. With contributions from many other users. R package version 3.3-1.

[133] H. O. Hartley and D. L. Harris. Monte Carlo computations in normal correlation problems. *Journal of Association for Computing Machinery*, 10:301–306, 1963.

[134] T. Hastie. *gam: Generalized Additive Models*, 2006. R package version 0.98.

[135] T. Hastie and R. Tibshirani. Generalized additive models. *Statistical Science*, 1(3):297–318, 1986. With discussion.

[136] T. Hastie, R. Tibshirani, and J. Friedman. *The Elements of Statistical Learning*. Springer Series in Statistics. Springer-Verlag, New York, 2001. Data mining, inference, and prediction.

[137] T. J. Hastie and R. J. Tibshirani. *Generalized Additive Models*. Chapman and Hall Ltd., London, 1990.

[138] W. K. Hastings. Monte Carlo sampling methods using Markov chains and their applications. *Biometrika*, 57:97–109, 1970.

[139] N. Henze. A multivariate two-sample test based on the number of nearest neighbor coincidences. *Annals of Statistics*, 16:772–783, 1988.

[140] N. Henze. On Mardia's kurtosis test for multivariate normality. *Communications in Statistics: Theory and Methods*, 23:1031–1045, 1994.

[141] N. Henze. Extreme smoothing and testing for multivariate normality. *Statistics and Probability Letters*, 35:203–213, 1997.

[142] N. J. Higham. *Accuracy and Stability of Numerical Algorithms*. SIAM Publications, Philadelphia, 1996.

[143] J. S. U. Hjorth. *Computer Intensive Statistical Methods: Validation, Model Selection and Bootstrap*. Chapman and Hall, London, 1992.

[144] W. Hoeffding. A class of statistics with asymptotically normal distribution. *Annals of Mathematical Statistics*, 19:293–325, 1948.

[145] W. Hoeffding. A non-parametric test of independence. *Annals of Mathematical Statistics*, 19:546–547, 1948.

[146] R. V. Hogg, J. W. McKean, and A. T. Craig. *Introduction to Mathematical Statistics*. Prentice Hall, Upper Saddle River, New Jersey, sixth edition, 2005.

[147] K. Hornik. *The R FAQ*. R Foundation for Statistical Computing, Vienna, Austria, 2007. ISBN 3-900051-08-9.

[148] R. J. Hyndman and Y. Fan. Sample quantiles in statistical packages. *The American Statistician*, 50:361–365, 1996.

[149] S. M. Iacus. *sde: Simulation and Inference for Stochastic Differential Equations*, 2006. R package version 1.9.5.

[150] J. P. Imhof. Computing the distribution of quadratic forms in normal variables. *Biometrika*, 48(3/4):419–426, 1961.

[151] J. P. Imhof. Corrigenda: Computing the distribution of quadratic forms in normal variables. *Biometrika*, 49(1/2):284, 1962.

[152] A. Inselberg. The plane with parallel coordinates. *The Visual Computer*, 1:69–91, 1985.

[153] R. G. Jarrett. A note on the intervals between coal-mining disasters. *Biometrika*, 66:191–193, 1979.

[154] M. E. Johnson. *Multivariate Statistical Simulation*. Wiley, New York, 1987.

[155] M. E. Johnson, W. Chiang, and J. S. Ramberg. Generation of continuous multivariate distributions for statistical applications. *American Journal of Mathematical Management Science*, 4:225–248, 1984.

[156] N. L. Johnson, S. Kotz, and N. Balakrishnan. *Continuous Univariate Distributions*, volume 1. Wiley, New York, Second edition, 1994.

[157] N. L. Johnson, S. Kotz, and N. Balakrishnan. *Continuous Univariate Distributions*, volume 2. Wiley, New York, Second edition, 1995.

[158] N. L. Johnson, S. Kotz, and A. W. Kemp. *Univariate Discrete Distributions*. Wiley, New York, Second edition, 1992.

[159] A. W. Kemp. Efficient generation of logarithmically distributed pseudorandom variables. *Applied Statistics*, 30:249–253, 1981.

[160] S. E. Kemp. *knnFinder: Fast Near Neighbour Search*. R package version 1.0.

[161] W. J. Kennedy, Jr. and J. E. Gentle. *Statistical Computing*. Marcel Dekker, New York, 1980.

[162] D. A. King and J. H. Maindonald. Tree architecture in relation to leaf dimensions and tree stature in temperate and tropical rain forests. *Journal of Ecology*, 87:1012–1024, 1999.

[163] C. Kleiber and S. Kotz. *Statistical Size Distributions in Economics and Actuarial Sciences*. Wiley, 2003.

[164] D. B. Knuth. *The Art of Computer Programming (Vol. 2: Seminumerical Algorithms)*. Addison-Wesley, Reading, third edition, 1997.

[165] D. Kundu and A. Basu, editors. *Statistical Computing: Existing Methods and Recent Developments*. Alpha Science International Ltd., Harrow, U.K., 2004.

[166] D. Kuonen. Saddlepoint approximations for distributions of quadratic forms in normal variables. *Biometrika*, 86(4):929–935, 1999.

[167] D. T. Lang, D. Swayne, H. Wickham, and M. Lawrence. *rggobi: Interface between R and GGobi*, 2006. R package version 2.1.4-4.

[168] K. Lange. *Numerical Analysis for Statisticians*. Springer-Verlag, New York, 1998.

[169] K. Lange. *Optimization*. Springer-Verlag, New York, 2004.

[170] R. J. Larsen and M. L. Marx. *An Introduction to Mathematical Statistics and Its Applications*. Prentice-Hall, Inc., New Jersey, fourth edition, 2006.

[171] P. M. Lee. *Bayesian Statistics*. Oxford University Press, New York, third edition, 2004.

[172] E. L. Lehmann. *Testing Statistical Hypotheses*. Springer, New York, second edition, 1986. Originally published New York: Wiley c1986.

[173] E. L. Lehmann and G. Casella. *Theory of Point Estimation*. Springer, New York, second edition, 1998.

[174] F. Leisch and E. Dimitriadou. *mlbench: Machine Learning Benchmark Problems*, 2007. R package version 1.1-3.

[175] R. V. Lenth. Algorithm AS 243 – Cumulative distribution function of the non-central t distribution. *Applied Statistics*, 38:185–189, 1989.

[176] A. Liaw and M. Wiener. Classification and regression by randomForest. *R News*, 2(3):18–22, 2002.

[177] U. Ligges. R-WinEdt. In K. Hornik, F. Leisch, and A. Zeileis, editors, *Proceedings of the 3rd International Workshop on Distributed Statistical Computing (DSC 2003)*, TU Wien, Vienna, Austria, 2003. ISSN 1609-395X.

[178] R. J. A. Little and D. B. Rubin. *Statistical Analysis with Missing Data.* Wiley, Hoboken, NJ, second edition, 2002.

[179] J. S. Liu. *Monte Carlo Strategies in Scientific Computing.* Springer, New York, 2001.

[180] C. Loader. *locfit: Local Regression, Likelihood and Density Estimation,* 2006. R package version 1.5-3.

[181] C. R. Loader. Bandwidth selection: Classical or plug-in? *The Annals of Statistics,* 27:415–438, 1999.

[182] J. Ludbrook and H. Dudley. Why permutation tests are superior to t and F tests in biomedical research. *The American Statistician,* 52(2):127–132, 1998.

[183] M. Maechler and many others. *sfsmisc: Utilities from Seminar fuer Statistik ETH Zurich,* 2007. R package version 0.95-9.

[184] J. Maindonald and J. Braun. *Data Analysis and Graphics Using R – an Example-based Approach.* Cambridge University Press, Cambridge, 2003.

[185] J. Maindonald and J. Braun. *DAAG: Data Analysis and Graphics,* 2007. R package version 0.95.

[186] E. Mammen. *When Does Bootstrap Work?* Springer, New York, 1992.

[187] K. V. Mardia. Measures of multivariate skewness and kurtosis with applications. *Biometrika,* 57:519–530, 1970.

[188] K. V. Mardia, J. T. Kent, and J. M. Bibby. *Multivariate Analysis.* Academic Press, San Diego, 1979.

[189] J. S. Marron and M. P. Wand. Exact mean integrated squared error. *The Annals of Statistics,* 20:712–736, 1992.

[190] G. Marsaglia, W. W. Tsang, and J. Wang. Fast generation of discrete random variables. *Journal of Statistical Software,* 11, 2004.

[191] A. D. Martin and K. M. Quinn. *MCMCpack: Markov Chain Monte Carlo (MCMC) Package,* 2007. R package version 0.8-1.

[192] W. L. Martinez and A. R. Martinez. *Computational Statistics Handbook with MATLAB.* Chapman & Hall/CRC, Boca Raton, FL, 2002.

[193] R. N. McGrath and B. Y. Yeh. Count Five test for equal variance. *The American Statistician,* 59:47–53, 2005.

[194] G. J. McLachlan and T. Krishnan. *The EM Algorithm and Extensions.* Wiley, New York, 1997.

[195] N. Metropolis. The beginning of the Monte Carlo method. *Los Alamos Science,* (15, Special Issue):125–130, 1987. Stanislaw Ulam 1909–1984.

[196] N. Metropolis. The Los Alamos experience, 1943–1954. In *A History of Scientific Computing (Princeton, NJ, 1987)*, ACM Press History Series, pages 237–250. ACM, New York, 1990.

[197] N. Metropolis, A. W. Rosenbluth, M. N. Rosenbluth, A. H. Teller, and E. Teller. Equations of state calculations by fast computing machine. *Journal of Chemical Physics*, 21:1087–1091, 1953.

[198] N. Metropolis and S. Ulam. The Monte Carlo method. *Journal of the American Statistical Association*, 44:335–341, 1949.

[199] D. Meyer, A. Zeileis, and K. Hornik. *vcd: Visualizing Categorical Data*, 2007. R package version 1.0.5.

[200] P. W. Mielke, Jr. and K. J. Berry. Permutation tests for common locations among samples with unequal variances. *Journal of Educational and Behavioral Statistics*, 19(3):217–236, 1994.

[201] I. Miller and M. Miller. *John E. Freund's Mathematical Statistics with Applications*. Prentice Hall, New Jersey, seventh edition, 2004.

[202] J. F. Monahan. *Numerical Methods of Statistics*. Cambridge University Press, Cambridge, 2001.

[203] H. G. Müller. *Nonparametric Regression Analysis of Longitudinal Data*. Springer-Verlag, Berlin, 1988.

[204] P. Murrell. *R Graphics*. Chapman & Hall/CRC, Boca Raton, FL, 2005.

[205] J. A. Nelder and R. Mead. A simplex algorithm for function minimization. *Computer Journal*, 7:308, 1965.

[206] J. Nocedal and S. J. Wright. *Numerical Optimization*. Springer, New York, 1999.

[207] P. L. Odell and A. H. Feiveson. A numerical procedure to generate a sample covariance matrix. *Journal of the American Statistical Association*, 61:199–203.

[208] G. Owen. *Game Theory*. Academic Press, New York, third edition, 1995.

[209] W. M. Patefield. Algorithm AS159. An efficient method of generating $r \times c$ tables with given row and column totals. *Applied Statistics*, 30:91–97, 1981.

[210] J. K. Patel and C. B. Read. *Handbook of the Normal Distribution*. Marcel Dekker, New York, second edition, 1996.

[211] J. C. Pinheiro and D. M. Bates. *Mixed-Effects Models in S and S-Plus*. Springer, 2000.

[212] M. Plummer, N. Best, K. Cowles, and K. Vines. *coda: Output analysis and diagnostics for MCMC*, 2007. R package version 0.11-2.

[213] W. H. Press, S. A. Teukolsky, W. T. Vetterling, and B. P. Flannery. *Numerical Recipes in C: The Art of Scientific Computing*. Cambridge University Press, New York, Second edition, 1992.

[214] M. L. Puri and P. K. Sen. *Nonparametric Methods in Multivariate Analysis*. Wiley, New York, 1971.

[215] M. H. Quenouille. Approximate tests of correlation in time series. *Journal of the Royal Statistical Society, Series B*, 11:68–84, 1949.

[216] M. H. Quenouille. Notes on bias in estimation. *Biometrika*, 43(3/4):353–360, 1956.

[217] R Development Core Team. *R: A Language and Environment for Statistical Computing*. R Foundation for Statistical Computing, Vienna, Austria, 2007. ISBN 3-900051-07-0.

[218] R Development Core Team. *R Installation and Administration*. R Foundation for Statistical Computing, Vienna, Austria, 2007. ISBN 3-900051-09-07.

[219] A. E. Raftery and S. M. Lewis. How many iterations in the Gibbs sampler? In J. M. Bernardo, J. O. Berger, O. P. Dawid, and A. F. M. Smith, editors, *Bayesian Statistics 4*, pages 763–773. Oxford University Press, Oxford, 1992.

[220] C. R. Rao, editor. *Linear Statistical Inference and Its Applications*. Wiley, New York, second edition, 1973.

[221] C. R. Rao, editor. *Computational Statistics*. Elsevier, The Netherlands, 1993.

[222] C. R. Rao, E. J. Wegman, and J. L. Solka, editors. *Handbook of Statistics, Volume 24: Data Mining and Data Visualization*.

[223] B. D. Ripley. *Stochastic Simulation*. Cambridge University Press, Cambridge, 1987.

[224] B. D. Ripley. *Pattern Recognition and Neural Networks*. Cambridge University Press, Cambridge, 1996.

[225] B. D. Ripley and D. J. Murdoch. *The R for Windows FAQ*. R Foundation for Statistical Computing, Vienna, Austria, 2007.

[226] M. L. Rizzo and G. J. Székely. *energy: E-statistics (energy statistics) tests of fit, independence, clustering*, 2007. R package version 1.0-6.

[227] C. P. Robert. Convergence control methods for Markov Chain Monte Carlo algorithms. *Statistical Science*, 10(3):231–253, 1995.

[228] C. P. Robert and G. Casella. *Monte Carlo Statistical Methods*. Springer, New York, Second edition, 2004.

[229] G. O. Roberts. Markov chain concepts related to sampling algorithms. In W. R. Gilks, S. Richardson, and D. J. Spiegelhalter, editors, *Markov Chain Monte Carlo in Practice*, pages 45–58. Chapman & Hall, Boca Raton, FL, 1996.

[230] G. O. Roberts, A. Gelman, and W. R. Gilks. Weak convergence and optimal scaling of random walk Metropolis algorithms. *Annals of Applied Probability*, 7:110–120, 1997.

[231] V. K. Rohatgi. *An Introduction to Probability Theory and Mathematical Statistics*. Wiley, New York, 1976.

[232] S. M. Ross. *A First Course in Probability*. Prentice-Hall, New Jersey, seventh edition, 2006.

[233] S. M. Ross. *Simulation*. Academic Press, San Diego, fourth edition, 2006.

[234] S. M. Ross. *An Introduction to Probability Models*. Academic Press, San Diego, ninth edition, 2007.

[235] J. P. Royston. Algorithm AS 181. The W test for normality. *Applied Statistics*, 31:176–180, 1982.

[236] J. P. Royston. An extension of Shapiro and Wilk's W test for normality to large samples. *Applied Statistics*, 31(2):115–124, 1982.

[237] P. Royston. Approximating the Shapiro-Wilk W-test for non-normality. *Statistical Computing*, 2:117–119, 1992.

[238] R. Y. Rubinstein. *Simulation and the Monte Carlo Method*. Wiley, New York, 1981.

[239] D. Sarkar. *lattice: Lattice Graphics*, 2007. R package version 0.15-8.

[240] M. F. Schilling. Multivariate two-sample tests based on nearest neighbors. *Journal of the American Statistical Association*, 81:799–806, 1986.

[241] D. W. Scott. On optimal and data-based algorithms. *Biometrika*, 66:605–610, 1979.

[242] D. W. Scott. Averaged shifted histograms: Effective nonparametric density estimators in several dimensions. *Annals of Statistics*, 13:1024–1040, 1985.

[243] D. W. Scott. Frequency polygons: Theory and application. *Journal of the American Statistical Association*, 80:348–354, 1985.

[244] D. W. Scott. *Multivariate Density Estimation. Theory, Practice, and Visualization*. John Wiley, New York, 1992.

[245] D. W. Scott and A. Gebhardt. *ash: David Scott's ASH routines.* S original by David W. Scott, R port by Albrecht Gebhardt. R package version 1.0-9.

[246] P. K. Sen. On some multisample permutation tests based on a class of U-statistics. *Journal of the American Statistical Association*, 62(320):1201–1213, 1967.

[247] J. Shao and D. Tu. *The Jackknife and Bootstrap.* Springer-Verlag, New York, 1995.

[248] S. S. Shapiro and M. B. Wilk. An analysis of variance test for normality (complete samples). *Biometrika*, 52:591–611, 1965.

[249] S. S. Shapiro, M. B. Wilk, and H. J. Chen. A comparative study of various tests of normality. *Journal of the American Statistical Association*, 63:1343–1372, 1968.

[250] B. W. Silverman. Choosing the window width when estimating a density. *Biometrika*, 65(1):1–11, 1978.

[251] B. W. Silverman. Some properties of a test for multimodality based on kernel density estimates. In *Probability, Statistics and Analysis*, volume 79 of *London Mathematical Society Lecture Note Series*, pages 248–259. Cambridge University Press, Cambridge, 1983.

[252] B. W. Silverman. *Density Estimation for Statistics and Data Analysis.* Chapman & Hall, London, 1986.

[253] P. W. F. Smith, J. J. Forster, and J. W. McDonald. Monte Carlo exact tests for square contingency tables. *Journal of the Royal Statistical Society. Series A (Statistics in Society)*, 159(2):309–321, 1996.

[254] G. Snow. *TeachingDemos: Demonstrations for teaching and learning*, 2005. R package version 1.5.

[255] C. Spearman. The proof and measurement of association between two things. *American Journal of Psychology*, 1904.

[256] Student. The probable error of a mean. *Biometrika*, 6:1–25, 1908.

[257] H. A. Sturges. The choice of a class interval. *Journal of the American Statistical Association*, 21:65–66, 1926.

[258] G. J. Székely. Potential and kinetic energy in statistics. Lecture notes, Budapest Institute of Technology (Technical University), 1989.

[259] G. J. Székely. E-statistics: Energy of statistical samples. Technical Report 03-05, Bowling Green State University, Department of Mathematics and Statistics, 2000.

[260] G. J. Székely. Student's t-test for scale mixture errors. In J. Rojo, editor, *Optimality, The Second Lehmann Symposium*, volume 49 of *IMS*

Lecture Notes – Monograph Series, pages 9–15. Institute of Mathematical Statistics, 2006.

[261] G. J. Székely and M. L. Rizzo. Testing for equal distributions in high dimension. *InterStat*, 11(5), 2004.

[262] G. J. Székely and M. L. Rizzo. Hierarchical clustering via joint between-within distances: extending Ward's minimum variance method. *Journal of Classification*, 22(2):151–183, 2005.

[263] G. J. Székely and M. L. Rizzo. A new test for multivariate normality. *Journal of Multivariate Analysis*, 93(1):58–80, 2005.

[264] G. J. Székely and M. L. Rizzo. The uncertainty principle of game theory. *The American Mathematical Monthly*, 8:688–702, October 2007.

[265] G. J. Székely, M. L. Rizzo, and N. K. Bakirov. Measuring and testing dependence by correlation of distances. *Annals of Statistics*, 35(6), December 2007.

[266] M. A. Tanner. *Tools for Statistical Inference: Methods for the Exploration of Posterior Distributions and Likelihood Functions*. Third edition, 1993.

[267] M. A. Tanner and W. H. Wong. The calculation of posterior distributions by data augmentation. *Journal of the American Statistical Association*, 82:528–549, 1987.

[268] T. M. Therneau and B. Atkinson. *rpart: Recursive Partitioning*, 2007. R port by Brian Ripley. R package version 3.1-36.

[269] R. A. Thisted. *Elements of Statistical Computing*. Chapman and Hall, New York, 1988.

[270] H. C. Thode, Jr. *Testing for Normality*. Marcel Dekker, Inc., New York, 2002.

[271] R. Tibshirani and F. Leisch. *bootstrap*, 2006. Functions for the book "An Introduction to the Bootstrap." S original by Rob Tibshirani, R port by Friedrich Leisch. R package version 1.0-20.

[272] L. Tierney. Markov chains for exploring posterior distributions (with discussion). *Annals of Statistics*, 22:1701–1762, 1994.

[273] Y. L. Tong. *The Multivariate Normal Distribution*. Springer, New York, 1990.

[274] J. Tukey. Bias and confidence in not quite large samples (abstract). *Annals of Mathematical Statistics*, 29:614, 1958.

[275] J. W. Tukey. *Exploratory Data Analysis*. Addison-Wesley, New York, 1977.

[276] S. Ulam, R. D. Richtmyer, and J. von Neumann. Statistical methods in neutron diffusion. *Los Alamos Scientific Laboratory*, report LAMS-551, 1947.

[277] W. N. Venables and B. D. Ripley. *S Programming*. Springer, New York, 2000.

[278] W. N. Venables and B. D. Ripley. *Modern Applied Statistics with S*. Springer, New York, fourth edition, 2002. ISBN 0-387-95457-0.

[279] W. N. Venables, D. M. Smith, and the R Development Core Team. *An Introduction to R*. R Foundation for Statistical Computing, Vienna, Austria, 2007. ISBN 3-900051-12-7.

[280] J. Verzani. *Using R for Introductory Statistics*. Chapman & Hall/CRC, Boca Raton, FL, 2005.

[281] R. von Mises. *Wahrscheinlichkeitsrechnung und Ihre Anwendung in der Statistik und Theoretischen Physik*. Deuticke, Leipzig, Germany, 1931.

[282] J. von Neumann. Zur Theorie der Gesellschaftsspiele. *Mathematische Annalen*, 100:295–320, 1928.

[283] J. von Neumann. Various techniques used in connection with random digits. *National Bureau of Standards Applied Mathematics Series*, 12:36–38, 1951.

[284] G. Wahba. *Spline Models for Observational Data*. SIAM, Philadelphia, 1990.

[285] G. G. Walter and X. Shen. *Wavelets and other Orthogonal Systems*. Chapman & Hall/CRC, Boca Raton, FL, second edition, 2001.

[286] M. Wand and B. Ripley. *KernSmooth: Functions for kernel smoothing for Wand & Jones (1995)*, 2007. S original by Matt Wand. R port by Brian Ripley. R package version 2.22-20.

[287] M. P. Wand. Frequency polygons: Theory and applications. *Journal of the American Statistical Association*, 80:348–354, 1985.

[288] M. P. Wand. Data-based choice of histogram bin width. *The American Statistician*, 51:59–64, 1997.

[289] M. P. Wand and M. C. Jones. *Kernel Smoothing*, volume 60 of *Monographs on Statistics and Applied Probability*. Chapman and Hall Ltd., London, 1995.

[290] G. R. Warnes. *gplots: Various R programming tools for plotting data*. Includes R source code and/or documentation contributed by Ben Bolker and Thomas Lumley. R package version 2.3.2.

[291] G. R. Warnes. *mcgibbsit: Warnes and Raftery's MCGibbsit MCMC diagnostic*, 2005. R package version 1.0.5.

[292] M. Watanabe and K. Yamaguchi. *The EM Algorithm and Related Statistical Models*. Marcel Dekker, New York, 2004.

[293] E. Wegman. Nonparametric probability density estimation: I. A summary of available methods. *Technometrics*, 14:533–546, 1972.

[294] E. Wegman. Hyper dimensional data analysis using parallel coordinates. *Journal of the American Statistical Association*, 85:664–675, 1990.

[295] E. J. Wegman. Computational statistics: A new agenda for statistical theory and practice. *Journal of the Washington Academy of Sciences*, 78:310–322, 1988.

[296] S. S. Wilks. On the independence of k sets of normally distributed statistical variables. *Econometrica*, 3:309–326, 1935.

[297] M. Wiper, D. R. Insua, and F. Ruggeri. Mixtures of gamma distributions with applications. *Journal of Computational and Graphical Statistics*, 10(3):440–454, 2001.

[298] P. Wolf and U. Bielefeld. *aplpack: Another Plot PACKage: stem.leaf, bagplot, faces, spin3R, ...*, 2006. R package version 1.0.

[299] D. Wuertz, many others, and see the source file. *fSeries: Rmetrics – The Dynamical Process Behind Markets*, 2006. R package version 240.10068.

Index